ECONOMIC SYSTEMS ANALYSIS AND ASSESSMENT

WILEY SERIES IN SYSTEMS ENGINEERING AND MANAGEMENT

Andrew P. Sage, Editor

A complete list of the titles in this series appears at the end of this volume.

ECONOMIC SYSTEMS ANALYSIS AND ASSESSMENT

Cost, Value, and Competition in Information and Knowledge Intensive Systems, Organizations, and Enterprises

Andrew P. Sage
Department of Systems Engineering and Operations Research,
George Mason University,
Fairfax, VA 22030-4444

William B. Rouse
Tennenbaum Institute
School of Industrial and Systems Engineering,
College of Computing,
Georgia Institute of Technology,
Atlanta, GA 30332

JOHN WILEY & SONS, INC.

Published by John Wiley & Sons, Inc., Hoboken, New Jersey
Published simultaneously in Canada

Library of Congress Cataloging-in-Publication Data:

Sage, Andrew P.
Economic systems analysis and assessment / Andrew P. Sage, William B. Rouse.
p. cm.– (Wiley series in systems engineering and management ; 54)
ISBN 978-0-470-13795-6 (hardback)
1. Microeconomics. 2. System analysis. I. Rouse, William B. II. Title.
HB172.S127 2011
338.501—dc22 2010042299

Printed in Singapore

10 9 8 7 6 5 4 3 2 1

CONTENTS

PREFACE

The purpose of this book is to provide a background in the fundamentals of economic systems analysis and assessment that is appropriate for engineers and managers concerned with the systems engineering and management of systems that are generally information technology intensive. It is assumed that readers of this book will have previously studied mathematics through calculus and differential equations, and that they have some background in linear algebra. No prior background in mathematical programming or economics is assumed, although a modest exposure to undergraduate microeconomics will be very helpful. The objectives of this book include a salient discussion of engineering economic systems that will be relevant for those who need or desire to use the subject matter in their professional practice. This book will also support those who must communicate and broker the results of engineering economic systems analyses and assessments between the many professionals having a stake in definition, development, and deployment of information technology intensive systems. Finally, this book provides a thorough grounding in investment analysis and assessment, particularly for technology portfolios, capacity improvement and expansion, and mergers and acquisitions to acquire technologies and/or capacity.

The book itself is comprised of five major parts as follows:

1. ***Microeconomics.*** We provide a concise overview of classic microeconomics including production and the theory of the firm; theory of the consumer; market equilibria and market imperfections; and normative or welfare economics, including imperfect competition effects and consumer and producer surplus. Chapters 2 to 5 contain this presentation. We also discuss some behavioral economics issues in this part, particularly in Chapter 5. These chapters are as follows:
   - **Chapter 1:** Introduction to Economic Systems Analysis and Assessment
   - **Chapter 2:** Production and the Theory of the Firm
   - **Chapter 3:** The Theory of the Consumer
   - **Chapter 4:** Supply–Demand Equilibria and Microeconomic Systems Analysis and Assessment Models
   - **Chapter 5:** Normative or Welfare Economics, Decisions and Games, and Behavioral Economics
2. ***Program Management Economics.*** We discuss economic valuation of programs and projects including investment rates of return,

cost–benefit and cost–effectiveness analysis, earned value management, cost structures and estimation of program costs and schedules, strategic and tactical pricing issues, and capital investment and options. There is one lengthy chapter in this part:

- **Chapter 6:** Cost–Benefit and Cost–Effectiveness Analyses and Assessments

3. ***Cost Estimation.*** Cost estimation technologies involve precedented and unprecedented development, commercial off-the-shelf (COTS) software, software reuse, application generators, and fourth-generation languages. Contemporary cost estimation methods are evaluated in terms of openness of underlying models, platform requirements, data required as inputs, output, and accuracy of estimates provided by the models. COCOMO I and II, and COSYSMO are examples of a cost model, function point cost estimation models. Cost is estimated for systems of systems engineering. There is a single chapter in this part:
   - **Chapter 7:** Cost Assessment
4. ***Strategic Investments in an Uncertain World.*** The final part of our economic systems analysis and assessment efforts is concerned with valuation of major investments such as technology portfolios and large-scale capacity expansions, as well as mergers and acquisitions. Here we provide a chapter that addresses alternative methods for valuation of firms including Stern–Stewart's EVA, Holt's CFROI, and various competing methodologies. Chapter **9** considers option-based valuation models including classic real option models (Black–Scholes) and extensions for multistage options with more robust portfolio assumptions. Valuation of information technology intensive enterprises is also addressed. Overall, this part provides a discussion of valuation methods for managing strategic investments in an uncertain world:
   - **Chapter 8:** Approaches to Investment Valuation
   - **Chapter 9:** Real Options for Investment Valuation
5. ***Extensions to the Work.*** There are many extensions possible to economic systems analysis and assessment. There are needed extensions to the classic microeconomics of economic systems analysis and assessment to enable satisfactory treatment of the increasing returns to scale, network effects, and path-dependent issues generally associated with contemporary ultra-large-scale telecommunications and information networks. Investing in the training and education, safety and health, and work productivity of humans is another very important issue. In our concluding chapter of this work, we present a very brief discussion of these issues:
   - **Chapter 10:** Contemporary Perspectives

We sincerely hope that readers find our discussions of economic systems analysis and assessment of value to their work in systems and software engineering, systems and enterprise management, and related areas.

Andrew P. Sage
Department of Systems Engineering and Operations Research
George Mason University
Fairfax, VA 22030-4444

William B. Rouse
Tennenbaum Institute
School of Industrial and Systems Engineering
College of Computing
Georgia Institute of Technology
Atlanta, GA 30332

CHAPTER 1

INTRODUCTION TO ECONOMIC SYSTEMS ANALYSIS AND ASSESSMENT: COST, VALUE, AND COMPETITION IN INFORMATION AND KNOWLEDGE INTENSIVE SYSTEMS, ORGANIZATIONS, AND ENTERPRISES

1.1 INTRODUCTION

This book is about one of the fundamental concerns in the engineering and management of systems of all types, and especially those with a major telecommunications and information network focus: the economic behavior of these systems. We discuss the very important role of economics in shaping our lives and designing our activities and institutions to achieve economic (and other) objectives. The purpose of this book is to present those fundamentals of classic and modern microeconomic systems analysis and assessment that are most necessary in the engineering and management of systems of machines, humans, and organizations that are effective and efficient, and equitable as well. We desire to equip ourselves to answer three fundamental questions:

1. What should be produced and how much of it should be produced?
2. How should the goods be produced?
3. Who should get the goods and services that are produced?

Economic Systems Analysis and Assessment,
by Andrew P. Sage and William B. Rouse

The first of these questions relates to *effectiveness*, the second to *efficiency*, and the third to *equity* concerns. There are a number of related concerns. Many other questions, and their answers, are also important. We are generally concerned with why, where, and when artifacts as well as what, how, and who. For example, we surely wish to ensure sustainability, by preserving the natural resource basis to enable continued satisfaction of human needs in an equitable manner over time. There are also issues that affect marketing of our products, as well as with research and development to enable the production of innovative products (and services). Thus, we wish to examine a plethora of issues associated with the engineering of economic systems.

This chapter will provide an overview of our undertakings. We will first summarize a framework for systems engineering and illustrate the important role of the economics of a firm in maximizing profits and that of the economics of the consumer in maximizing satisfaction by allocating resources, all within the constraints of finite resources. Then we will provide an introductory discussion of the microeconomics of firms and consumers operating together in various markets. Our presentation will stress the information base and other conditions necessary to ensure what we will call a perfectly competitive economy.

These conditions will, as will be apparent, typically not prevail. Various distortions from perfect competition will then result. Our discussions will concern normative economics—how individuals and organizations should ideally behave from an axiomatic perspective to best achieve identified objectives. We will also discuss descriptive economics—how individuals and organizations actually behave. Finally, we will discuss prescriptive economics—how individuals and organizations should behave in realistic settings. This chapter provides a relatively detailed outline of this work and our objectives in writing it.

1.2 A FRAMEWORK FOR SYSTEMS ENGINEERING AND MANAGEMENT

A central purpose of systems engineering is to assist clients in organizing knowledge that contributes to the efficiency, effectiveness, equity, and explicability of decisions and associated resource allocations. Systems engineering methodology provides a framework for the formulation, analysis, and interpretation of issues and problems that lead to the resolution of issues of large scale and scope. Within this framework, content, concepts, and methods are selected. The systems process, in which client(s) and analyst(s) cooperate to establish useful policies, plans, or designs, involves three fundamental steps:

1. *Formulation* of the issue or problem,
2. *Analysis* of the (impacts of) alternatives, and
3. *Interpretation* of results for the value systems of relevant stakeholders, thereby leading to the *evaluation* and *prioritization* of alternatives as well as the *selection* and *implementation* of selected alternative(s).

The systems engineering process is typically characterized by

1. a systematic, rational, and purposeful course of action;
2. a holistic approach in which issues or problems are generally examined in relation to their environment, as well as to due attention to the causal or symptomatic, institutional, and value aspects of the issue under consideration; and
3. the eclectic use of methods and knowledge based on the normative theory of systems science and operations research, as well as the behavioral theory of systems and organizational management.

The typical product of a systems engineering study is a plan to implement a decision, or a plan to implement another phase of a systems study that will ultimately result in such a plan. Economic concerns are vital in developing appropriate plans. It is the study of engineering economic systems analysis that is of interest here. This study is all the more valuable if we first embed it within a discussion of the entire systems process.

A very important fundamental concept of systems engineering is that all systems are associated with life cycles. These are of several types: we have a life cycle for the engineering of the system, and another life cycle for the use of the system. Similar to all natural systems that exhibit a birth–growth–aging–death lifecycle, human-made systems also have a life cycle. Generally, this life cycle consists of three essential phases: *definition* of the requirements for a system, *development* of the system itself, and *deployment* of the system in an operating environment. Each of these may be described by a larger number of more fine-grained phases. These three phases are found in all intentional systems evolutionary efforts. Most realistic life-cycle processes comprise more than three phases. One of the major contributions of systems engineering is in adopting an appropriate perspective for the life cycles associated with engineering the system.

This life-cycle perspective should also be associated with a long-term view toward planning for systems evolution, research to bring about any new and emerging technologies needed for this evolution, and a number of activities associated with actual systems evolution, or acquisition. Thus, we see that the efforts involved in the life-cycle phases of definition, development, and deployment need to be implemented across three life cycles that comprise:

- systems planning and marketing;
- research, development, test, and evaluation (RDT&E); and
- systems acquisition or procurement.

We briefly examine these life-cycle phases here. Discussions of the methods for systems engineering are very important. Here we emphasize economic systems analysis and its application to telecommunications and information networks. We emphasize that these discussions would be incomplete if they are not associated with some discussion of systems engineering life cycles, processes, or methodology and the systems management efforts that lead to selection of appropriate processes.

Systems engineering is a management technology to assist and support policy making, planning, decision making, and associated resource allocation or action deployment. Systems engineers accomplish this by quantitative and qualitative formulation, analysis and assessment, and interpretation of the impacts of action alternatives on the needs perspectives, the institutional perspectives, and the value perspectives of their clients or customers.

The key words in this definition are formulation, analysis and assessment, and interpretation, which form an integral part of systems engineering. We may exercise these in a formal sense, or in an experientially based intuitive sense. These are the components comprising a structural framework for systems methodology and design. We need a guide to formulation, analysis and assessment, and interpretation efforts, and systems engineering provides this through embedding these three steps into life cycles, or processes, for systems evolution.

Systems management and integration issues are of major importance in determining the effectiveness, efficiency, and overall functionality of systems designs. To achieve a high measure of functionality, it must be possible for a systems design to be efficiently and effectively produced, used, maintained, retrofitted, and modified throughout all phases of a life cycle. This life cycle begins with need conceptualization and identification, through specification of systems requirements and architectures, to ultimate systems installation, operational implementation, evaluation, and maintenance throughout a productive lifetime.

For our purposes, we may also define systems engineering as the definition, design, development, production, and maintenance of functional, reliable, and trustworthy systems within cost and time constraints. It is generally accepted that we may define things according to

- structure,
- function, or
- purpose.

Often, definitions are incomplete if they do not address structure, function, and purpose. Our continued discussion of systems engineering will be assisted by the provision of a structural, functional, and purposeful definition of systems engineering as follows:

Structure. Systems engineering is an appropriate combination of methods and tools, made possible through a suitable methodology and systems management procedures, in a useful process-oriented setting that is appropriate for the resolution of real-world problems, often of large scale and scope.

Function. Systems engineering is a management technology to assist clients through the formulation, analysis and assessment, and interpretation of the impacts of proposed policies, controls, or complete systems on the need perspectives, institutional perspectives, and value perspectives of stakeholders to issues under consideration.

Purpose. The purpose of systems engineering is information and knowledge organization that will assist clients who desire to define, develop, and deploy total systems to achieve a high standard of overall quality, integrity, and integration as related to performance, trustworthiness, reliability, availability, and maintainability of the resulting system.

Each of these definitions is important and an understanding of all three is generally needed, as we have noted. In our three-level hierarchy of systems engineering there is generally a nonmutually exclusive correspondence between function and tools, structure and methodology, and purpose and management, as illustrated in Fig. 1.1. A systems engineering process results from efforts at the level of systems management to pick an appropriate methodology, or appropriate set of procedures, or a process for engineering a system. A systems engineering product, or service, results from this process, or product line, together with an appropriate set of methods and metrics. These are illustrated in Fig. 1.2.

We have illustrated three hierarchical levels of systems engineering in Fig. 1.1. These are associated with structure, function, and purpose, as also indicated in Fig. 1.1. The evolution of a systems engineering product, or service, from the chosen systems engineering process is illustrated in Fig. 1.2. The systems engineering process is driven by systems management, and there are a number of drivers for systems management, such as the competitive strategy of the organization. The basic activities of systems engineers are usually concentrated on the evolution of an appropriate process to enable the definition, development, or deployment of a system or on the formulation, analysis, and interpretation of issues associated with one of these phases. Figure 1.3 illustrates the basic systems engineering process phases and steps.

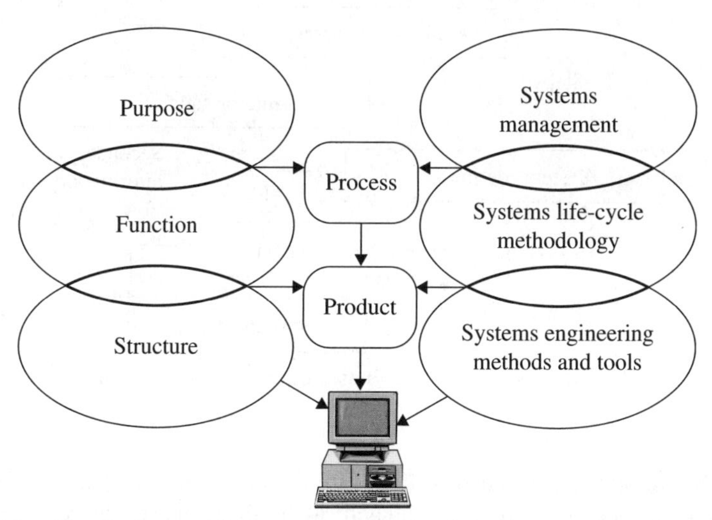

Figure 1.1. The Evolution of Process and Product from Purpose, Function, and Structure and the Three Levels of Systems Engineering: Systems Management, Methodology, and Methods and Tools.

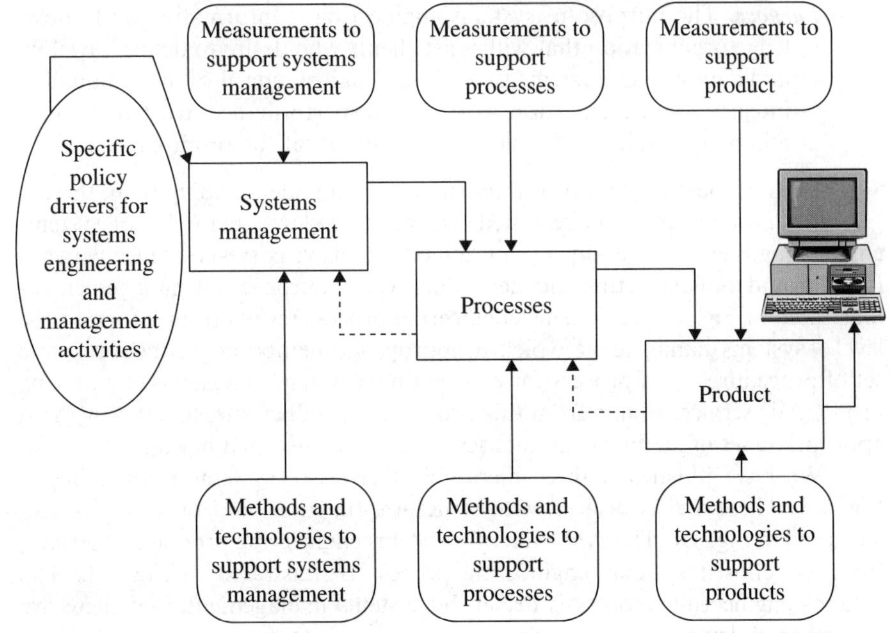

Figure 1.2. Three-Level Systems Engineering and Management Perspective on the Engineering of Systems.

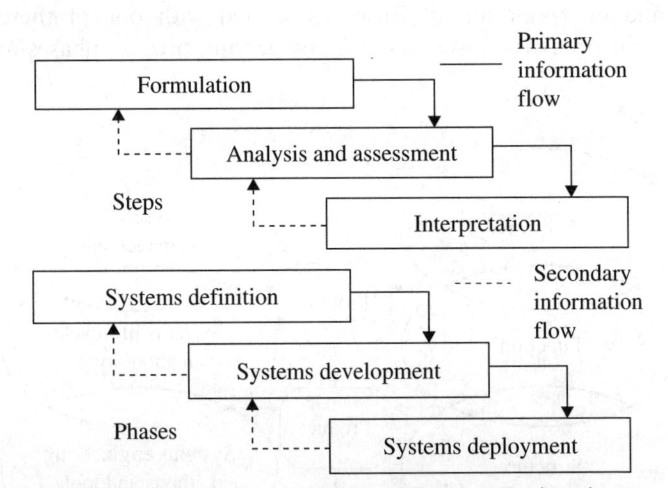

Figure 1.3. The Three Basic Steps and Phases of Systems Engineering.

Generally, these are combined to illustrate the occurrence of each of the three steps of systems engineering within each of the three phases, as represented in Fig. 1.4. A three-element-by-three-element matrix structure representation of a systems engineering framework is also possible as shown in Fig. 1.5.

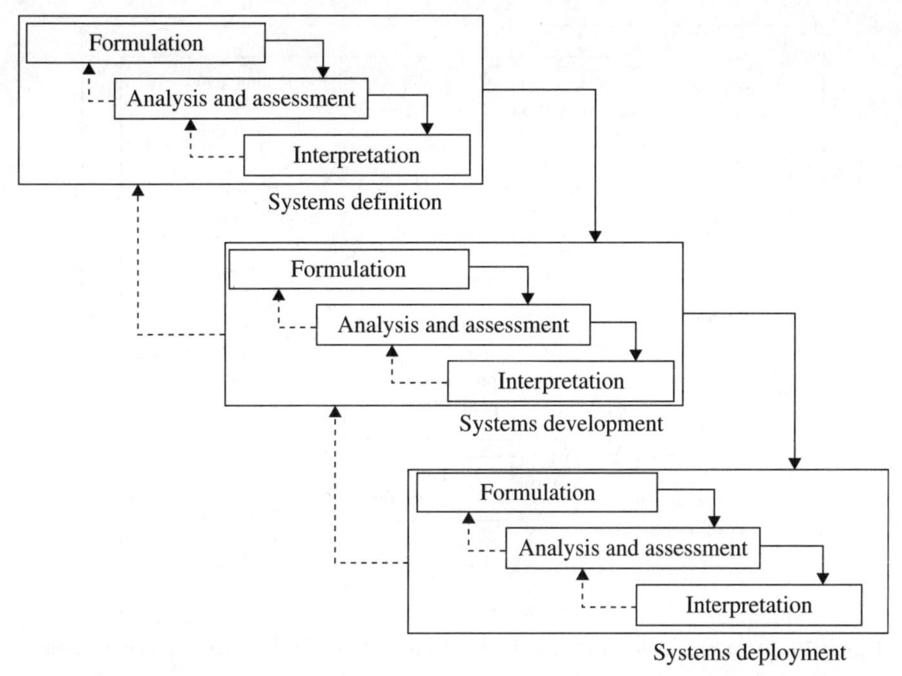

Figure 1.4. A Systems Engineering Framework Comprised of Three Phases and Three Steps Per Phase.

	Formulation	Analysis and assessment	Interpretation
Definition	Activity 1	Activity 2	Activity 3
Development			
Deployment			Activity 9

Figure 1.5. Illustration of Nine Activity Cells for a Simple Two-Dimensional Systems Engineering Framework.

A systems engineering framework, from a formal perspective at least, consists of three fundamental steps: issue formulation, issue analysis, and issue interpretation. These are conducted at each of the life-cycle phases that have been chosen to implement the basic life-cycle phased efforts of definition, development, and deployment. There are three general systems life cycles, as suggested by Fig. 1.6:

- research, development, test, and evaluation (RDT&E);
- acquisition (or production, or manufacturing, or fielding);
- planning and marketing.

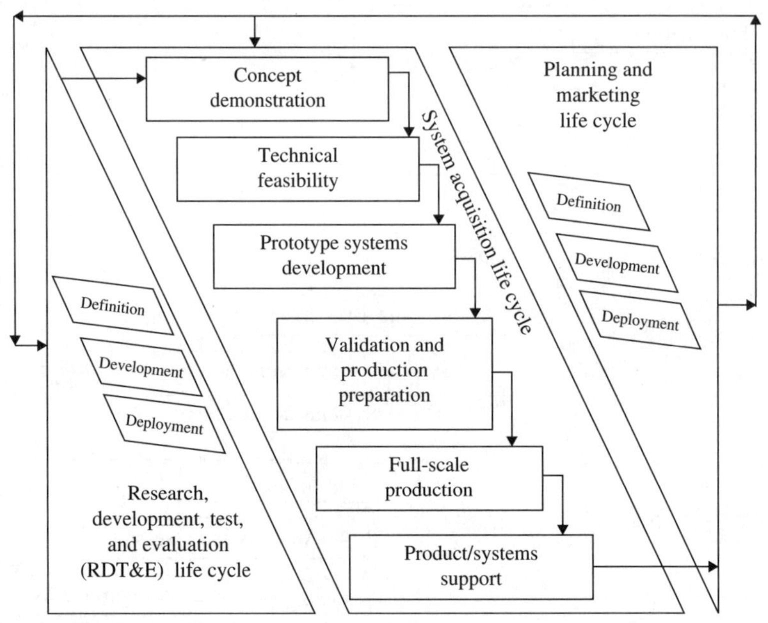

Figure 1.6. Interactions across the Three Primary Systems Engineering Life Cycles.

Systems engineers are involved in efforts associated with each of these life cycles and the associated functions, often in a technical direction or systems management capacity. The detailed life-cycle phases are shown only in the systems acquisition life cycle in the figure. Only the three basic phases are shown for the RDT&E life cycle, and the planning and marketing life cycle. An objective in this is to engineer trustworthy and sustainable systems that have such desirable attributes as those shown in Fig. 1.7.

There are a number of frameworks that we might use to characterize systems engineering and management efforts. Without a sound and well-understood process for the acquisition or production of large systems, it is very likely that there will be a number of flaws in the resulting system itself. Thus, the definition, development, and deployment of an appropriate process, or a set of processes, for the engineering of systems are very important. To undertake a study of systems engineering methods only and their potential use to support the engineering of trustworthy systems, without some understanding of systems engineering processes, is likely to lead to very unsatisfactory results.

Systems engineers provide a needed interface between the client or stakeholder group, or enterprise, to which an operational system will ultimately be delivered, and a detailed design and implementation group, which is responsible for specific systems production and implementation. Figure 1.8 illustrates this view of a systems engineering team as an interface group that provides conceptual design and technical direction to enable the products of a detailed design group to be responsive to client needs. Thus, systems engineers

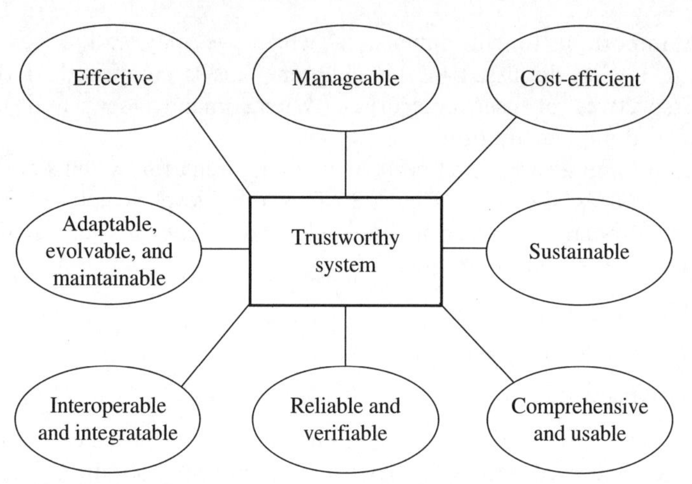

Figure 1.7. Attributes of a Trustworthy Systems Engineering Product or Service.

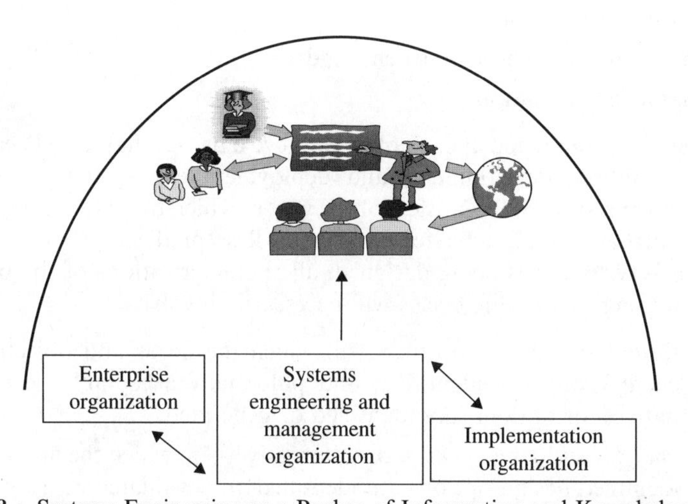

Figure 1.8. Systems Engineering as a Broker of Information and Knowledge.

act, in part, as brokers of information between a client group and those responsible for detailed design and systems production. Knowledge and use of the principles of economic systems analysis are especially important in achieving this needed brokerage.

Systems engineering processes are to a very large extent based on frameworks for systems methodology and design. The framework chosen here consists of three dimensions:

- logic dimension, which consists of three fundamental steps;
- time dimension, which consists of three basic life-cycle phases; and
- life-cycle dimension, which consists of three stages or life cycles.

An important fourth dimension, which we may call a perspectives dimension, is also discussed in this chapter. This is comprised of the three basic perspectives of user enterprise, systems management and technical direction, and implementation.

We envision a three-level performance hierarchy for systems engineering phased efforts, as shown in Fig. 1.3. This three-level structured hierarchy comprises a systems engineering life cycle and is one of the ingredients of systems engineering methodology. It involves

- systems definition,
- systems development, and
- systems deployment.

The structural definition of systems engineering we posed earlier indicates that we are concerned with a framework for problem resolution that, from a formal perspective at least, consists of three fundamental steps for a systems engineering activity:

- issue formulation,
- issue analysis and assessment, and
- issue interpretation.

These are conducted at each of the life-cycle phases that have been chosen for the definition, development, and deployment efforts that lead to the engineering of a system. Regardless of the way in which the systems engineering life-cycle process is characterized, and regardless of the type of product or system or service that is being designed, all characterizations of the phases of the systems engineering life cycles will necessarily involve

1. *formulation* of the problem—in which the needs and objectives of a client group are identified, and potentially acceptable design alternatives, or options, are identified or generated;
2. *analysis and assessment* of the alternatives—in which the impacts of the identified design options are identified and evaluated or assessed; and
3. *interpretation* and selection—in which the options, or alternative courses of action, are compared by means of interpretation and comparison of the assessed impacts of the alternatives and how the client group values these. The needs and objectives of the client group are necessarily used as a basis of this selection. The most acceptable alternative is selected for implementation or further study in a subsequent phase of systems engineering.

Our model of the steps of the logic structure of the systems process is based on this conceptualization. These three steps can be, and generally are, disaggregated into a number of other more detailed steps. Each of these steps of systems engineering is accomplished for each of the life-cycle phases. As is the case with respect to the life-cycle phases, it is generally needed to have iteration and learning associated with these steps. This strongly suggests that evolutionary life-cycle approaches, as extensions of the waterfall models illustrated here, will generally be very desirable. In a later Chapter 10, we explicitly consider this in the form of evolutionary economic analysis.

As we have noted, there are generally three different systems engineering life cycles. These relate to the three different stages of effort that are needed to deliver a competitive product or service to the marketplace:

- research, development, test, and evaluation (RDT&E);
- system acquisition or production; and
- systems planning and marketing.

Thus we may imagine a three-dimensional model of systems engineering that is comprised of steps associated with each phase of a life cycle, the phases in the life cycle, and the life cycles that comprise the coarse structure or stages of systems engineering. Figure 1.9 illustrates this across three distinct but interrelated life cycles, for the three steps, and the three phases that we have described here. This is one morphological framework for systems engineering. As we have noted, it will generally be necessary to expand the three steps and three phases we indicate here into a larger number of steps and phases. Often,

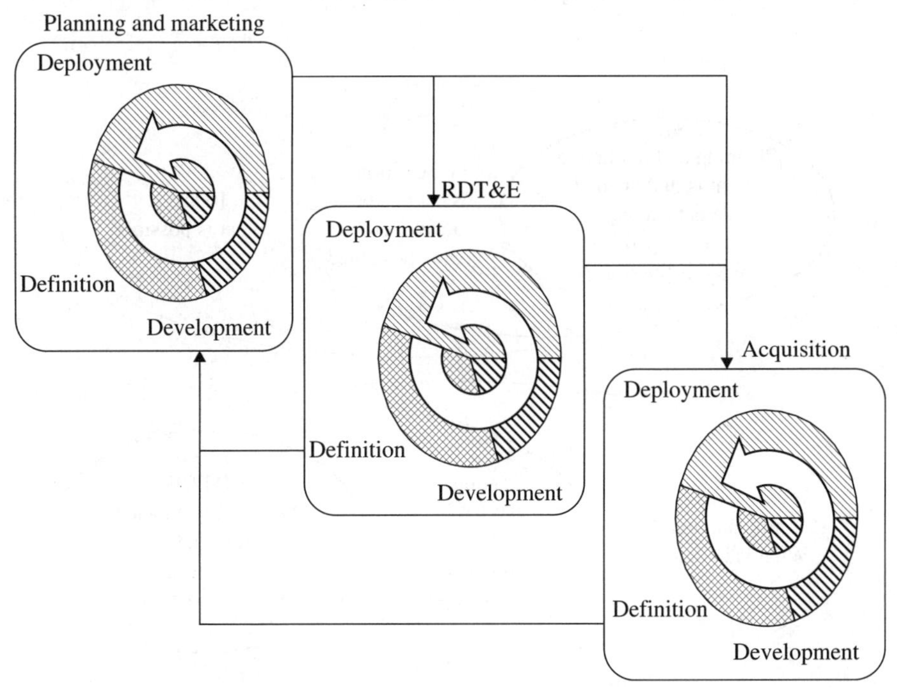

Figure 1.9. Major Systems Engineering Life Cycles with Three Phases within Each Life Cycle.

also, there will be a number of concurrent RDT&E and systems acquisition efforts that will be needed to ultimately bring about a large-scale system.

It is necessary that efforts across these three life cycles be well integrated and coordinated, or else difficulties can ensue. Figure 1.10 represents the relationships across these life cycles. The systems planning and marketing life cycle yields answers to the question: What is in demand? The RDT&E life cycle yields answers to the question: What is (technologically) possible (within reasonable economic and other considerations)? The acquisition life cycle yields answers to the question: What can be developed (from an efficiency, effectiveness, trustworthiness, and sustainability perspective)? It is only in the region where there is overlap, in an *n*-dimensional space, that responsible actions should be implemented to bring about programs for all three life cycles. This suggests that the needs of one life cycle should not be considered independently of the other two. Figure 1.10 represents this conceptually in a two-dimensional Venn-diagram-like representation of the possibility space for each life cycle. Effort should be undertaken to address only the issues within the ellipse represented by the thicker exterior.

Each of the logical steps of systems engineering is accomplished for each of the life-cycle phases. There are generally three different systems engineering life cycles or stages for a complete systems engineering effort, as we have indicated. Thus we may imagine a three-dimensional model of systems engineering that is comprised of steps associated with each phase of a life cycle, the

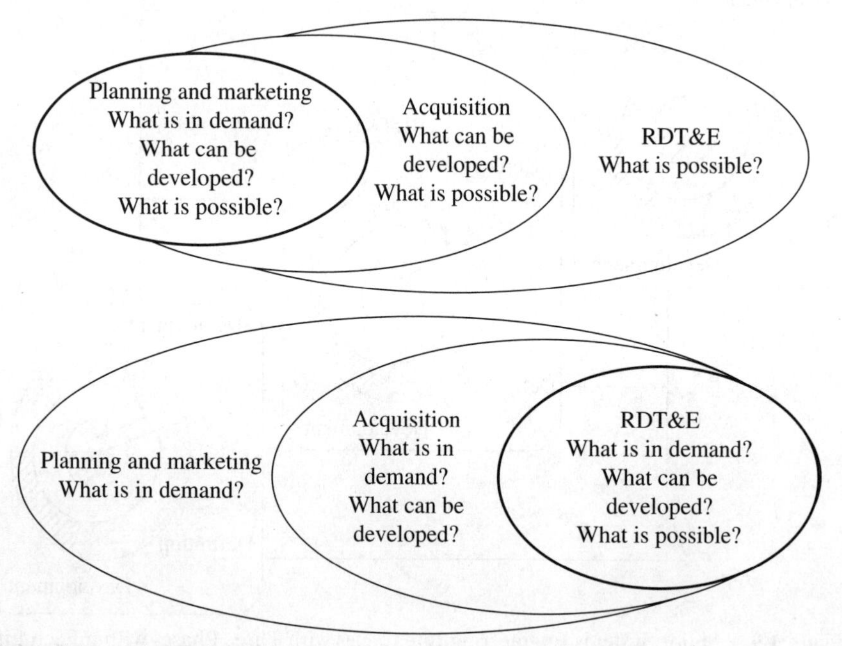

Figure 1.10. Illustrations of the Need for Coordination and Integration across Life Cycles.

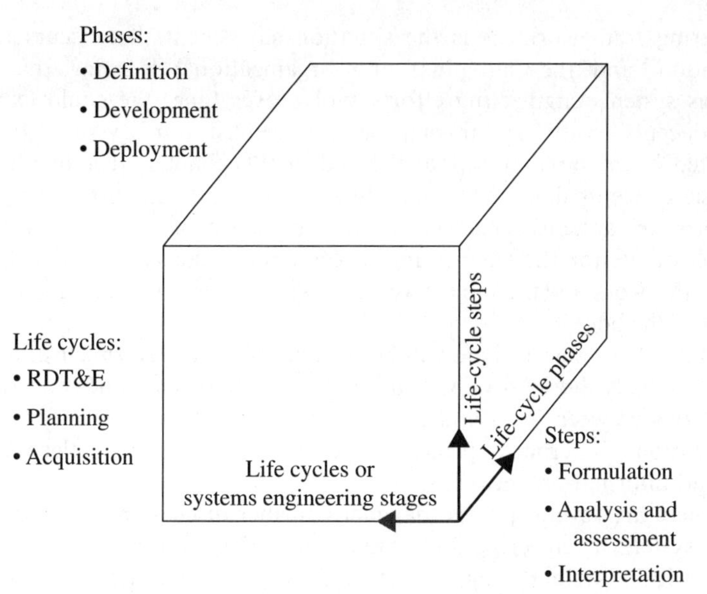

Figure 1.11. Three-Dimensional Framework for Systems Engineering.

phases in the life cycle, and the life cycles or stages of a complete systems engineering effort. Figure 1.10 illustrates this framework of steps, phases, and stages as a three-dimensional cube. This is one three-dimensional framework, in the form of a morphological box, for systems engineering. The word morphology is adapted from biology and means a study of form. As we use it, a methodology is an open set of procedures for problem solving. Consequently, a methodology involves a set of methods, a set of activities, and a set of relations between the methods and the activities. To use a methodology we must have an appropriate set of methods. Generally, these include a variety of qualitative and quantitative approaches from a number of disciplines that enable formulation, analysis, and interpretation of the phased efforts that are associated with the definition, development, and deployment of both an appropriate process and the product that results from this process. Associated with a methodology is a structured framework with which particular methods are associated for the resolution of a specific issue.

Of course, systems engineering is comprised of much more than just a methodological framework, or frameworks. In an earlier three-level view of systems engineering, we indicated that we can consider systems engineering efforts at the levels of

- systems engineering methods and tools, and associated metrics;
- systems methodology, or life-cycle processes; and
- systems management.

We suggested Figs. 1.1 and 1.2 as illustrative of this representation of systems engineering. This is also an important dimension to a systems engineering framework, as is the situation assessment that occurs for issue recognition, and the individual and organizational learning that should occur as systems engineering efforts evolve over time. We could expand on these concepts greatly, and this has been accomplished in several of the references cited in the bibliography at the end of this chapter. It is our hope that the basic concepts illustrated here will serve as a suitable introduction to the principles of systems engineering and management for appreciation of its implications for the engineering of economic systems, which is the major focus of this work and the subject we address in the remainder of this chapter and the rest of the book.

Our major concern here will be the analysis and assessment of a systems effort, especially those portions that involve the microeconomic concerns of the interactions between firms and consumers in markets and the evaluation and prioritization of alternative projects, especially as they concern telecommunications and information networks.

There are several points that merit further discussion here. First of all, neither systems engineering and management nor economic systems analysis efforts within systems engineering and management are processed in a sequenced linear way. They involve a process in which iteration plays a central part. Insights obtained from one part of the effort might lead to a revision of approaches taken earlier, making iteration and feedback necessary. Second, the steps and phases outlined are helpful as a guide, not as a restrictive format. Flexibility in the procedures and methods used is a central feature of systems engineering and management. It should be noted, however, that each of the steps and phases outlined above represents an important ingredient in a systems engineering effort, and omission or neglect of any step increases the risks of failure. Third, since systems engineering is a process in which people work together to realize the various steps and phases of the effort, the selection of an appropriate combination of capable analysts, experts, or other participants, and methods or aids in the process, is at least as important as adherence to the several steps of the systems engineering framework. Figure 1.12 presents a conceptual flowchart of the steps in a typical systems engineering process. These steps are conducted across each of the phases in the process.

In this section, we have presented an essential introduction to systems engineering. Our purpose here is primarily to present some of the essential underlying concepts. The references related to systems engineering in the selected bibliography at the end of this chapter provide much supporting detail.

1.3 THEORY OF THE FIRM

Chapter 2 will be concerned with the classic theory of the firm. We will adopt as a fundamental hypothesis the assumption that the goal of a firm is to maximize profit. To do this, the firm will need to know the costs of production. These costs will depend on the market conditions extant for the three fundamental

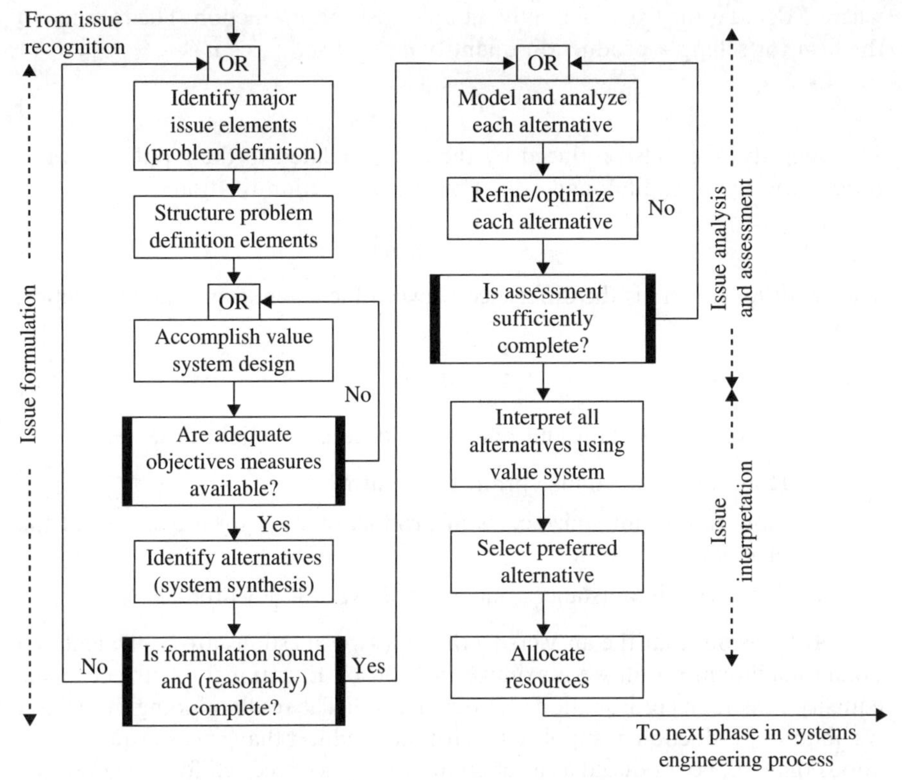

Figure 1.12. Prototypical Flowchart of Steps in the Systems Engineering Process.

types of classic economic resources: land,[1] capital, and labor. In much of the literature on modern economic systems, information and knowledge are considered to be the fourth fundamental resource. We will develop the classic theory of the firm in Chapter 2, and introduce newer concepts in the later chapters.

If we know the way in which the quantity of a product depends on the input of land, capital, and labor to production, then it becomes possible to determine the cost of a given quantity of produced goods in terms of the unit costs of the inputs to production and the quantity of these inputs that are used. If we use amounts T, K, and L of land, capital, and labor, respectively, and if the known wages[2] of these factor input to production are given by w_T, w_K, and w_L, then the costs of production are

$$C(T, K, L) = w_T T + w_K K + w_L L + F \tag{1.1}$$

[1]"Land" is the classic economic term that implies all raw materials and natural resources.

[2]It is interesting to conjecture on the meaning of the wages for capital. It turns out that the difference between the wages for capital and capital represents interest.

where F denotes the fixed—initially set up—costs of production. The revenue to the firm for selling a production quantity q at a fixed price p is

$$R = pq \tag{1.2}$$

The quantity of goods produced by the firm is related to the input factors of production (land, capital, and labor) by the production relation

$$q = f(T, K, L) \tag{1.3}$$

The profit to the firm is the difference between the revenue and the production costs, or

$$\Pi = R - C \tag{1.4}$$

There are several important and relevant questions we might pose here:

1. How can we maximize profit to the firm?
2. How can we minimize costs of production of a given quantity of the product?
3. Are there circumstances under which we will not produce?

It turns out that the answers to the first two questions are equivalent. To obtain maximum profit, we maximize Π given by Equation 1.4, subject to the equality constraints of Equations 1.1 through 1.3. The result of doing this is that we obtain a production or supply curve for the producer that gives the quantity of goods that will be produced as a function of the price received for the goods (or services). To minimize production costs we minimize the costs of production, which is given by C in Equation 1.1 subject to the equality constraints of Equations 1.2 and 1.3. Doing this results in a relation for the minimum production cost $C(q)$ for producing a quantity of goods q. The answer to the third question is that we should produce as long as we can obtain a nonnegative profit.

We will explore issues such as these in considerably greater detail in Chapter 2. A number of extensions will be undertaken. In particular, we will consider the case where there is a sole producer of a given product who has perfect information about consumer demand for the product. This situation is known as a monopoly. This will be the first of several situations that we will examine in which one or more of the conditions for "perfect economic competition" are violated.

One very important notion is that of return to scale in production. If increasing all factors of production by some amount $\lambda > 1$ increases the quantity produced by the same amount, we say that the production function possesses constant returns to scale (CRS):

$$\text{CRS} : \lambda q = \lambda f(T, K, L) = f(\lambda T, \lambda K, \lambda L), \quad \lambda > 1 \tag{1.5}$$

In the case where increasing all factors of production by the same positive amount $\lambda > 1$ results in a produced amount that is less than λ times the initial amount, we have decreasing returns to scale (DRS):

$$\text{DRS}: \lambda q = \lambda f(T, K, L) > f(\lambda T, \lambda K, \lambda L), \quad \lambda > 1 \tag{1.6}$$

The majority of classic production functions possess either DRS or, in a very few cases, CRS. When we examine information and knowledge intensive products, such as software, in our later work, we will generally find that they possess increasing returns to scale (IRS), such that increasing all factors of production by an amount $\lambda > 1$ results in a production quantity that is greater than λ times the amount initially produced:

$$\text{IRS}: f(\lambda T, \lambda K, \lambda L) > \lambda f(T, K, L) = \lambda q, \quad \lambda > 1$$

There are many very interesting properties of these information intensive products that are often called *network effects*: consumption externalities, switching costs and lock-in, and many issues that affect compatibility, standards, and complementarity of these products. These are generally a result of these positive or increasing economies of scale in production.

1.4 THEORY OF THE CONSUMER

Why should a firm produce a product or service? One answer is that there is a *demand* for the product or service, because the firm is *effective* in fulfilling some (perceived) need, and that the firm is *efficient* in producing it and can make a profit by doing so. In Chapter 3 we examine various aspects of the economic theory of the consumer. We assume that the consumer has a utility function that expresses the satisfaction received from the possession or consumption, a term used by economists to also include savings or investment, of a bundle of goods and services. The consumer is assumed to have a utility function

$$U = U(x_1, x_2, \ldots, x_N) = U(\mathbf{x}) \tag{1.7}$$

where $\mathbf{x} = [x_1, x_2, \ldots, x_N]^{\mathrm{T}}$ is a bundle of goods and services or commodity bundle. There is a price vector $\mathbf{p} = [p_1, p_2, \ldots, p_N]^{\mathrm{T}}$ that represents the fixed price that has to be paid for a unit of each of the N goods and services.

The consumer is assumed to be greedy and selfish, in that "more" of any given good or service is always better than "less." Sadly, the consumer has limited resources and cannot pay more than some fixed "income" I for these. The fundamental problem of the consumer is to maximize utility, given by Equation 1.7, subject to the resource constraint

$$I \geq \sum_{i=1}^{N} p_i x_i = \mathbf{p}^{\mathrm{T}} \mathbf{x} \tag{1.8}$$

We will explore various facets of consumer behavior in attempting to maximize the effectiveness of limited resources in maximizing satisfaction. The result of resolving the maximization of utility with a constraint on disposable income is the demand curve for a consumer. Figure 1.13 shows six supply–demand

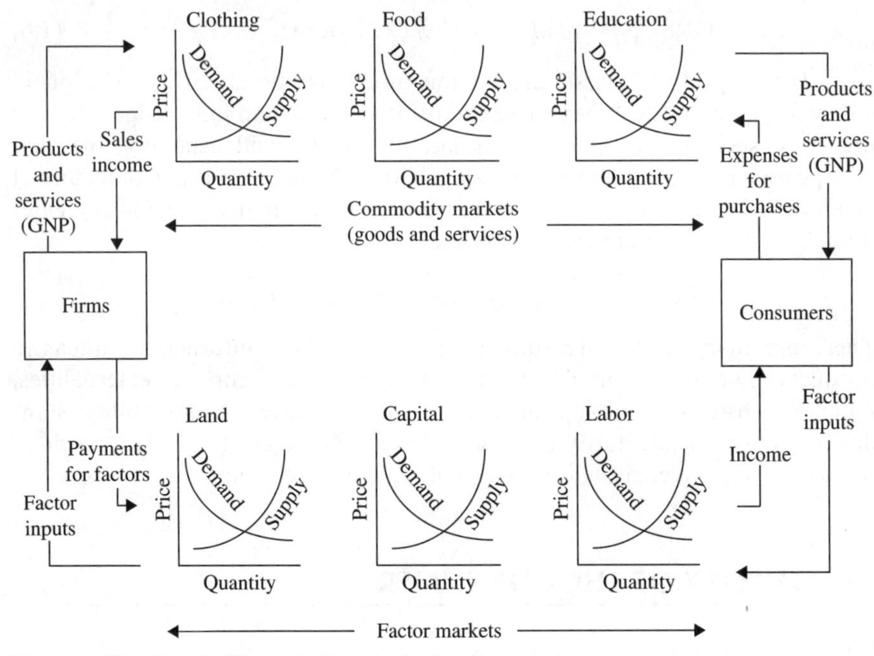

Figure 1.13. Simple Flows in Economic Systems.

curves for various factors and consumer goods and services in a classic representation of a (free-market) economic system where there are DRS, which is the classic case. We will have much more to say about these relationships in Chapter 3. In our later chapters, we will also explore the many issues associated with information and knowledge intensive networks and products where there are IRS.

1.5 THE INTERACTION OF THE THEORIES OF FIRMS AND CONSUMERS: MICROECONOMIC MODELS OF ECONOMIC ACTIVITY

Chapters 2 and 3 discuss the theories of firms and consumers. In Chapter 4 we will extend these concepts to microeconomic models that describe the behavior of economic agents such as firms, consumers, and resource owners in a (free) market economy. We will be primarily concerned with the conditions that prevail in a market system that is in equilibrium and in which no imperfections (monopolies, externalities, etc.) exist. Microeconomic models such as these serve primarily as guides to the behavior that will result in the greatest satisfaction for each economic agent.

The foundation for a microeconomic model is a set of relations that describe

1. the price and quantity of goods and services that will be desired by a consumer who is maximizing their utility;
2. the price and quantity of goods and services that will be desired by a firm that is maximizing its profits; and
3. the general conditions characterizing the markets in which firms and consumers interact.

These relations are combined to determine the equilibrium market conditions that will result in the greatest mutual satisfaction for all firms and consumers in the economic system. This equilibrium is, mathematically, the intersection of the supply and demand curves for products and the intersection of the supply and demand curves for the factor inputs to production.

Microeconomic models provide insight into the workings and effects of "ideal" market systems and can be used to evaluate alternative policies designed to regulate economic behavior or alter economic conditions. They can be used to investigate the effects of changes in such elements as preferences of consumers, technologies of firms, and the availability and costs of the various factor resource inputs to production.

Typical final results or products of the use of supply–demand models of microeconomic activity include

1. a quantitative model describing the interaction of some set of economic agents, including firms and consumers, in a market economy;
2. a determination of the market conditions that will exist in equilibrium when all economic agents are deriving maximum satisfaction;
3. increased understanding of the workings and effects of a free-market system;
4. a set of relations describing the quantity of commodities and resources that each economic agent will desire for a given price; and
5. a determination of those economic decisions that will result in maximum utility for the consumers and maximum profit for the firms.

The first step in building a microeconomic model of the supply–demand relations describing economic activity is to identify the basic components of the economic system under consideration. These components will generally include

1. a "consumption sector," generally represented by a set of consumers or households;
2. a "production sector," generally represented by a set of firms;
3. a set of final goods, commodities, or services; and
4. a set of economic resources that are the factor inputs to production.

Generally these consist of capital (K), land (T), and labor (L). The set of relations that provide the foundation for a microeconomic model are derived from theoretical considerations of

1. the economic behavior of firms (Chapter 2),
2. the economic behavior of households (Chapter 3), and
3. the equilibrium conditions that prevail in the markets where households and firms exchange resources and commodities (Chapter 4).

Let us provide some more perspective on each of these.

The Economic Behavior of Productive Units. The role of a firm in a classic economic system generally consists in buying factor inputs in the form of land, capital, and labor; producing goods and services from these resources; and finally selling these goods to households and other firms for consumption. In economic systems, it will be necessary for some firms to use goods and services produced by other firms as factor inputs to their own production. For example, a firm that manufactures television sets may purchase components from another firm that manufactures electronic parts. It is assumed that associated with each firm is a *production function* that describes the maximum amount of final goods a firm can produce for a given quantity of factor inputs. The form of this production function will depend on the firm's technology and productive capacity. The *profit* of a particular firm is the total value of goods and services sold (revenues) minus the cost of producing these commodities. It is assumed that the basic goal of the firm is to maximize profits subject to the constraints on its technology and capacity as reflected in its production function. The solution to this optimization problem will result in the quantity of commodities produced and sold for consumption for given and assumed fixed prices.[3] This will be the supply curve for the firm in question. The solution to this profit maximization equation will also result in a factor demand equation. We will obtain a relation that gives the demand for factor inputs to production in terms of their prices. Of course, we could postulate a supply equation for a firm with several unspecified parameters. Regression techniques could, in principle, be used to identify parameters such that we identify a supply curve for a given firm as it operates in practice.

The Economic Behavior of Consumers. The role of the household or consumer in an economic system generally consists in selling resources, such as labor, or factor inputs for production to firms and in buying commodities (final goods) for consumption with the income received from selling these factors. The utility function describes the amount of satisfaction a household derives from the possession of a given set of commodities and a given set of factor inputs. The form of this utility function will depend on the tastes and preferences of each consumer or household. Each consumer will also have a budget constraint, which reflects the total amount of income a household has to spend on final

[3]The firm is assumed to be a price taker in the sense that it accepts prices, which may vary over time, of course, as given. The firm is "too small" to attempt to control prices.

goods and services. The income of a household comes from the sales of factor inputs and from the ownership of firms. The basic goal of a household is to maximize its utility function subject to the budget constraint.[4] The solution to this optimization problem identifies an "optimum" quantity of commodities demanded for consumption and an "optimum" quantity of factor inputs supplied to firms for given commodity and factor input prices. The quantities that are optimum for a particular household will depend on its resources and preferences as reflected in its utility function and budget constraint. For example, consumers will sell factor inputs, such as labor for wages, to increase their incomes. Consumers will also derive some utility from unsold factor inputs. Labor hours not supplied by a consumer can be interpreted, for example, as leisure time.

Market Equilibrium Conditions. An economic market is in *equilibrium* when the quantity of all the goods and services demanded is equal to that which is supplied. In microeconomic models in which firms, consumers, factor inputs, and commodities are all involved, equilibrium conditions require that the following two conditions hold:

1. For each final good or commodity produced, the total quantity supplied by all firms is equal to the total quantity demanded by all households. This gives rise to a set of commodity market clearing equations.
2. For each economic factor input to production, the total quantity supplied by all households is equal to the total quantity demanded by all firms. This gives rise to a set of factor market clearing equations.

After the basic structure of the model has been identified according to the issue formulation guidelines presented earlier, data need to be collected and used to determine the precise forms of the production function and consumers' utility functions. Utility functions might be reconstructed from observed past behavior or elicited in a more direct form. Statistical techniques, such as regression analysis and estimation theory, are also useful in constructing models of production functions for firms and utility curves for consumers.

The next step in the construction of a microeconomic supply–demand model is to actually solve the equations describing economic equilibrium. In theory, this requires the solution to the optimization problem for each household and each firm, subject to constraints and the market clearing conditions. The resulting solution will give the quantities of commodities and factor inputs that will, theoretically, be exchanged in equilibrium, together with the market prices at which they will be exchanged. This information can then be used to

[4]The household also accepts prices as given; it needs such a small fraction of any available product or service that it cannot control price.

investigate the effects of changes in such factors as consumer preference, firm technology, income distribution, market structure, and resource availability on equilibrium market conditions. It can be used to evaluate the consequences of alternative policies designed to regulate or improve existing economic conditions. Figure 1.13 illustrates, conceptually, the results that we might obtain from the construction of a simple microeconomic model. In Fig. 1.13 there are two fundamental feedback loops: one involving the flow of products and the other involving the flow of capital. We will expand on this diagram in Chapter 4. Here it is especially important to note the flows into and out of the firms and consumer sectors. The input to the firms is the income due to sales and the factors for production; the output from the firms is the products of goods and services, the aggregate total of which is generally termed the gross national product (GNP), although the term gross domestic product (GDP) is now more common, and payments for the factor inputs to production.

Microeconomic models can be used to gain insight into the behavior of economic systems, as well as to assess the impact of alternative policies designed to alter or regulate behavior. They can also be used to determine market structures that will result in some desired utilization of resources. Thus microeconomic models may be useful for identifying alternatives, as well as for the analysis of their impacts. A simple example will illustrate some of the concepts involved in constructing a microeconomic model.

Example 1.1: We shall construct a microeconomic model based on a very simple economy consisting of two firms and two consumers. We will present much more detailed and realistic models in our later chapters. Imagine that in this small market system firm 1 buys labor from the consumers to make clothes to sell to the consumers, while firm 2 buys oil from the consumers to produce food to sell to the consumers. The generic questions we wish to answer are: How much food and how much clothing will each consumer purchase, and how much labor will each firm use? To construct a model that can answer these questions, we make some definitions and assumptions.

We use the following terminology. We let

C = clothes produced by firm 1

F = food produced by firm 2

L_1 = labor bought by firm 1

L_2 = labor bought by firm 2

C_1 = clothes bought by consumer 1

F_1 = food bought by consumer 1

C_2 = clothes bought by consumer 2

F_2 = food bought by consumer 2

P_C = price of one unit of clothing

p_F = price of one unit of food

w_L = wage per unit of labor

Π_1 = profit of firm 1
Π_2 = profit of firm 2
U_1 = utility of consumer 1
U_2 = utility of consumer 2

The question naturally arises as to the time interval or horizon over which these prices, quantities, and profits are valid. The answer is that a specified planning interval over which incomes are obtained, goods produced, etc., must be identified. The models of firms and consumers should be valid for that interval. Since this is intended to be a simple illustration of economic analysis, we do not need to specify these here.

We also need a brief description of the production, consumption, and market mechanisms.

Production. We assume that firm 1 can make units of clothing for units of labor purchased in such a way that its production function is given by $C = 8(5L_1)^{0.5}$. The profits of firm 1 are the value of the clothes sold minus the cost of the labor required to produce the clothing. Thus, the profit for this firm is given by the relationship

$$\Pi_1 = p_C C - w_L L_1$$

Similarly, firm 2 can produce units of food from each unit of labor in such a way that its production function is given by the relationship $F = 8(10L_2)^{0.5}$ and its profits are given by the relationship

$$\Pi_2 = p_F F - w_L L_2$$

Consumption. We assume that each consumer initially owns 40 units of labor, *all* of which is sold to the firms, so that the income of each consumer is given by $I = 40w_L$. We assume that consumer 1 is partial to clothes and derives satisfaction from the consumption of clothes and food according to the utility function for consumer 1 given by

$$U_1 = (C_1)^{0.75}(F_1)^{0.25}$$

Also, assuming that consumer 1 spends all of the income earned, the purchase of clothing and food will be subject to the budget constraint

$$I = 40w_L = p_C C_1 + p_F F_1$$

Consumer 2 prefers food and has utility and budget equations given by

$$U_2 = (C_2)^{0.25}(F_2)^{0.75}$$
$$I = 40w_L = p_C C_2 + p_F F_2$$

Market Equilibrium. In equilibrium, the quantities of labor, clothing, and food supplied must equal the respective quantities demanded. In the market for labor, then, the total quantity of labor purchased by the two firms must be equal to the quantity of labor provided by the consumers. This is represented by the factor market clearing equation:

$$L_1 + L_2 = 40 + 40 = 80$$

Similarly, the quantities of clothes and food purchased by the households must equal the quantities produced by firms 1 and 2, respectively. Thus the commodity market clearing equations are

$$C = C_1 + C_2$$
$$F = F_1 + F_2$$

We can now set up the optimization problems that must be solved to determine the quantities of goods and resources that will be exchanged when each firm and household is maximizing its own satisfaction function, profit for the firms, and utility for the consumers. Each firm must maximize profit subject to the production constraint, whereas each consumer must maximize utility subject to the budget constraint. We have four optimization problems:

Firm 1	*Firm* 2
Maximize $\Pi_1 = p_C C - w_L L_1$	Maximize $\Pi_2 = + p_F F - w_L L_2$
subject to $C = 8(5L_1)^{0.5}$	subject to $F = 16(5L_1)^{0.5}$
Consumer 1	*Consumer* 2
Maximize $U_1 = (C_1)^{0.75}(F_1)^{0.25}$	Maximize $U_2 = (C_2)^{0.25}(F_2)^{0.75}$
subject to $p_C C_1 + p_F F_1 = 40w_L$	subject to $p_C C_2 + p_F F_2 = 40w_L$

Also, we have two sets of clearing equations, one each for the commodity and factor markets.

Commodity market clearing equations:

$$C = C_1 + C_2$$
$$F = F_1 + F_2$$

Factor market clearing equation:

$$L_1 + L_2 = 80$$ ■

These are the basic equations for this economic model. Attempts to indicate the full details concerning the solution to this problem would take us into subjects we will explore thoroughly in Chapter 4; we will therefore only indicate the highlights of our solution procedure here. It is easy to find the commodity bundles that maximize the utility of consumers 1 and 2. We obtain

$$\hat{F}_1 = \frac{10w_L}{p_F}, \hat{C}_1 = \frac{30w_L}{p_C}$$

$$\hat{F}_2 = \frac{30w_L}{p_F}, \hat{C}_2 = \frac{10w_L}{p_C}$$

where we use the symbol ^ to indicate optima.

By using the commodity market clearing equations, we obtain from the foregoing

$$\hat{C} = \hat{C}_1 + \hat{C}_2 = \frac{40w_L}{p_C}$$

$$\hat{F} = \hat{F}_1 + \hat{F}_2 = \frac{40w_L}{p_F}$$

As we should intuitively expect, these commodity demand relations show that the quantity of each product demanded decreases with increasing price of the commodity. To obtain maximum profit for each firm we find the quantity produced such that we have the necessary condition for profit maximization: $\partial\Pi_1/\partial C = \partial\Pi_2/\partial F = 0$. We easily obtain

$$\hat{C} = \frac{160p_C}{w_L}, \quad \hat{F} = \frac{160p_F}{w_L}$$

As we would intuitively expect, the quantity of the commodity that is produced increases with the price the firm can get for it.

We note that the first two relations for $\hat{C}$ and $\hat{F}$ are demand relations, whereas the latter two relations are supply relations. Economic equilibrium requires that these two be equal. We obtain as conditions for economic equilibrium $w_L = 2p_C = 4p_F$. This simply says that one of the prices, or the labor wage, can serve as a "numeraire" or "anchor" on which all other prices are based. We should not expect otherwise, as a little thought will show. Thus we may as well let $w_L = 1$ and get $p_C = 0.5$ and $p_F = 0.25$.

We already incorporated the factor market clearing equation when we wrote the budget constraint for each consumer, as this constraint included the clearing equation. Thus, all the necessary relations

1. the equations for optimality of the firm,
2. the equations for optimality of the consumers, and
3. the market clearing equations

have now been obtained. We obtain as the equilibrium relations the results $\hat{C}_1 = 60$, $\hat{C}_2 = 20$, $\hat{F}_1 = 40$, and $\hat{F}_2 = 120$.

In Chapter 4, we will provide a relatively general discussion of market equilibria and the effects that market imperfections, such as due to monopolist firms, may produce on equilibria.

1.6 WELFARE OR NORMATIVE ECONOMICS

Welfare or normative economics is that branch of microeconomics that is primarily concerned with the allocation of resources among competing uses. The purpose of welfare or normative economics is to determine patterns of resource allocation, commodity distribution, and economic organization that will result in maximum economic welfare for the economic system as a whole. As such, it is a part of what may be called *normative economics*. In deciding what distribution of resources among the members of a society is most desirable or equitable, ethical value judgments must be made. In Chapters 2 to 4 we are concerned primarily with efficiency and effectiveness; in Chapter 5 we are also concerned with equity. There are a number of issues associated with equity, the notion of sustainable development being a very important one of these. We will provide an introductory discussion of these generalized equity concerns in Chapter 5.

The basis for normative or welfare economics consists of a set of mathematical relations that describe

1. the satisfaction derived by each consumer from the possession or consumption of given quantities of goods and resources;
2. the productive capacity of the economic system for transforming economic resources into consumable goods or services; and
3. the level of general economic welfare associated with each distribution of goods and resources among the individual economic agents in the system.

These relations can be embedded into an optimization problem, the solution of which will determine a resource allocation and commodity distribution pattern yielding maximum economic *efficiency*, maximum economic *effectiveness*, and maximum *equity*, all according to the specified welfare function. We will show in Chapter 5 that it is not possible to optimize equity independently of efficiency and effectiveness concerns. The converse is also true.

The intent of a welfare economics application is often to determine policies that will enhance or increase the general economic welfare. Examples of application areas include policies pertaining to the income distribution among citizens, the allocation of raw materials among alternative productive uses, the distribution of consumable energy, government funding of public projects, and issues that concern sustainability.

The typical final results or product of a welfare or normative economic study may include

1. a quantitative model describing the allocation, utilization, and consumption of economic resources and commodities by consumers and firms;
2. a mathematical relation that assigns an index of social desirability or general economic welfare to each resource allocation and commodity distribution pattern in the economic system;

3. a determination of the economic configuration for which the social welfare function will attain a maximum value;
4. a determination of economic decisions that will result in maximum satisfaction for individual consumers;
5. a determination of resource allocations and utilizations to best ensure sustainable development;
6. a determination of how each consumer's satisfaction contributes to the overall social welfare; and
7. an increased understanding of the workings of the economic system under consideration.

All of this seems very desirable; however, there are major requirements for the large-scale application of welfare economics concepts. These include

1. a set of utility functions that represent each consumer's tastes and preferences with respect to the possession or consumption of given quantities of resources and commodities;
2. an aggregate production function representing the total productive capacity of the economic system such that we know the maximum quantity of consumable goods that can be produced for given amounts of resources available for productive use;
3. a social welfare function that maps levels of utility attained by individual consumers into an overall measure of social economic welfare;
4. a set of relations that describe the sustainable development concepts to be incorporated; and
5. a specification of the information and knowledge sources on which the welfare economics model is based.

Not all normative or welfare economics models will fully incorporate all of these issues. In particular, sustainable development issues are not incorporated in many classic welfare functions.

The first step in building a welfare economics model is to identify the components of the economic system under consideration. These components will generally include

1. a set of commodities or final (consumable) goods,
2. a set of economic resources used as factor inputs to the production of final goods,
3. a set of consumers who sell resources (e.g., labor) to firms and who buy final goods for consumption, and
4. a set of firms that produce final goods using factor inputs.

The construction of a welfare economics model will require the determination of

a. consumer preferences,
b. production functions, and
c. an overall economic welfare function.

We describe these three concepts in more detail here.

Consumer Preferences. Associated with each consumer is a utility function that describes the level of satisfaction derived from the consumption of goods and services. The form for this function depends on the relative preferences and tastes of the consumer, the amount of resources with which the consumer is endowed, and the preferences of the consumer with respect to the factor inputs that they supply to firms and those which are retained for other uses, such as hours devoted to leisure. Generally, the utility function describes the level of satisfaction a consumer attains for a given state of the economy.

Productive Capacity of Firms. Each firm produces final goods from factor inputs subject to the constraints on its technology and productive capacity. The production functions of all the firms in the economic system can, in principle, be combined to form an aggregate production function that describes the total productive capacity of the system. This aggregate production function gives the maximum quantity of consumable goods that the system can produce for given amounts of available resources.

General Economic Welfare. The utility function of all consumers in the economic system is used to determine a *social welfare function* that describes the overall desirability of a given economic state. The form of this function depends on the relative contribution of each consumer's satisfaction level to the level of economic welfare of the whole system, and thus depends on very subjective value judgments. The premise is that the maximum value of an appropriately constructed social welfare function results from an economic state in which the distribution of resources and goods among consumers is the most *equitable* one.

After the major components of the model have been identified, the next step is to obtain data or other information that enables issue formulation. Production functions may be determined on the basis of past economic or technological information; consumer utility functions might be derived either by reconstruction of past behavior or by direct elicitation, while, generally, a direct attempt will be made to elicit a social welfare function for those interested in the results of the analysis.

The next step is to solve the resulting mathematical optimization problem, namely, to maximize the social welfare function subject to the productive capacity constraint represented in the aggregate production function. The result is called the *welfare optimum*.

Solutions to problems of welfare economics systems analysis are generally closely associated with decision analysis issues, and we will examine both normative and behavioral issues in the concluding part of Chapter 5. It is essential that this be done to incorporate descriptive realities. Microeconomic models are often helpful in determining the structure of the economic system with which the welfare economics model is concerned. Welfare economics models often supply very useful inputs to a cost–benefit analysis effort in which a measure of the economic desirability of several alternative projects is to be evaluated. The conceptual value to be obtained from welfare or normative economics models is very great, even though it will be necessary to incorporate behavioral realities obtained through such approaches as decision regret and prospect theory to obtain realistic solutions.

1.7 PROGRAM AND PROJECT MANAGEMENT ECONOMICS

Cost–benefit analysis is fundamentally concerned with the evaluation of alternative projects that will enhance individual, institutional, national, or international welfare so as to enable the selection of one or more of them for implementation. It is natural to ask: How is this different from welfare or normative economics? An answer is that the goals of the two are much the same. Welfare economics is much more concerned with adjusting parameters associated with a single policy alternative such that they are the best in maximizing a given or identified (scalar) welfare function. In the cost–benefit approach it is desired to evaluate a number of proposed alternative courses of action. It is recognized that welfare economics and other microeconomic approaches will provide very useful conceptual frameworks for the evaluation of proposed alternatives, but that not all of their prescriptions can be followed precisely. Cost–benefit analysis is a method for evaluating the net balance of benefits and costs, as these evolve over time, that are associated with proposed plans or projects. Both quantifiable and nonquantifiable effects are taken into account, as well as various intangible and secondary effects of proposed activities. Indirect methods, based on concepts such as consumer surplus and shadow pricing, are used to infer benefits and costs when there is no direct indication that appears either realistic or valid.

In conducting a cost–benefit analysis, it is first necessary to identify the costs and benefits of the impacts of the proposed alternatives. These are then quantified and discounted over time to obtain the net present values of benefits and costs. An assessment of the costs and benefits of a project that are not easily converted into economic costs and benefits may be more appropriately treated through a cost–effectiveness analysis. In this approach, effectiveness indices are based on the noneconomic "costs" and the economic and noneconomic benefits of proposed projects. Cost–benefit and cost–effectiveness analyses are very practical methods to conduct project appraisals. As such, this is very appropriate material with which we can continue our study of economic systems analysis, and we present a rather detailed discussion of the many subjects related to systems engineering and program management economics. In particular, in Chapter 6 we discuss economic valuation of programs and projects, including investment rates of return, cost–benefit and cost–effectiveness analyses, earned value management, cost structures and estimation of program costs and schedules, strategic and tactical pricing issues, and capital investment and options.

1.8 CONTEMPORARY ISSUES CONCERNING INFORMATION AND INFORMATION TECHNOLOGY ECONOMICS

Fundamentally, information technology (IT) is concerned with improvements in a variety of human problem-solving endeavors through design, development, and use of technologically based systems and processes that enhance the efficiency and effectiveness of information and associated knowledge in a variety of strategic, tactical, and operational situations. Ideally, this is accomplished through critical attention to the information needs of humans in problem-solving tasks and in the provision of technological aids, including computer-based systems of hardware and software and associated processes, that assist in these tasks. Information technology activities and efforts complement and enhance, as well as transcend, the boundaries of traditional engineering through emphasis on the information basis for engineering.

Information technology is comprised of hardware and software that enable the acquisition, representation, storage, transmission, and use of information. Success in IT is dependent on being able to cope with the overall architecture of systems, their interfaces with humans and organizations, and their relations with external environments. It is also very critically dependent on the ability to successfully convert information into knowledge.

The initial efforts at provision of IT-based systems concerned implementation and use of new technologies to support office functions. These have evolved from electric typewriters and electronic accounting systems to include very advanced technological hardware, such as FAX machines and personal computers to perform functions as electronic file processing, accounting, and word processing. Now, networking is a major facet of IT.

In the early days of human civilization, development was made possible through the use of human effort, or labor, primarily. Human ability to use natural resources led to the ability to develop based not only on labor but also on the availability of natural resources, land being a classic example. At that time, most organizations were comprised of small proprietorships. The availability of financial capital during the Industrial Revolution led to this being a third fundamental economic resource, and also to the development of large, hierarchical corporations. This period is generally associated with centralization, mass production, and standardization.

In the later part of the Industrial Revolution, electricity was discovered and the semiconductor was invented. This has led to the information age, or the IT age. Among the many potentially critical IT-based tools are database machines, e-mail, artificial intelligence tools, facsimile transmission (FAX) devices, fourth-generation programming languages, local area networks (LAN), integrated service digital networks (ISDN), optical disk storage (CD-ROM) devices, personal computers, parallel processing algorithms, word processing software, computer-aided software engineering packages, word processing and accounting software, and a variety of algorithmically based software packages. There are many others, and virtually anything that supports information acquisition, representation, transmission, and use can be called an IT product. This includes the Internet.

Major availability of technologies for information capture, storage, and processing has led to information, as well as its product knowledge, becoming a fourth fundamental economic resource for development. This is the era of total quality management, mass customization of products and services, reengineering at the level of product and process, and decentralization and horizontalization of organizations, and systems management. While IT has enabled these changes, much more than just IT is needed to bring them about satisfactorily. In this book, we fill focus our attention primarily on IT and its use by individuals and organizations to improve their productivity.

Many commentators have long predicted the coming of the information age and its characteristics are described in a number of contemporary writings. For example, Alvin Toffler writes of three waves: the agriculture, industrial, and information or knowledge ages. Within these are numerous subdivisions. For example, the information age could be partitioned into the era of vertically integrated and stand-alone systems, process reengineering, total quality management, and knowledge and enterprise integration. Information and knowledge are now fundamental resources that augment the traditional economic resources: land and natural resources, human labor, and financial capital. Critical success factors for success in the third wave, or information age, have been identified and include strategy, customer value, knowledge management, business organization, market focus, management accounting, measurement and control, shareholder value, productivity, and transformation to the third wave model for success. There are numerous other methods of partition. The information age could be partitioned into the age of mainframe computers, minicomputers, microcomputers, networked and client-server computers, and the age of knowledge management.

Major growth in the power of computing and communicating, and associated networking is quite fundamental and has changed relationships among people, organizations, and technology. These capabilities allow us to study much more complex issues than was formerly possible. They provide a foundation for dramatic increases in learning and both individual and organizational effectiveness. In large part, this is due to the networking capability that enables enhanced coordination and communication among humans in organizations. It is also due to the vastly increased potential availability of knowledge to support individuals and organizations in their efforts. However, information technologies need to be appropriately integrated within organizational frameworks if they are to be broadly useful. This poses a transdisciplinary challenge of unprecedented magnitude if we are to move from high performance IT to high performance organizations.

In years past, broadly available capabilities never seemed to match the visions proffered, especially in terms of the time frame of their availability. Consequently, despite these compelling predictions, traditional methods of information access and utilization continued their dominance. In the past years, the pace has quickened quite substantially and the need for integration of IT issues with organizational issues has led to the creation of a field of study sometimes called "organizational informatics," or more recently "knowledge management," the objectives of which generally include

- capturing human information and knowledge needs in the form of systems requirements and specifications;
- developing and deploying systems that satisfy these requirements;
- supporting the role of cross-functional teams in work;
- overcoming behavioral and social impediments to the introduction of IT systems in organizations; and
- enhancing human communication and coordination for effective and efficient workflow through knowledge management.

The Internet, World Wide Web, and networks in general have become ubiquitous in supporting these endeavors. However, organizational productivity is not necessarily enhanced unless attention is paid to the human side of developing and managing technological innovation to assure that systems are designed for human interaction. These are, of course, major objectives for systems engineering and systems management.

There are several ways in which we can define IT. The U.S. Bureau of Economic Analysis appears to define it in terms of office, computing, and accounting machinery. Others consider IT as equivalent to information processing equipment, which includes communications equipment, computers, software, and related office automation equipment. Still others speak about the technologies of the information revolution and identify such technologies as advanced semiconductors, advanced computers, fiber optics, cellular technology, satellite technology, advanced networking, improved human computer interaction, and digital transmission and digital compression. We would not argue about the content in this list, although we would certainly add software and middleware technology to this list. It could be argued that software is intimately associated with advanced computers and communications. This is doubtlessly correct; however, there is still software associated with the integration of these various technologies of hardware and software to comprise the many IT-based systems in evidence today and which will be in use in the future.

The information revolution is driven by technology and market considerations and by market demand and pull for tools to support transaction processing, information warehousing, and knowledge formation. Market pull has been shown to exert a much stronger effect on the success of an emerging technology than technology push. There is hardly any conclusion that can be drawn other than that society shapes technology or, perhaps more accurately stated, technology and the modern world shape each other in that only those technologies that are appropriate for society will ultimately survive.

The potential result of this mutual shaping of IT and society is knowledge capital, and this creates needs for knowledge management. The costs of the IT needed to provide a given level of functionality have declined dramatically over the past decade—especially within the past few years—due to the use of such technologies as broadband fiber optics, spectrum management, and data compression. A transatlantic communication link today costs one-tenth of the price that it did a decade ago, and may well decline by another order of magnitude within the next three or four years. The power of computers continues to increase and the cost of computing has declined by a factor of 10,000 over the past 25 years. Large central mainframe computers have been augmented, and in many cases replaced, by smaller, more powerful, and much more user-friendly personal computers. There has, in effect, been a merger of the computer and telecommunications industries into the IT industry and it is now possible to store, manipulate, process, and transmit voice, digitized data, and images at very little cost.

Current industrial and management efforts are strongly dependent on access to information. The world economy is in a process of globalization and it is possible to detect several important changes. The contemporary and evolving world is much more service oriented, especially in the more developed nations. The service economy is much more information and knowledge dependent and much more competitive. Further, the necessary mix of job skills for high-level employment is changing. The geographic distance between manufacturers and consumers, and between buyers and sellers, is often of little concern today. Consequently, organizations from diverse locations compete to provide products and services. Consumers potentially benefit as economies become more transnational.

The IT revolution is associated with an explosive increase of data and information, with the potential for equally explosive growth of knowledge. Information technology and communication technology have the capacity to radically change production and distribution of products and services and, thereby, bring about fundamental socioeconomic changes. In part, this potential for change is due to progressively lowered costs of computer hardware. This is associated with reduction in the size of the hardware and, therefore, with dematerialization of systems. This results in the ability to use these systems in locations and under conditions that would have been impossible just a few years ago. Software developments are similarly astonishing. The capabilities of software increase steadily, the costs of production decrease, reliability increases, functional capabilities can be established and changed rapidly, and the resulting systems are ideally and often user friendly through systems integration and design for user interaction. The potential for change is also brought about due to the use of IT systems as virtual machines, and the almost unlimited potential for decentralization and global networking due to simultaneous progress in optical fiber and communication satellite technology.

The life cycle of IT development is quite short and the technology transfer time in the new "postindustrial," or knowledge-based, society brought about by the information revolution is usually much less than that in the Industrial Revolution. Information technology is used to aid problem-solving endeavors by using technologically based systems and processes and effective systems management. Ideally, this is accomplished through

- critical attention to the information needs of humans in problem-solving and decision-making tasks; and
- provision of technological aids, including computer-based systems of hardware and software and associated processes, to assist in these tasks.

Success in IT and engineering-based efforts depends on a broad understanding of the interactions and interrelations among the components of large systems of humans and machines. Moreover, a successful IT strategy also seeks to meaningfully evolve the overall architecture of systems, the systems' interfaces with humans and organizations, and their relations with external environments.

As discussed, the most dominant recent trend in IT has been more and more computer power in less and less space. Gordon Moore, a founder of Intel, noted that since the 1950s the density of transistors on processing chips has doubled every 18 to 24 months. This observation is often called "Moore's law." He projected that doubling would continue at this rate. Put differently, Moore's law projects a doubling of computer performance every 18 months within the same physical volume. The implication of this is that computers will provide increasingly impressive processing power. The key question, of course, is what we will be able to meaningfully accomplish with all of this power.

Advances in computer technology have been paralleled by trends in communications technology. The ARPAnet emerged in the 1960s, led to the Internet Protocol in the 1970s, and the Internet in the 1980s. Connectivity is now on most desktops, e-mail has become a "must have" business capability, and the World Wide Web is on the verge of becoming a thriving business channel. The result is an emerging networking market. Telecommunications companies are trying to both avoid the obsolescence that this technology portends and figure out how to generate revenues and profits from this channel. The result has been a flurry of mergers and acquisitions in this industry.

That the price of computing has dropped to half approximately every two years over the last two decades or so is nothing short of astounding. Had the rest of the economy matched this decline in prices, the price of an automobile would be in the vicinity of $10. Organizational investments in IT have increased dramatically and now account for approximately 10% of new capital equipment investments by U.S. organizations. Roughly half of the labor force is employed in information-related activities. On the other hand, productivity growth seems to have continually declined since the early 1970s, especially in the service sector that comprises about 80% of IT investments. This situation implies needs to effectively measure IT contributions to productivity, identify optimal investment strategies in IT, and enhance IT effectiveness through knowledge management for enhanced productivity.

Although IT does indeed potentially support improvement of the designs of existing organizations and systems, it also enables fundamentally new ones, such as virtual corporations and major expansions of organizational intelligence and knowledge. It does so not only by allowing for interactivity in working with clients to satisfy present needs, but also through proactivity in planning and plan execution. An ideal organizational knowledge strategy accounts for future technological, organizational, and human concerns, to support the graceful evolution of products and services that aid clients. Today, we realize that human and organizational considerations are vital to the success of IT. This is clearly the network age of information and knowledge. One of the major challenges of today is that of capturing value in the network age. Associated with these changes are a wide range of new organizational models. Distributed collaboration across organizations and time zones is becoming increasingly common. The motivation for such collaboration is the desire to access sources of knowledge and skills not usually available in one place. The result of such new developments in IT as network computing, open systems architectures, and major new software advances has been a paradigm shift that has prompted the reengineering of organizations; the development of high performance business teams, integrated organizations, and extended virtual enterprises; and the emergence of loosely structured organizations. These are the issues we examine in the later part of this book, specifically in Chapters 7 through 10.

1.9 ECONOMIC PITFALLS IN THE ENGINEERING OF SYSTEMS

In any problem-solving effort, we must be careful to state the critical assumptions on which our developments and solution procedures are based. If this is not done, or not done carefully, it is an easy task to obtain "solutions" that are false because they are based on a set of assumptions that are different from those characteristic of the original problem or issue. In this section, we will state some of the assumptions on which our developments will be based and present some critical points associated with them.

In Chapters 2 to 4 we develop a number of results based on a perfectly competitive economy. There are six critical assumptions:

1. In the market for each and every good or service, there are a large number of relatively small producers and consumers.
2. All firms in the same industry produce similar goods of comparable quality. No consumer has any reason to prefer the output of one firm over that of another; products and services are completely standardized among firms that make the product or service; and there is no brand loyalty among consumers.
3. Resources are completely mobile. Owners of productive resources (land, capital, labor) or factor inputs to production are free to put them to whatever use they find will yield them maximum return. People can work in or sell their resources to any industry they wish. No barriers exist to prevent the establishment of a firm in any industry, or to prevent any firm from leaving an industry.

4. Each economic agent has perfect information and knowledge. All firms and all consumers know with certainty all present and future prices; each knows their own characteristics (production functions and utilities). There are no uncertainties.
5. Each economic agent (producer or consumer) is an optimizer; each acts to maximize satisfaction. Thus each consumer acts to maximize its utility and each firm acts to maximize its profits.
6. There are no price controls. Prices may move up or down freely, subject only to market pressures.

If the above six conditions hold, and they are really rather idealistic, it is easy to show that

a. prices and the quantity of goods produced and consumed are determined by the market-demand equilibrium; and
b. all goods are produced and sold at the lowest possible price, at least in the long run after transients have died out.

Surely, not all of these six conditions will always exist. We examine cases where they do not exist in our detailed treatments to follow. When condition 1 does not exist, a monopoly or monopsony may well result: Firms will have knowledge of how prices will vary with demand and will be able to control the entire quantity produced (monopoly), or suppliers of labor will be able to exert similar control (monopsony). When condition 2 does not hold, we need to consider similar products as if they were separate and develop different supply and demand relationships for the different brands of a product. The non-existence of condition 2 gives rise to the profession of marketing.

When conditions 4 and 5 do not hold, as is often the case, we need to consider various aspects of game theory and behavioral theories of the firm and of the consumer. When conditions 3 and 6 do not hold, we can impose various controls and constraints on admissible solutions. We do this in several places in this book. We do, for example, consider the effects of taxes and price controls.

In some cases additional conditions hold; then additional simplifications result. These conditions include the following:

a. Individuals are "selfish." Each person's feelings are determined only by personal consumption. People are devoid of both sympathy and envy. The fortunes and misfortunes of others do not affect a person's feelings of satisfaction or dissatisfaction.
b. Individuals are "greedy." More is always better, at least never worse. A person never achieves satiation and always feels better off by consuming more.
c. Preferences are such that the rate at which additional amounts of one good, x, may be substituted for another, y, to retain the same level of utility diminishes with increasing amounts of the good x; that is, diminishing marginal rates of substitution between goods exist. Indifference curves of constant utility are convex to the origin.

d. There are no externalities, such as cases in which the increased production of one firm results in decreased production ability for another firm.

e. There are no production processes that exhibit IRS in the sense that increased production quantity will always result in increased profits.

f. All goods and services are exchanged in markets, and all markets are in steady-state equilibrium.

g. There are no public goods; there are only private goods that are consumed by a single individual or household.

h. There is neither government taxation nor government subsidization of the production or consumption of any good or service.

If conditions 1–6 and a–h are satisfied, then we may establish the very important result that all goods and services have market prices, and that the market prices are exactly equal to the corresponding shadow prices, such that the shadow price and the market price have true social values.

Even though some of these conditions may not hold, we can often predict, by determining which assumptions are violated, the direction in which the observed price will deviate from the "shadow price" or price that represents true value. When prices do not exist, which is often the case when there are no markets for goods and services like national defense, public schools, or parks, we can suggest measurement guidelines to assist the systems analyst in making approximations. A great deal of information is needed to determine concepts like "value" and "willingness to pay," information that is generally not readily available. All of this complicates the subject of economic systems analysis. This does not make it either an impossible or a sterile subject by any means, but it does suggest that we must be keenly aware of potential limitations, especially with respect to difficulties of observation and measurement and with respect to the fact that there are indices of performance other than utility and profit, and that there is much literature that shows that humans do not have the cognitive stamina to optimize in unaided situations. This also strongly suggests the need for parsimony as well as perception in the organization of the knowledge bases of economic and other activities that support systems engineering and management in both the private and public sectors.

In a seminal paper, now three decades old, Oskar Morgenstern identified 13 points in contemporary economic theory that, if ignored, act as pitfalls in economics and in sound systems engineering applications of economic concepts. Ten of Morgenstern's points are of special concern in this book. Slightly modified in some instances to conform to the thrust of our discussions, these ten points are as follows:

1. ***Control of Economic Variables.*** The maxima of profit and utility that we so ardently seek in much of Chapters 2 to 4 may make good sense and be truly attainable if we can identify and have control over all the variables on which the maxima, or minima, depend. Often no single economic agent "controls" all of the essential parameters in a study. Game theory and distributed information control studies could, in principle, resolve difficulties due to multiple agent control issues; but these subjects introduce their own rigidities.

2. ***Revealed Preference Theory.*** In principle, it is possible to observe the behavior of a consumer operating under budget constraints and to determine the priority order or ranking of the consumer's preferences. In practice, this is extraordinarily difficult. There is, for example, no way of being sure that an observed order is complete. The time sequences in which commodities are obtained are important, as there are questions of price changes over time, different expected lines of various commodities, changes in income streams over time, and complications due to purchasing "bundles" of commodities at one time rather than individual goods.
3. ***Pareto Optimality.*** Much of our discussion in Chapter 5 will be based on the assumption that an improvement or increase in one person's utility with no decrease in anyone else's utility is an improvement for all of society. In the simplest case, the utility of each individual is independent of everyone else's. A question of interest is: How do we find out that there has been an improvement or a decrease? We can ask people or we can make a judgment based on observation. But how do we know that people are not denying that they are better off in the hopes of getting still greater returns? Or perhaps people are not able to state truthfully whether they are better off or not under some changed resource allocation. If one infers that a person is better off with some change in allocation of resources, is that not making the interpersonal comparison of utilities the concept of Pareto optimality seeks to avoid? Our view here is that we are, and we resolve this dilemma by asserting that interpersonal comparisons of utility are necessary for any practically successful group decision-making effort.
4. ***Tatonnement.*** In the assumed behavior that leads to general economic equilibrium, initially assumed strategies for buyers and sellers will either lead to equilibrium or they will be modified. This mechanism of groping for stable prices is known as tatonnement. This involves the use of various production rules, heuristics, or standard operating policies. But will the equilibrium be stable and unique? Even if the equilibrium is stable and unique, will it be reached in any finite time? We will not be able to develop dynamic economic models to the extent needed to answer these questions here. Generally accepted answers seem to be that any realistic economic model will exhibit stable equilibria. The question of multiple equilibria is an interesting one, as it offers the opportunity to alter social and national welfare for the better by shifting the economy from one stable equilibrium point to another. The existence of multiple equilibria would require a sudden jump in economic variables such as price. This seems to have occurred, for example, with respect to gold and oil prices over the

past decade. Does this indicate a jump-type phenomenon indicating multiple equilibria, such as could be modeled with catastrophe theory? Or does it represent the faster than normal change in a single stable equilibrium point over time? Either view is possible, although the latter seems more so in keeping with most economic thought. And what about the time required to reach a stable equilibrium? Is this within a human lifetime? Will the equilibrium that is reached be a perfectly competitive one, or will coalitions of individuals be formed who agree on prices at which they will sell or purchase goods? Questions such as these need, in effect, to be answered in the process of issue formulation and impact analysis, which are essential parts of economic systems analysis.

5. ***Fixation on Free or Perfect Competition.*** Most of our work is based on what has traditionally been called "perfect competition." The popular use of the word *competition* implies struggle, maneuvering, bargaining, negotiation, compromise, and conciliation. But free competition, as used in the traditional economic sense, implies that no one really has any influence on anything; everyone accepts prices as given and adapts to maximize utility and/or profit. Surely, a lesson in this is that we are dealing with what we may call *economic rationality* only in our work here. We do this with no shame at all, since economic rationality is very important. But we must remember that while economic rationality presumes technological rationality, there are other forms of rationality—social, political, and legal, for example—that are noncommensurate or at least not necessarily commensurate with economic rationality. Each must be considered in effective systems engineering practice.
6. ***Resource Allocation.*** Most of the resource allocation efforts that we will discuss assume perfect competition conditions. But there are monopolies, oligopolies, monopsonies, and oligopsonies, as well as governments, that use bargaining, negotiation, and finally voting to allocate resources, rather than the economic "market mechanism." These influences are surely important and we cannot regard answers obtained using the methods of economic systems analysis only as necessarily *the* answers, for we do not live in an economic system, but in a socio-legal-political economic system.
7. ***Substitution.*** A good that is a substitute for another good is said to have the same (economic) value. We must be careful with this concept, for while it is often valid, a particular context may make it not valid. In particular, the relationship between one good and others that are present in the commodity bundle is important, as is the form of the utility function. If a utility function is given by $U = x_1^{0.5} x_2^2$, then the utility of $\mathbf{x}^T = [1 \;\; 2]$ is the same as that of $\mathbf{x}^T = [16 \;\; 1]$. This does not say that 15 units of x_1 are worth 1 unit of x_2, but, rather, at 2 units of x_2

and 1 unit of x_1, a decrease of 1 unit of x_2 is balanced out by a gain of 15 units of x_1. It would seem that only in the additive form $U = a_1x_1 + a_2x_2$ could we talk about so many units of x_1 being worth 1 unit of x_1.

8. ***Supply and Demand.*** We shall spend much time in the beginning chapters deriving supply curves of products for firms and factor inputs to production for consumers, and demand curves of products for consumers and factor inputs to production for firms. The notion of time is curiously absent, yet obviously time is present. We desire food and newspapers every day, whereas we purchase automobiles and stereos only every few years. Only if we are careful with respect to the consideration of time periods will we be able to obtain the correct interpretation of our supply–demand curves. Suppose, for example, that the demand curve for product x is given by $p = a - bx$. If the price is initially p_1, we buy a quantity $x_1 = (a - p_1)/b$. Suppose the price suddenly drops to p_2. Will we buy quantity $x_2 = (a - p_2)/b$? The answer is generally no, since we have just bought quantity x_1 at price p_1. We should expect to, perhaps, purchase some additional amount $0 \leq \overline{x_2} \leq x_2 - x_1$ such that the total amount purchased will be $x_1 \leq x_1 + \overline{x_2} \leq x_2$. Only if we purchase quantity x_1 at price p_1 in one demand period and our need for the good is reconstituted, and if our income remains the same such that the demand remains the same, will we purchase amount x_2 at price p_2 when the time shifts to a different time period. Different people have different reconstitution time periods. The important notion of consumer surplus, which we will find very useful for cost–benefit studies, is very affected by this notion. What we have said here is equally applicable to supply curves and producer surplus concepts.
9. ***Indifference Curve Concepts and Uncertainty.*** Generally we base indifference curves on concepts of value and preference for goods that are received with certainty. This is a fundamental part of the theory of the consumer. These notions of value have been shown to be correct under conditions of certainty, but need to be modified somewhat to allow the consideration of risk. Basically, there is a relative risk aversion function that can be used to transform the value function for outcomes received with certainty to outcomes received with some known probability. Attempts to fully explore these notions would take us into the very interesting and important subject of decision analysis, which we will explore later in this book.
10. ***Theory of the Firm.*** Our theory of the firm is based on productive units that maximize their profit. This concept fits well for many producers of "hardware." But what about the service sector? There it may fit well or hardly at all. Maximizing profit is the primary objective of many consulting firms, but surely not all of them. Is the primary objective of a ballet company or theater group to maximize profit? Surely not! And even if the foremost desire of the owner of a firm making hardware is

to maximize profit, is it the objective of all of the managers of the firm? Probably not! There are behavioral theories of the firm and satisficing theories of human behavior that add a greater amount of reality and rationality to the behavior of the firm than that exclusively contained in the economically rational theory that we develop here. It is reasonable for the reader to ask, then why are we studying all of this economic rationality? The answer is that what we develop here is one viewpoint on rationality and a very important one. It is more developed at present than political rationality concepts, social rationality concepts, legal rationality concepts, and organizational rationality concepts. These other rationality concepts can and must, and do, augment economic rationality concepts; they cannot replace it.

In Chapters 2 through 5, we will discuss many of these pitfalls from the vantage point of classic microeconomic systems analysis. Chapter 6 is concerned with many of these issues from the perspective of systems engineering program and project management. In a real sense, Chapter 6 is a transition chapter that also brings in some of the issues associated with information and knowledge economics as driven by the IT revolution. The last three chapters are concerned with information and knowledge as a fourth fundamental economic resource that now affects the engineering and management of systems of all types. Chapter 10 briefly discusses some of the contemporary issues we have not been able to explore in this book.

1.10 SUMMARY

In this chapter we described some of the rationale behind this book and provided an introduction to many of the concepts to follow. We will hopefully have an interesting journey that is relevant to the engineering of economic and productive systems of all types, especially those that are information and knowledge intensive.

PROBLEM

1. Pick a contemporary issue of interest to you that has strong economic components. Describe the concerns associated with this issue from the perspectives discussed in this chapter.

BIBLIOGRAPHY AND REFERENCES

Many references to contemporary works in economic systems analysis could be cited here. Since these are cited in the chapters to follow, they will not be repeated here.

Discussions of systems engineering may be found in

Sage AP. Systems engineering. New York: Wiley; 1992.

Sage AP. Systems management for information technology and software engineering. New York: Wiley; 1995.

Sage AP, Armstrong JE, Jr. An introduction to systems engineering. New York: Wiley; 2000.

Sage AP, Palmer JD. Software systems engineering. New York: Wiley; 1990.

Sage AP, Rouse WB, editors. Handbook of systems engineering and management. 2nd ed. New York: Wiley; 2009.

Section 1.9 is based in part on the seminal paper

Morgenstern O. Thirteen critical points in contemporary economic theory: an interpretation. J Econ Lit 1972;10:1163–1189.

Other classic references that discuss contemporary issues of relevance to our book include the following. We will provide a variety of contemporary references in the chapters to follow.

Bell D. The coming of post industrial society. New York: Basic Books; 1973.

Bradley SP, Nolan RL. Sense and respond: capturing value in the network age. Boston, MA: Harvard Business School Press; 1998.

Brynjolfsson E, Yang S. Information technology and productivity: a review of the literature. In: Yovitz M, editor. Advances in computers, vol. 43. New York: Academic Press; 1996. pp. 179–214.

Dertouzos M. What will be: how the new world of information will change our lives. New York: Harper Collins; 1997.

Hope J, Hope T. Competing in the third wave: the ten key management issues of the information age. Boston, MA: Harvard Business School Press; 1997.

Shapiro C, Varian HR. Information rules: a strategic guide to the network economy. Boston, MA: Harvard Business School Press; 1999.

Toffler A. The third wave. New York: Morrow, Bantam Books; 1980 and 1991.

Toffler A, Toffler H. Creating a new civilization. Atlanta, GA: Andrews and McNeel; 1995.

CHAPTER 2

PRODUCTION AND THE THEORY OF THE FIRM

2.1 INTRODUCTION

We will use the word *production* or *productivity* in a very general sense to denote all activities that satisfy consumer demand for goods and services. In the classic use of the term, production is associated with the use of natural resource inputs to produce finished products that are *subsequently* marketed to consumers through various service efforts. In the initial part of this book, we will be concerned with this classic interpretation. In the later part, we will also consider production of inherently information and knowledge intensive products, such as software. Production is therefore one of the basic components of micro-economic theory. A (classic) *firm* is a unit that uses natural resource inputs, capital, and labor to produce output goods or services for purchase by consumers. The economic problem faced by a production unit or a firm is that of determining the quantity of output to produce and the amounts of the various input factors to be used in the production process. This will depend, of course, on technological relationships between the prices or costs of input factors, or input supplies, and the price that can be obtained for the output quantity, which is a function of the demand for the product of the firm. This is a problem in resource allocation that involves

1. the technology of the production process or the production function,
2. the price, or equivalently costs, of the input resource quantities or input factors such as labor, natural resources, and capital, and
3. the price of and demand for the output quantity.

Figure 2.1 is a block diagram of the firm as an economic institution and it shows the role of the three elements above. To obtain an economic theory of the firm we will need to discuss each of these elements, which is the objective of this chapter.

We will use the term *producer* to denote a decision-making entity that converts commodity bundles or inputs, by means of production, into other commodity bundles or outputs. We will generally assume that the producer

Economic Systems Analysis and Assessment,
by Andrew P. Sage and William B. Rouse

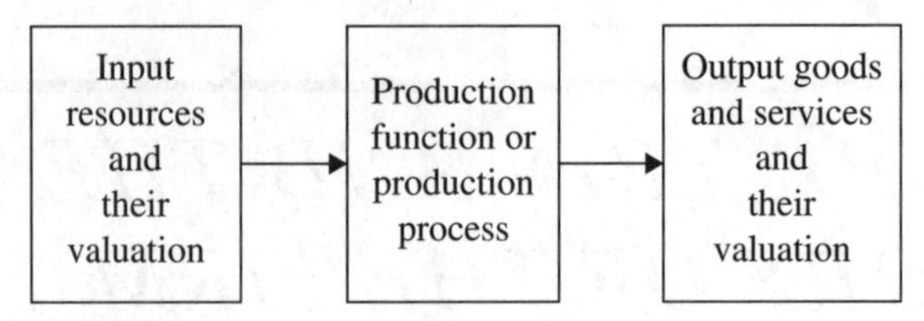

Figure 2.1. Block Diagram of the Production Process.

desires to maximize the profits of the firm. We will define *profit* to be the difference between the revenue that the firm obtains for the products it produces minus all of the costs of producing the products. The particular decision variables that are subject to adjustment by the producer depend, of course, on the type of production issue that is being considered. In the simplest problem, the decision involves setting the output level; in a slightly more complicated problem both the output level and the price may be set by the firm. This setting of the price by the firm violates assumptions concerning "perfect competition." It introduces the need for imperfect competition—in this case monopoly—considerations.

2.2 THE PRODUCTION FUNCTION

We begin our discussion of production with the simplest case of a single-product firm. We will then briefly mention how the results of a single-product firm can be extended to include the multiproduct firm, where the same firm generally produces a number of goods.

We assume that our single-product firm produces an output from several inputs. We will use the symbol q to denote the (scalar) output quantity. We assume that x_i is the ith input quantity to the production process. Thus x_i represents the various input resources or input factors to the production process. These resources will generally be of three distinguishable types

1. raw materials or natural resources (traditionally often denoted by M),
2. capital, such as investment in production machinery (traditionally often denoted by K or C), and
3. labor (traditionally often denoted by L).

In many economic studies, particularly traditional ones, only an economic and technological valuation of these inputs is attempted. However, it is becoming increasingly recognized that psychological, social, and political valuations are of great importance. In our later chapters we will consider these to some extent, although a full development of a behavioral theory of the firm is, while very important, beyond the scope of this book.

The production function, traditionally denoted by f, is a specific mapping or technological relation between the input variables x_i and the production process and the output quantity produced, denoted by q. The specific mapping chosen is that presumably unique number that represents the maximum output that can be produced for a given set of input quantities.

Example 2.1: If we have two input factors x_1 and x_2 to the production process, then we say that the product output level or quantity of production q is, for the ith production technique P_i,

$$q = P_i(x_1, x_2) \tag{2.1}$$

We define the following: the output product q is the number of suits of clothing manufactured; x_i are the input factors, such that x_1 is the number of hours of input labor to the manufacturing process and x_2 the number of square meters of input fabric; and a_i are the coefficients of production, such that a_1 is the number of hours of labor needed to make a suit and a_2 the number of square meters of fabric necessary to make a suit. Then we see that the output production level is the quantity x_1/a_1 or x_2/a_2, which is the smallest. We have for our specific production process or technique

$$q = \min\left(\frac{x_1}{a_1}, \frac{x_2}{a_2}\right) \tag{2.2}$$

This is a very restrictive production process. Normally it would be possible to trade off clothing for labor such that if labor costs rose, we could cut the amount of labor, say, by a less restrictive process, which could increase the amount of fabric used per suit. We might then obtain a production function like

$$q = f(x_1, x_2) = cx_1 + \mathrm{d}x_2$$

Here we could maintain a constant amount of production by decreasing x_1 and increasing x_2. We would obtain this production function from a class of production techniques by finding the one that yields the maximum output for a given set of inputs. This would be of the form

$$q = f(x_1, x_2) = \max_j [P_j(x_1, x_2)]$$

We will generally assume a production function of the form

$$q = f(\mathbf{x}) = f(x_1, x_2, \ldots, x_n) \tag{2.3}$$

in which there are n production input factors defined by the n-vector

$$\mathbf{x} = \begin{bmatrix} x_1 \\ x_2 \\ \vdots \\ x_n \end{bmatrix}$$

These factors are determined such that there is no wastage of the input production factors. We do this by defining the production function f such that the following two conditions are satisfied.

1. For any given fixed level of input factors, the technique of production that gives the maximum output production is selected.
2. At any given fixed output production level q, the consumption of each input factor is at a minimum.

We ensure condition 1 by the requirement that, for fixed $\mathbf{x}$,

$$q = \max_j [P_j(\mathbf{x})] = f(\mathbf{x}) \tag{2.4}$$

We accomplish this by selecting the best production technique or process. We ensure condition 2 by the further requirement that the output production must increase if we increase the resource input to or factors of production. Thus, we require that $q^1 \geq q^2$, or in equivalent terminology $f(x^1) \geq f(x^2)$, whenever[1] $x^1 \geq x^2$.

The desirability of these nonwaste conditions is physically apparent. In most physical production processes it is possible that further technological progress may reduce the amount of input products required for a given output or improve production efficiency or reduce waste. Also, there may be constraints, such as those imposed by worker safety, that will not allow complete freedom in process selection. All of these can, should, and will be incorporated into our theory of production. All that we require by our nonwaste condition is that under the existing production constraints there be no waste production factors.

Example 2.2: We return to our previous example, in which the production function for a particular technology was selected as

$$q = \min\left(\frac{x_1}{a_1}, \frac{x_2}{a_2}\right) \tag{2.5}$$

where, as before,

q = output production level
x_1 = input factor of labor
x_2 = input factor of fabric
a_1 = labor coefficient of production
a_2 = fabric coefficient of production

For this example, our nonwaste requirement says that we would be foolish to use more than $a_1 q$ units of labor or $a_2 q$ units of fabric, where q is determined from Equation 2.5 for a specific x_1 and x_2.

[1]We say that vector $\mathbf{x}^1$ is greater than or equal to vector $\mathbf{x}^2$ when each component of $\mathbf{x}^1$ is greater than or equal to each component of $\mathbf{x}^2$, i.e., $x_1^1 \geq x_1^2, x_2^1 \geq x_2^2, \ldots, x_n^1 \geq x_n^2$.

To expand on this discussion further, let us suppose that

$a_1 = 4\,\text{h}$ of labor per suit
$a_2 = 3\,\text{m}^2$ of fabric per suit.

We can now use Equation 2.5 to draw a production *isoquant*, or curve showing the values of x_1 and x_2 that produce a constant output q. We do this for several constant values of q and easily obtain the set of isoquants shown in Fig. 2.2. Every value of x_1, x_2 that lies on one side of a given isoquant must necessarily result in a higher output production, whereas every value of x_1, x_2 on the other side must result in a lower production. The line $q = x_1/4 = x_2/3$ defines the minimum input factors for this production technique. Clearly, we should operate on this line to satisfy the nonwaste conditions. ■

Two important axioms are assumed to be satisfied by a production function. The first axiom is that there is an *economic region* in which a decreased output cannot result from an increased input. This axiom is entirely equivalent to and follows from condition 2 of noninput factor waste. This leads to an important quantity called the marginal productivity, **MP(x)**, which is the first partial derivative or gradient of the production function. We have

$$\mathbf{MP}(\mathbf{x}) = \frac{\partial f(\mathbf{x})}{\partial \mathbf{x}} = \begin{bmatrix} \text{MP}_1(\mathbf{x}) \\ \text{MP}_2(\mathbf{x}) \\ \vdots \\ \text{MP}_n(\mathbf{x}) \end{bmatrix} = \begin{bmatrix} \dfrac{\partial f(\mathbf{x})}{\partial x_1} \\ \dfrac{\partial f(\mathbf{x})}{\partial x_2} \\ \vdots \\ \dfrac{\partial f(\mathbf{x})}{\partial x_n} \end{bmatrix} \tag{2.6}$$

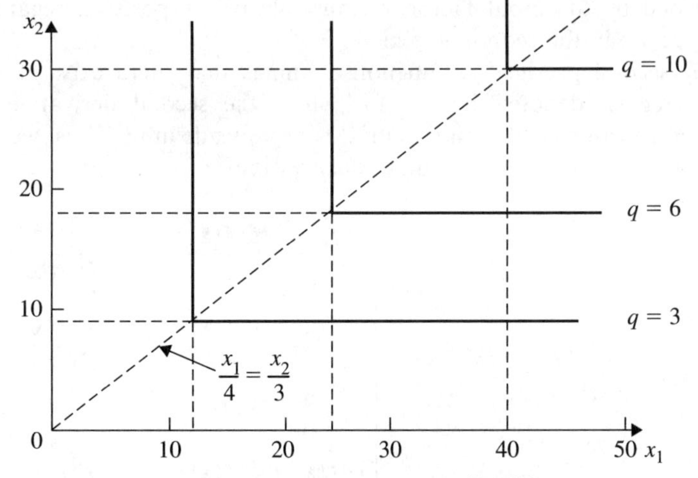

Figure 2.2. Isoquants for Example 2.2, with $q = \min\left(\frac{x_1}{4}, \frac{x_2}{3}\right)$.

An important feature of each component of the marginal productivity or marginal return that follows from condition 2 is that it is positive, or surely nonnegative to be more precise, in the economic region. We can easily illustrate this by any of several procedures. One particular approach that is very insightful is to consider a graph of the output quantity produced for various values of x_1 with all the other factor inputs held constant. We will use the symbol $\mathbf{x}_{\bar{i}}$ to denote all the components of the vector $\mathbf{x}$ except x_i. We may write, using this definition,

$$q = f(\mathbf{x}) = f(x_i, \mathbf{x}_{\bar{i}}) \tag{2.7}$$

Now suppose that we fix the value of the vector $\mathbf{x}_{\bar{i}}$ at some particular value of $\mathbf{x}_{\bar{i}}$. By increasing x_i to $x_i + \Delta x_i$, for Δx_i positive, we necessarily do not decrease the output q and thus have

$$f(x_i + \Delta x_i, \mathbf{x}_{\bar{i}}) \geq f(x_i, \mathbf{x}_{\bar{i}}), \quad \Delta x_i \geq 0$$

The definition of marginal productivity is equivalent to

$$\frac{\partial \mathrm{MP}_1(\mathbf{x})}{\partial x_1} = \lim_{\Delta x_1 \to 0} \left(\frac{f(x_1 + \Delta x_1, \mathbf{x}_{\bar{i}}) - f(x_1, \mathbf{x}_{\bar{i}})}{\Delta x_1} \right)$$

Thus, we immediately see that each component of the marginal productivity is always nonnegative, and generally always positive, in the economic region.

A quantity similar to the marginal productivity that is of importance is the average productivity or average return, defined by the expression

$$\mathbf{AP}_i(\mathbf{x}) = \frac{q}{x_i} = \frac{f(x_i, \mathbf{x}_{\bar{i}})}{x_i} \tag{2.8}$$

This average productivity, the total productivity with all the input factors but x_i fixed divided by the input factor x_i, must always be positive, regardless of whether we are in the economic region.

Our second production function axiom is that there exists a *relevant economic region*, denoted by R_p, for which the second derivative of the production function, or Hessian matrix, is negative definite. Thus, we have in this relevant economic region a matrix expression

$$\mathbf{H}(\mathbf{x}) = \frac{\partial^2 f(\mathbf{x})}{\partial \mathbf{x}^2} = \frac{\partial \mathrm{MP}(\mathbf{x})}{\partial \mathbf{x}} = \begin{vmatrix} \frac{\partial^2 f(\mathbf{x})}{\partial x_1^2} & \frac{\partial^2 f(\mathbf{x})}{\partial x_1 \partial x_2} & \cdots & \frac{\partial^2 f(\mathbf{x})}{\partial x_1 \partial x_n} \\ \frac{\partial^2 f(\mathbf{x})}{\partial x_2 \partial x_1} & \frac{\partial^2 f(\mathbf{x})}{\partial x_2^2} & \cdots & \frac{\partial^2 f(\mathbf{x})}{\partial x_2 \partial x_n} \\ \vdots & \vdots & & \vdots \\ \frac{\partial^2 f(\mathbf{x})}{\partial x_n \partial x_1} & \frac{\partial^2 f(\mathbf{x})}{\partial x_n \partial x_2} & \cdots & \frac{\partial^2 f(\mathbf{x})}{\partial x_n^2} \end{vmatrix}$$

that is negative definite.[2] The production function $q = f(\mathbf{x})$ is therefore strictly concave (for every nonnegative output q) in the relevant economic region.

In the relevant economic region for a production function, we can show that the main diagonal components of this Hessian matrix are negative, in that

$$\frac{\partial^2 q}{\partial x_i^2} = \frac{\partial^2 f(\mathbf{x})}{\partial x_i^2} = \frac{\partial \mathrm{MP}_i(\mathbf{x})}{\partial x_i} < 0 \tag{2.9}$$

for all i in the input factor set. This is physically a very significant result. It is a statement of the *law of diminishing marginal productivity* or the *law of diminishing marginal returns.* As we add more and more input x_i to a process with all the other input factors, $\mathbf{x}_{\bar{i}}$, fixed, we must eventually reach a point at which the marginal productivity for that input x_i drops to zero and we reach the boundary of the relevant economic region.

We illustrate these concepts graphically in Fig. 2.3, which shows a typical production curve and the variation of production with a single input factor, as in Equation 2.7. Figure 2.3 also illustrates the marginal productivity $\mathbf{MP_1(x)}$

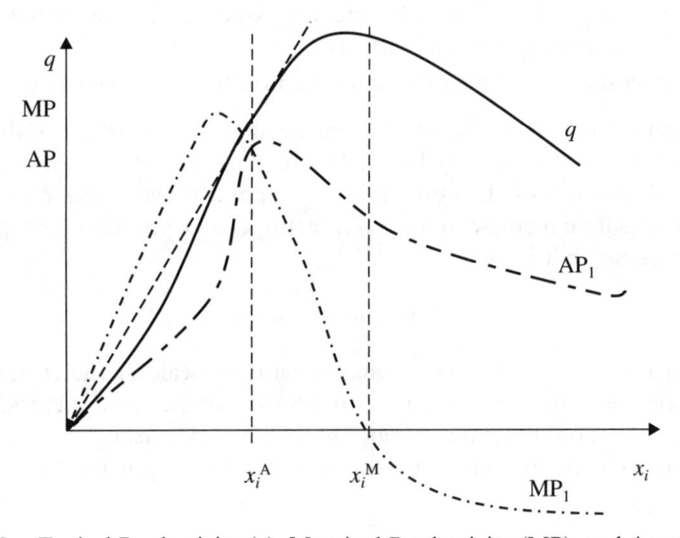

Figure 2.3. Typical Productivity (q), Marginal Productivity (MP), and Average Productivity (AP) Curves.

[2]If $\mathbf{H}(x)$ is negative definite, $f(\mathbf{x})$ is strictly concave. If $\mathbf{H}(\mathbf{x})$ is negative semidefinite, then $f(\mathbf{x})$ is (quasi)concave. Normally, concavity is all that is required in the relevant economic region. We recall that a convex function $f(\mathbf{x})$ is one in which for $0 \le a \le 1$, we have $af(x_1) + (1-a)f(x_2) \ge f[a(x_1) + (1-a)x_2]$. A function $f(\mathbf{x})$ is *concave* if for $0 \le a \le 1$, we have $f[ax + (1-a)y] \ge a f(x) + (1-a)f(y)$. The concept of a convex set is also of interest. If S is convex and if $x \in S$, $y \in S$, then we have $ax + (1-a)y \in S$, $\forall \boldsymbol{a} \in (0, 1)$. The graph of a concave function will always lie below all tangent lines for the function. A function is quasi-concave if $A - \{x{:}f(x) \ge t\}$ is a convex set.

and the average productivity $\mathbf{AP}_1(\mathbf{x})$ for the hypothetical production function chosen. We note several interesting items concerning Fig. 2.3:

1. For small x_1 factor inputs the marginal and average productivities increase with increasing factor input.
2. There is a maximum factor input x_1^M, and if x_1 is increased above this value, we move outside of the relevant economic region and production is decreased.
3. Above x_1^M the marginal productivity is negative. This is another symptom of our being outside the relevant economic region.
4. For values of x_1 between zero and x_1^A the marginal productivity is greater than the average productivity. Further increases in x_1 result in the marginal productivity becoming less than the average productivity. Often this will be undesirable. The value of x_1 at which the maximum average productivity occurs is the value beyond which the marginal productivity is less than the average productivity.
5. There are three distinct productivity regions for this input factor x_1:

   a. a region in which the marginal productivity is greater than the average productivity,
   b. a region in which the marginal productivity is positive but less than the average productivity, and
   c. a region in which the marginal productivity is negative.

When all the factor inputs change by the same constant multiple, the concept of *return to scale* may be used to further characterize the production function. A production function is said to possess *constant return to scale* characteristics if an increase in the input factor $\mathbf{x}$ to $a\mathbf{x}$ produces a proportional output increase, that is, if

$$f(a\mathbf{x}) = af(\mathbf{x}), \quad a > 1 \tag{2.10}$$

If a production function possesses constant return to scale characteristics, then a 50% increase in all the inputs will result in a 50% increase in output productivity.

A production function is said to possess *decreasing return to scale* characteristics if the output increases by a smaller proportion than does the input, that is, if

$$f(a\mathbf{x}) < af(\mathbf{x}), \quad a > 1 \tag{2.11}$$

Similarly, a production function possesses *increasing return to scale* characteristics if changing all the inputs by a constant multiple a changes the output by more than this proportion, that is, if

$$f(a\mathbf{x}) > af(\mathbf{x}), \quad a > 1 \tag{2.12}$$

Production functions may possess increasing returns to scale for some points in input factor space, constant returns to scale for other points in input factor space, and decreasing returns to scale for still other points in input factor space. Generally, for sufficiently large $\mathbf{x}$, production of all realistic classical natural physical products will exhibit decreasing return to scale. In the very important case of goods strongly dependent on information, which we will examine in the later chapters, this is not the case and increasing returns to scale may well be present.

The case of constant return to scale is particularly important, as production functions with this property are mathematically convenient. Fortunately, a great many production functions do possess constant or nearly constant return to scale characteristics. Two things that will often prohibit constant return to scale are limited input factor supplies that cannot be greatly increased to accommodate greatly expanded outputs and the fact that some inputs only occur in integer units.

Example 2.3: It is convenient to discuss return to scale in terms of the *homogeneous production function of degree s* for which

$$q = f(\mathbf{x})$$

$$a^s q = f(a\mathbf{x})$$

Here, a is greater than 1. When $s > 1$, we have increasing returns to scale; when $s < 1$, we have decreasing returns to scale; and when $s = 1$, we have the constant return to scale case. We may be tempted to conclude that a linear production function is the only one that has the constant return to scale property. It is the case that any linear function that has zero output for zero input

$$q = \sum_{i=1}^{n} b_i x_i$$

certainly does have constant return to scale, in that

$$f(a\mathbf{x}) = \sum_{i=1}^{n} b_i(ax_i) = c\sum_{i=1}^{n} b_i x_i = af(\mathbf{x})$$

However, the nonlinear production function

$$f(\mathbf{x}) = x_1^{b_1} x_2^{b_2}$$

is also a homogeneous production function, in that

$$f(a\mathbf{x}) = a^{b_1 + b_2} f(\mathbf{x})$$

This is a linear homogeneous production function if $b_1 + b_2 = 1$. Thus we see that functions that are not linear may also possess constant return to scale. ■

Elasticity of production or production elasticity or output elasticity is a measure of the return to scale as a function of the input factor vector $\mathbf{x}$. It is defined as the ratio of the marginal productivity to the average productivity with respect to a change in the ith input. This factor elasticity[3] of production is written as

$$\varepsilon_i(\mathbf{x}) = \frac{\mathrm{MP}_i(\mathbf{x})}{\mathrm{AP}_i(\mathbf{x})} \tag{2.13}$$

We can use the definitions of marginal and average productivity, Equations 2.6 and 2.8, such that the foregoing becomes

$$\varepsilon_i(\mathbf{x}) = \frac{\partial f(\mathbf{x})/\partial x_i}{f(x_i, x_{\bar{i}})/x_i} = \frac{\partial f(\mathbf{x})/f(\mathbf{x})}{\partial x_i/x_i} = \frac{\partial(\ln q)}{\partial(\ln x_i)} \tag{2.14}$$

Factor elasticity has a number of interpretations. Perhaps the most interesting one is that it is the proportional variation in the output divided by the proportional variation in the factor input. Thus if x_i increases by 2% and the output increases by 6%, the elasticity is 3. Of course, the definition we have used of elasticity, that is, the ratio of the marginal productivity to the average productivity, is important and has much physical meaning.

For the vector case, we can formally define the scalar elasticity as the inner or scalar product of the marginal productivity and the reciprocal average product vector defined by

$$\mathbf{RAP}(\mathbf{x}) = \frac{\mathbf{x}}{f(\mathbf{x})} \tag{2.15}$$

We thus have for the elasticity

$$\varepsilon(\mathbf{x}) = \mathbf{MP}^{\mathrm{T}}(\mathbf{x})\mathbf{RAP}(\mathbf{x}) = \left[\frac{\partial f(\mathbf{x})}{\partial \mathbf{x}}\right]^{\mathrm{T}} \frac{\mathbf{x}}{f(\mathbf{x})} \tag{2.16}$$

which becomes

$$\varepsilon(\mathbf{x}) = \sum_{i=1}^{n} \frac{\partial f(\mathbf{x})}{\partial x_i} \frac{x_i}{f(\mathbf{x})} \tag{2.17}$$

Thus we see that the total elasticity is just the sum of the local factor input elasticities and given by

$$\varepsilon(\mathbf{x}) = \sum_{i=1}^{n} \varepsilon_i(\mathbf{x}) \tag{2.18}$$

where $\varepsilon_i(\mathbf{x})$ is defined by Equation 2.14.

[3]We will encounter several types of elasticity in this book. In all cases, elasticity will be the ratio of a marginal quantity to an average quantity. We will generally use subscripts and superscripts to denote the type of elasticity when this distinction is needed.

Example 2.4: We consider the linear production function of the form

$$q = \sum_{i=1}^{n} b_i x_i = f(\mathbf{x})$$

The individual production factor elasticity is given by its fundamental definition, which is the ratio of the marginal productivity to the average productivity, as

$$\varepsilon_j(\mathbf{x}) = \frac{\mathrm{MP}_j(\mathbf{x})}{\mathrm{AP}_j(\mathbf{x})} = \frac{\partial f(\mathbf{x})/\partial x_j}{f(\mathbf{x})/x_j} = \frac{b_j x_j}{\sum_{i=1}^{n} b_i x_i}$$

and the total factor elasticity of production is

$$\varepsilon(\mathbf{x}) = \sum_{j=1}^{n} \varepsilon_j(\mathbf{x}) = 1$$

This is as we should expect, since the linear production function possesses constant return to scale and is a homogeneous function of degree 1.

For the homogeneous production function of degree $b_1 + b_2$

$$q = f(\mathbf{x}) = x_1^{b_1} + x_2^{b_2}$$

we have for the factor elasticity

$$\varepsilon_j(\mathbf{x}) = b_j, \quad j = 1, 2$$

and we have for the total elasticity

$$\varepsilon(\mathbf{x}) = \sum_{j=1}^{2} \varepsilon_j(\mathbf{x}) = b_1 + b_2$$

The total elasticity is 1 if $b_1 + b_2 = 1$, and this is the requirement we had obtained in our previous example for constant return to scale. ■

The *marginal rate of substitution* or the rate of technical substitution of one input product for another is an important concept, as it allows insight into the alternate factor input possibilities that result in the same output. Along a specific isoquant of constant production q^k, given by

$$q^k = f(\mathbf{x})$$

we must have $dq^k = 0$. We will constrain all factor inputs but x_i and x_j to be constant. Thus we have from the general relation along an isoquant

$$\mathbf{MP}^{\mathrm{T}}(\mathbf{x})d\mathbf{x} = 0$$

the relation

$$dq^k = 0 = \frac{\partial f(\mathbf{x})}{\partial x_i} dx_i + \frac{\partial f(\mathbf{x})}{\partial x_j} dx_j = \text{MP}_i(\mathbf{x}) dx_i + \text{MP}_j(\mathbf{x}) dx_j \tag{2.19}$$

We rearrange this relation and obtain for the marginal rate of substitution along an isoquant

$$\text{MRS}_{ij}(\mathbf{x}) = -\frac{dx_j}{dx_i} = \frac{\partial f(\mathbf{x})/\partial x_i}{\partial f(\mathbf{x})/\partial x_j} = \frac{\text{MP}_i(\mathbf{x})}{\text{MP}_j(\mathbf{x})} \tag{2.20}$$

An interesting property of the marginal rate of substitution $\text{MRS}_{ij}(\mathbf{x})$ is the fact that it is the ratio of the marginal productivities of factors i and j. This is of considerable value, as it allows us to determine the optimum economic production factor inputs.

We may also define an *elasticity of substitution* as the expression

$$\begin{aligned}\varepsilon_{ij}(\mathbf{x}) &= -\frac{d[\ln(x_j/x_i)]}{d\{\ln[\text{M}_{pi}(\mathbf{x})/\text{MP}_j(\mathbf{x})]\}} \\ &= \frac{d(x_j/x_i)/x_j/x_i}{d\text{MRS}_{ij}(\mathbf{x})/\text{MRS}_{ij}(\mathbf{x})} = \frac{\text{MRS}_{ij}(\mathbf{x})}{x_j/x_i}\frac{d(x_j/x_i)}{d\text{MRS}_{ij}(\mathbf{x})}\end{aligned} \tag{2.21}$$

This elasticity represents the proportional change in the ratio of the inputs i and j divided by the proportional change in the ratio of the marginal products or in the marginal rate of substitution. We will be able to show, using the results of the next section, that the elasticity of substitution is also the ratio of the change in the relative factor inputs to the associated change in their relative wages or prices. This is given by

$$\varepsilon_{ij}(\mathbf{x}) = \frac{\partial(x_j/x_i)}{\partial(w_j/w_i)}\frac{w_j/w_i}{x_j/x_i}$$

Example 2.5: For the linear production function

$$q = \sum_{i=1}^{n} b_i x_i = f(\mathbf{x})$$

we obtain the marginal rate of substitution, Equation 2.20, from the marginal productivities $\text{MP}_i(\mathbf{x}) = b_i$ as

$$\text{MRS}_{ij}(x) = \frac{\text{MP}_i(\mathbf{x})}{\text{MP}_j(\mathbf{x})} = \frac{b_i}{b_j}$$

This relation tells us that to maintain a constant output production q we can substitute input products x_i and x_j according to the relationship

$$\mathrm{d}x_j = -\frac{b_i \mathrm{d}x_i}{b_j}$$

This relation is, of course, valid only for "small enough" changes. In Examples 2.3 and 2.4, we found that a linear production function has constant return to scale and a production elasticity of 1. In this example, we have found that the marginal rate of substitution $\mathrm{MRS}_{ij}(\mathbf{x})$ is b_i/b_j. Since this is a constant, the elasticity of substitution is infinite. Such a large elasticity means that even the smallest change in the rate of marginal substitution, which is constant everywhere, will produce an extraordinarily large change in the ratio of input products that determines the proportional mixes of the input factors.

For the homogeneous production function of degree $b_1 + b_2$,

$$q = f(\mathbf{x}) = x_1^{b_1} x_2^{b_2}$$

we have for the marginal rate of substitution

$$\mathrm{MRS}_{12}(x_1, x_2) = \frac{b_1 x_2}{b_2 x_1}$$

which indicates that input factor substitutability does depend on the present factor input level.

The elasticity of substitution is obtained from Equation 2.21 as $\varepsilon_{12}(x_1, x_2) = 1$. The fact that the elasticity of substitution is 1 throughout the complete range of input factors is desirable, because it ensures reasonable substitutability of one input factor for another over a wide range of inputs. This particular production function is a special case of the Cobb–Douglas production function, which is frequently used in microeconomic analysis. ■

In this section, we have introduced the subject of production and the theory of the firm. A number of examples have been presented that deal with two frequently used types of production functions: the linear production function

$$q = f(\mathbf{x}) = \mathbf{b}^{\mathrm{T}}\mathbf{x} = \sum_{i=1}^{n} b_i x_i \tag{2.22}$$

and the Cobb–Douglas production function

$$q = f(\mathbf{x}) = b_0 \prod_{i=1}^{n} x_i^{b_i} = b_0 x_1^{b_1} x_2^{b_2} \ldots x_n^{b_n} \tag{2.23}$$

There are many theoretical advantages to linear functions. Fortunately a function that is log linear, such as the Cobb–Douglas production function

$$\log q = \log b_0 + \sum_{i=1}^{n} b_i \log x_i \tag{2.24}$$

has many of these advantages. For example, linear regression analysis may be used to estimate the b_i coefficients in Equation 2.24 and we may use it to estimate the b_i coefficients in Equation 2.22.

Figures 2.4 and 2.5, respectively, present curves of the linear and Cobb–Douglas production functions for the case of two factor inputs. Various characteristics of these two useful production functions are summarized in Table 2.1 for the general case.

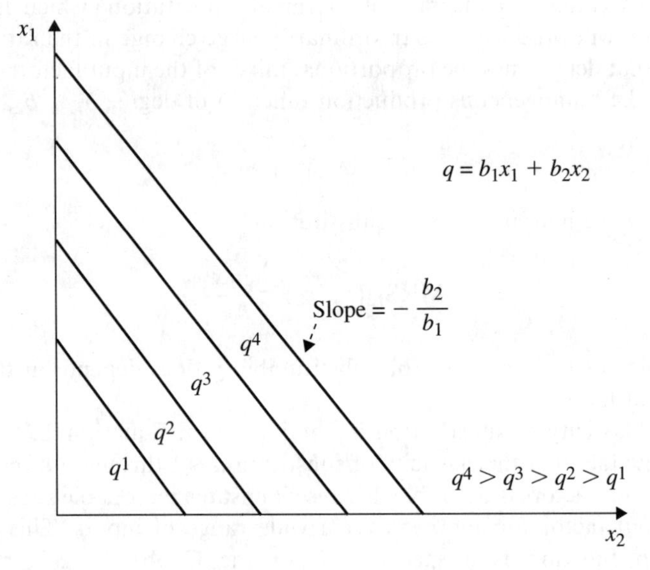

Figure 2.4. Linear Production Function.

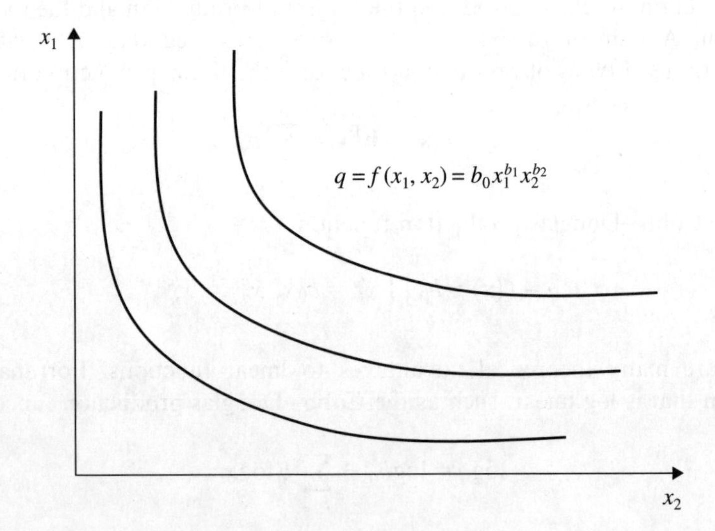

Figure 2.5. Cobb–Douglas Production Function.

TABLE 2.1. Properties of Linear and Cobb–Douglas Production Functions

Property	Linear	Cobb–Douglas
Defining Equations (2.22) and (2.23)	$q = f(\mathbf{x}) = \mathbf{b}^{\mathrm{T}}\mathbf{x} = \sum_{i=1}^{n} b_i x_i$	$q = f(\mathbf{x}) = b_0 \prod_{i=1}^{n} x_i^{b_i} = b_0 x_1^{b_1} x_2^{b_2} \cdots x_n^{b_n}$
Marginal productivity $\mathrm{MP}_j(\mathbf{x})$—Equation 2.6	b_j	$b_0 b_j x_j^{-1} \prod_{i=1}^{n} x_1^{b_1}$
Average productivity $\mathrm{AP}_j(\mathbf{x})$—Equation 2.8	$\frac{\sum_{i=1}^{n} b_i x_i}{x_j}$	$b_0 x_j^{-1} \prod_{i=1}^{n} x_1^{b_1}$
Relevant economic region—Equations 2.8 and 2.9	$0 < x_j < \infty$	$0 < x_j < \infty$
Condition or region where $\mathrm{MP}_j(\mathbf{x}) > \mathrm{AP}_j(\mathbf{x})$	No condition, no region	All regions if $b_j > 1$
Factor elasticity of production $\varepsilon_j(\mathbf{x})$—Equations 2.13 and 2.14	$\frac{b_j x_j}{\sum_{i=1}^{n} b_i x_i}$	b_j
Total elasticity of production $\varepsilon(\mathbf{x})$—Equations 2.16 and 2.18	1	$\sum_{i=1}^{n} b_i$
Marginal rate of substitution $\mathrm{MRS}_{ij}(\mathbf{x})$—Equation 2.20	$\frac{b_i}{b_j}$	$\frac{b_i x_j}{b_j x_i}$
Elasticity of substitution $\sigma_{ij}(\mathbf{x})$—Equation 2.21	∞	1

A summary of the central results of this section, in the context of a profit maximization example, yields much that is of importance to our future efforts. We will suppose that there is a single factor input to production, perhaps labor. The production function is

$$q = f(L) \tag{2.25}$$

We assume that the firm maximizes profits subject to constraints involving the production function and market realities concerning marketing products and purchasing inputs. The firm is assumed to be too small to affect the wages w of labor or the price of its product—it is a "price taker in all markets."

We now examine some issues concerning profit of a firm. For simplicity, we will initially assume that profit is not a function of time or, equivalently, that we are considering a fixed and known time interval and all data are for that known time interval. Thus, the profit of the firm is given by the expression

$$\Pi = \text{revenue minus costs} = pq - wL = pf(L) - wL \tag{2.26}$$

Here the assumed variable is the amount of labor to use in the production process. To maximize profit we must have

$$\frac{d\Pi}{dL} = p\frac{df(L)}{dL} - w = 0 \tag{2.27}$$

and

$$\frac{d^2\Pi}{dL^2} = p\frac{d^2f(L)}{dL^2} < 0 \tag{2.28}$$

The first-order condition is a necessary condition for optimality; it represents both the output supply function and the input factor–demand function. This second order condition is also very important, as it represents the sufficient condition for a maximum. From it we see that the *marginal product* (of labor in this case) must be decreasing at the optimum production level for maximum profit. This is the same as the requirement that the production function be concave in the region of maximum profit.

Example 2.6: We can see the optimality requirements graphically and physically from examination of the first-order condition Equation 2.27, and a typical profit. This curve, given by Equation 2.26, is illustrated in Fig. 2.6 for the production function

$$q = f(L) = 24.74L^2 - 1.5L^3$$

with $w = 45$, $p = 1$.

We obtain two possible values of labor, $L = 1$ and 10, which cause the marginal profit $d\Pi/dL$ to be zero. But $L = 1$ is a point of minimum, and

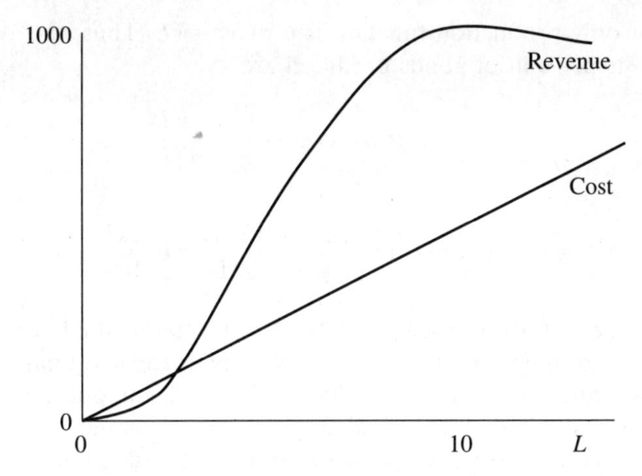

Figure 2.6. Revenue and Cost Curves for Example 2.6.

negative ($\Pi=-21.75$), profit. At $L=10$ we obtain the true maximum profit of $\Pi=525$. ■

The major conclusion that we obtain from this last example is that the production function must be concave at the (local) optimum where profit-maximizing behavior occurs. We also know that the maximum profit is nonnegative, since the firm can always obtain zero profit by doing nothing at all. We use these observations to obtain some interesting relations concerning sensitivities. From Equation 2.26 we obtain

$$\frac{f(L)}{L} \geq \frac{w}{p} \tag{2.29}$$

Using the wage–price ratio obtained from Equation 2.27, we have

$$\frac{L}{f(L)} \frac{\mathrm{d}f(L)}{\mathrm{d}L} \leq 1$$

We can rewrite this relation as the equivalent sensitivity expression

$$\frac{\mathrm{d}f(L)/f}{\mathrm{d}L/L} \leq 1 \tag{2.30}$$

Thus, we see that the "sensitivity" or "elasticity" of the output with respect to the input is less than or equal to 1. Again this requires nonincreasing return to scale. We can show, in general, that a profit-maximizing firm will only produce a finite nonzero amount under conditions of nonincreasing returns to scale.

Often it is more meaningful to interpret these results in terms of the average and marginal costs of production. The cost of production for our simple labor-only production function is C=cost=wL. Thus, the average and marginal costs per unit of goods produced are

$$\text{AC} = \text{average cost} = \frac{C}{q} = \frac{wL}{f(L)} \tag{2.31}$$

$$\text{MC} = \text{marginal cost} = \frac{\mathrm{d}C}{\mathrm{d}q} = \frac{\mathrm{d}C}{\mathrm{d}L}\frac{\mathrm{d}L}{\mathrm{d}q} = w\frac{\mathrm{d}L}{\mathrm{d}f(L)} = p \tag{2.32}$$

Here we use the fact, from Equation 2.27, that the marginal labor productivity is just the ratio of the wages to the prices. We divide the marginal costs by the average costs and use Equation 2.30 to obtain the important result that AC=average costs ≤ MC=marginal costs, or that average costs cannot be greater than marginal costs. At the factor input that leads to profit optimality, the marginal cost is at least as great as the average cost. If this were not the case, we could change the operating point and this would have led to lower average costs and hence greater profits. We will now examine these results in greater detail and with greater generality.

Example 2.7: The concept of a production possibility frontier is often used in representing trade-offs between produced items that result from constrained inputs to production. The production frontier allows us to represent trade-offs between goods (and services and investments) that compete for scarce resources and may be of assistance in determining an operating point that allocates resources to production in an efficient, effective, and equitable way. We will need this concept in our discussions of welfare, or normative, economics in Chapter 5.

Suppose that we have a very simple economy that produces food and clothing according to the production functions

$$F = \ln(1 + L_F) \tag{2.33}$$

$$C = (L_C)^{0.5} \tag{2.34}$$

where L_F and L_C are the labor devoted to production of food and clothing. For simplicity, we assume that the price of food and the price of clothing are each 1. We also assume that the wages for labor are the same in the two markets. We may obtain the conditions for maximum profit of the two firms from the profit equations

$$\Pi_F = \ln(1 + L_F) - w_L L_F \tag{2.35}$$

$$\Pi_C = (L_C)^{0.5} - w_L L_C \tag{2.36}$$

The optimum operating conditions and associated profits are obtained from $\mathrm{d}\Pi_i/\mathrm{d}L_i=0$ as

$$L_F = \frac{1}{w_L} - 1, \quad \Pi_F = \ln\left(\frac{1}{w_L}\right) - (1 - w_L)$$

$$L_C = \frac{1}{4w_L^2}, \quad \Pi_C = \frac{1}{4w_L}$$

The optimization conditions assume a potentially unlimited supply of potential labor. If the wages w_L are between 0 and 1, there will be a profit to both food and clothing producers. For labor wages in excess of 1 the profit in food production is negative and so no food will be produced. It is interesting to note that the maximum amount of labor available L_M may well be less than that demanded by the two firms, which is $L_D = L_F + L_C = \frac{1}{4w_L^2} + \frac{1}{w_L} - 1$.

If we have for the wages of labor w_L=0.1, then we obtain, for the unrestricted labor case, the results F=2.30, L_F=9, Π_F=1.4, L_C=25, Π_C=2.5, and C=5. It turns out here that if labor is restricted but there are more than 34 units available, the firms will not use as much labor as is available and, assuming perfect competition, each firm can maximize its profit equation as given by Equations 2.35 and 2.36. If there are less than 34 units of labor available, then some sort of rationing of labor will be needed, since we do not allow a firm to raise its wage rate to attract more labor. Firm F will not want any more than 9 units of labor, and firm C will not need any more than 25.

If L_M=9, then we see in this case that F_M will vary from 2.30 to 0 as C_M varies from 0 to 3.0. The profits of the two firms will vary from Π_F=1.4 to 0, and from Π_C=0 to 2.5, as F_M varies from 2.30 to 0 and L_{F_M} varies from 9 to 0. If the object were to maximize total profit $\Pi=\Pi_F + \Pi_C$, then the best operating point on the production frontier is L_{F_M}=0, L_{C_M}=9, F_M=0, C_M=3.0, Π_F=0, Π_C=2.5. But society may well not desire a situation in which all labor is devoted to clothing manufacture and none to food production. We see that there may be concerns other than profit maximization that are introduced into even a simple economic model. We will examine a number of these later as we explore such subjects as normative economics. ■

2.3 MULTIPRODUCT FIRMS AND MULTIPRODUCT PRODUCTION FUNCTIONS

Extensions of our production function concepts from the case of the single-product firm are relatively straightforward conceptually. We now briefly consider the case where there are a number of goods and services produced from the input factors. We will let q_j denote the output level of good or service j produced. In general, it will turn out that we will need output good or service q_i^i as an input to the production process to produce good or service q_j. Also, there is no fundamental reason why good or service q_i^i is not needed in the process of producing q_j. Thus we see that our generalization to the multiproduct case results in the need for production functions of the form

$$q_i = f_i(q_1^i, q_2^i, \ldots, q_i^i, \ldots, q_m^i, x_1^i, x_2^i, \ldots, x_n^i) \tag{2.37}$$

for $i=1, 2, \ldots, m$. Using vector notation, we can write this as

$$\mathbf{q} = \mathbf{f}(\mathbf{q}, \mathbf{x}) \tag{2.38}$$

or in an equivalent form

$$\varphi(\mathbf{q}, \mathbf{x}) = \mathbf{0} \tag{2.39}$$

It will sometimes be convenient to define an augmented vector **z** consisting of **q** and **x**,

$$\mathbf{z} = \begin{bmatrix} \mathbf{q} \\ \mathbf{x} \end{bmatrix} \tag{2.40}$$

With this notion, we have for the generalized vector production function of a multiproduct firm

$$\mathbf{g}(\mathbf{z}) = \mathbf{0} \tag{2.41}$$

In Chapter 4, concerning supply–demand equilibrium and input–output analysis, we will need to consider the multiproduct firm and will therefore delay further discussion of this topic at this point.

2.4 CLASSIC THEORY OF THE FIRM

There are four fundamental economic concerns of a firm and the microeconomic decision problem for a firm involves each of these:

1. What total amount of factor inputs x_i should be purchased?
2. What is the best allocation of the factor purchase resources among the input factors?
3. What is the best level of total output production?
4. What is the best allocation of the input factors to the various output production components?

For the single-product firm, three basic economic problems emerge from these considerations.

1. We wish to minimize the total cost or value of the input factors or input resources used in the production process to produce a given output product.
2. We wish to maximize the profit of the firm.

3. We wish to maximize the profit of the firm subject to resource constraints on input factors.

Often small-scale projects resolve the first problem. There is no fixed total budget constraint; however, we wish to turn out a product that has the lowest production cost. If we wish to optimize the production of soft drink containers, we might thus wish to solve problem 1. Problem 1 is equivalent to the problem of maximizing output production subject to a budget constraint on the input factors. Problem 2 is generally spoken of as the problem of the firm in the long term. When input resource constraints, such as those due to physical plant size or input factor availability, are present, we generally speak of the resulting problem as a problem of the firm in the short term. This is problem 3.

In all cases we wish to maximize the output q, or the profit Π, generally through minimizing production costs C by choosing appropriate inputs to the production process. The production function is assumed to be known, as is the output product price p. The input prices, or wages per unit of factor input, and denoted by the vector

$$\mathbf{w} = \begin{bmatrix} w_1 \\ w_2 \\ \vdots \\ w_n \end{bmatrix}$$

are also assumed to be known. The revenue R to the firm is a product of the output level and the output price and is

$$R(\mathbf{x}) = pq = pf(\mathbf{x}) \tag{2.42}$$

The *production cost* PC is equal to the total price or payments for all factor inputs and is

$$\mathrm{PC} = \mathbf{w}^{\mathrm{T}}\mathbf{x} = \sum_{i=1}^{n} w_i x_i \tag{2.43}$$

The fixed cost FC is generally assumed to be independent of production costs.[4] The total cost C is the sum of the fixed cost and the production cost and is given by the expression

$$C(\mathbf{x}) = \mathrm{PC} + \mathrm{FC} = \mathbf{w}^{\mathrm{T}}\mathbf{x} + \mathrm{FC} \tag{2.44}$$

The profit of the firm, Π, is the difference between the total revenue and the total cost and is

[4]Generally, we can assume that the fixed costs are equal to zero, since we obtain optimality by differentiating total cost. The constant fixed costs will not influence the conditions for optimality. This simply says that no input represents a true economic cost unless it is possible to change it by varying the input. Fixed costs influence profit, of course, and thus influence whether or not a firm will produce at all.

$$\Pi(\mathbf{x}) = R(\mathbf{x}) - C(\mathbf{x}) = pf(\mathbf{x}) - \mathbf{w}^{\mathrm{T}}\mathbf{x} - \mathrm{FC} \tag{2.45}$$

We are now in a position to state the classic microeconomic optimization problem of the firm in a relatively general form.

The resource allocation problem for each firm is that of minimizing its $C(\mathbf{x})$, given by Equation 2.44, subject to the equality constraint $q=f(\mathbf{x})$. In other words, we minimize the production costs associated with producing a fixed output quantity. We assume that all factor inputs will be nonnegative and thus do not impose an inequality constraint on the factor inputs. This general problem may be resolved by using the well-known Lagrange multiplier method of optimization theory. We adjoin the general equality constraint to be satisfied,

$$\mathbf{0} = \mathbf{b} - \mathbf{g}(\mathbf{x}) \tag{2.46}$$

to the cost function to be minimized,

$$J = C(\mathbf{x}) \tag{2.47}$$

by means of a vector Lagrange multiplier $\boldsymbol{\lambda}$ to obtain

$$J' = C(\mathbf{x}) + \boldsymbol{\lambda}^{\mathrm{T}}[\mathbf{b} - \mathbf{g}(\mathbf{x})] \tag{2.48}$$

This function J' is often called the Lagrangian and is denoted by $\mathcal{L}$. We note that we have added a special form of nothing to the original cost function J, since if Equation 2.46 is satisfied, J' will surely become equal to J of Equation 2.47. Our procedure will be to minimize J' of Equation 2.48 and then to determine $\boldsymbol{\lambda}$ such that Equation 2.46 is satisfied.

Use of basic calculus leads to the necessary requirement to minimize J' in Equation 2.48. We obtain

$$\frac{\partial J'}{\partial \mathbf{x}} = \mathbf{0} = \frac{\partial C(\mathbf{x})}{\partial \mathbf{x}} - \frac{\partial g^{\mathrm{T}}(\mathbf{x})}{\partial \mathbf{x}}\boldsymbol{\lambda} \tag{2.49}$$

$$\frac{\partial J'}{\partial \boldsymbol{\lambda}} = \mathbf{0} = \mathbf{b} - \mathbf{g}(\mathbf{x}) \tag{2.50}$$

and these are the two vector equations that must be solved to determine the solution to the original problem. Table 2.2 summarizes the results of optimization, with an equality constraint.

For our particular resource allocation economic problem, Equations 2.49 and 2.50 become

$$\mathbf{w} - \frac{\partial f(\mathbf{x})}{\partial \mathbf{x}}\boldsymbol{\lambda} = \mathbf{0} \tag{2.51}$$

$$q = f(\mathbf{x}) \tag{2.52}$$

TABLE 2.2. Basic Maximization (Minimization) Problem with Equality Constraints

Cost function to be maximized or minimized	$J = \Phi(\mathbf{x})$
Production equation equality constraint	$\mathbf{g}(\mathbf{x}) = \mathbf{b}$
Langrangian	$\mathcal{L} = \Phi(x) + \boldsymbol{\lambda}^T[b - g(x)]$
Necessary conditions	$\frac{\partial \mathcal{L}}{\partial \mathbf{x}} = \frac{\partial \Phi(\mathbf{x})}{\partial \mathbf{x}} - \frac{\partial g^T(\mathbf{x})}{\partial \mathbf{x}} \boldsymbol{\lambda} = \mathbf{0}$ $\frac{\partial \mathcal{L}}{\partial \boldsymbol{\lambda}} = \mathbf{b} - \mathbf{g}(\mathbf{x}) = \mathbf{0}$
Sufficient conditions for a maximum	$\frac{\partial^2 \mathcal{L}}{\partial \mathbf{x}^2}$ is negative definite along $\mathbf{g}(\mathbf{x}) = \mathbf{b}$
Sufficient conditions for a minimum	$\frac{\partial^2 \mathcal{L}}{\partial \mathbf{x}^2}$ is positive definite along $\mathbf{g}(\mathbf{x}) = \mathbf{b}$

since the Lagrangian for this problem is, in the single-product firm case,

$$\mathcal{L} = \mathbf{w}^T\mathbf{x} + \lambda[q - f(\mathbf{x})]$$

Example 2.8: We desire to ensure a given output level from a productive unit that has a Cobb–Douglas production function

$$q = f(\mathbf{x}) = b_0 x_1^{b_1} x_2^{b_2} \tag{2.53}$$

so as to minimize the total cost of production

$$J = C(\mathbf{x}) = w_1 x_1 + w_2 x_2 + \mathrm{FC} \tag{2.54}$$

We adjoin Equation 2.53 to Equation 2.54 by means of a Lagrange multiplier to obtain

$$J' = C(\mathbf{x}) + \lambda[q - f(\mathbf{x})] = w_1 x_1 + w_2 x_2 + \mathrm{FC} + \lambda(q - b_0 x_1^{b_1} x_2^{b_2}) \tag{2.55}$$

We use the basic calculus necessary conditions and obtain

$$\frac{\partial J'}{\partial x_1} = 0 = w_1 - \lambda b_0 b_1 x_1^{b_1 - 1} x_2^{b_2} \tag{2.56}$$

$$\frac{\partial J'}{\partial x_2} = 0 = w_2 - \lambda b_0 b_2 x_1^{b_1} x_2^{b_2 - 1} \tag{2.57}$$

$$\frac{\partial J'}{\partial \lambda} = 0 = q - b_0 x_1^{b_1} x_2^{b_2} \tag{2.58}$$

By combining Equations 2.56 and 2.57 we obtain

$$\frac{b_1}{b_2} = \frac{w_1 x_1}{w_2 x_2} \tag{2.59}$$

We will obtain a similar relation often in this chapter and will give it a very special and useful interpretation later. We may combine Equations 2.59 and 2.58 to obtain the desired solutions for the optimum factor inputs $\hat{x}_1$ and $\hat{x}_2$. There are

$$\hat{x}_1^{b_1+b_2} = \frac{q}{b_0}\left(\frac{b_2 w_1}{b_1 w_2}\right)^{-b_2}, \quad \hat{x}_2 = \frac{b_2 w_1}{b_1 w_2}\hat{x}_1$$ ■

Example 2.9: An interesting general resource allocation problem that has an analytic solution consists of minimizing the input factor costs

$$\mathbf{J} = \mathrm{C}(\mathbf{x}) = \mathbf{w}^{\mathrm{T}}\mathbf{x} \tag{2.60}$$

subject to the constraint of a given output level q from the productive unit, where the production function is assumed to be a general quadratic plus linear function of input factors

$$q = f(\mathbf{x}) = \mathbf{a}^{\mathrm{T}}\mathbf{x} - 0.5\mathbf{x}^{\mathrm{T}}\mathbf{B}\mathbf{x} \tag{2.61}$$

with $\mathbf{B}$ a symmetric negative definite matrix and $\mathbf{a} > \mathbf{0}$. For this production function, the marginal productivity is given by Equation 2.6 as

$$\mathbf{MP}(\mathbf{x}) = \frac{\partial f(\mathbf{x})}{\partial \mathbf{x}} = \mathbf{a} + \mathbf{B}\mathbf{x} \tag{2.62}$$

Since the marginal productivity is always positive in the economic region and since use of negative amounts of the input factors to production is not meaningful, we see that the economic region is given by $\mathbf{0} \le \mathbf{x} \le -\mathbf{B}^{-1}\mathbf{a}$. The elasticity of production is, from Equation 2.16,

$$\varepsilon(\mathbf{x}) = \mathbf{MP}^{\mathrm{T}}(\mathbf{x})\mathbf{RAP}(\mathbf{x}) = \left(\frac{\partial f(\mathbf{x})}{\partial \mathbf{x}}\right)^{\mathrm{T}} \frac{\mathbf{x}}{f(\mathbf{x})} = 1 + \frac{0.5\mathbf{x}^{\mathrm{T}}\mathbf{B}\mathbf{x}}{\mathbf{a}^{\mathrm{T}}\mathbf{x} + 0.5\mathbf{x}^{\mathrm{T}}\mathbf{B}\mathbf{x}} \tag{2.63}$$

We see that the elasticity is 1 for low values of $\mathbf{x}$ and decreases as $\mathbf{x}$ increases. At the end point $\mathbf{x} = -\mathbf{B}^{-1}\mathbf{a}$ of the economic region, the elasticity is zero as it should be, since the marginal productivity is zero at the boundary of the economic region.

To obtain the best resource allocation we adjoin Equation 2.61 to Equation 2.60, using a Lagrange multiplier λ to obtain

$$\mathbf{J}' = \mathbf{w}^{\mathrm{T}}\mathbf{x} + \lambda(q - \mathbf{a}^{\mathrm{T}}\mathbf{x} - 0.5\mathbf{x}^{\mathrm{T}}\mathbf{B}\mathbf{x}) \tag{2.64}$$

The necessary conditions for minimum cost of the input factors are, from Equations 2.49 and 2.50, given by the expressions

$$\frac{\partial J'}{\partial \mathbf{x}} = \mathbf{0} = \mathbf{w} - \lambda\mathbf{a} - \lambda\mathbf{B}\mathbf{x} \tag{2.65}$$

$$\frac{\partial J'}{\partial \lambda} = 0 = q - \mathbf{a}^{\mathrm{T}}\mathbf{x} - 0.5\mathbf{x}^{\mathrm{T}}\mathbf{B}\mathbf{x} \tag{2.66}$$

Solution of these two equations yields the optimum Lagrange multiplier $\hat{\lambda}$. From Equation 2.65 we obtain

$$\hat{\mathbf{x}} = \hat{\lambda}^{-1}\mathbf{B}^{-1}(\mathbf{w} - \hat{\lambda}\mathbf{a}) \tag{2.67}$$

and from Equation 2.66, using this value of $\hat{\mathbf{x}}$, we have

$$\hat{q} = -0.5\mathbf{a}^{\mathrm{T}}\mathbf{B}^{-1}\mathbf{a} + 0.5\lambda^{-2}\mathbf{w}^{\mathrm{T}}\mathbf{B}^{-1}\mathbf{w} \tag{2.68}$$

Thus, we see that the optimum Lagrange multiplier is given by

$$\hat{\lambda} = \left(\frac{\mathbf{w}^{\mathrm{T}}\mathbf{B}^{-1}\mathbf{w}}{2q + \mathbf{a}^{\mathrm{T}}\mathbf{B}^{-1}\mathbf{a}}\right)^{1/2} \tag{2.69}$$

The optimum factor input is then

$$\hat{\mathbf{x}} = -\mathbf{B}^{-1}\mathbf{a} + \left(\frac{2q + \mathbf{a}^{\mathrm{T}}\mathbf{B}^{-1}\mathbf{a}}{\mathbf{w}^{\mathrm{T}}\mathbf{B}^{-1}\mathbf{w}}\right)^{1/2}\mathbf{B}^{-1}\mathbf{w} \tag{2.70}$$

The minimum cost of the input factor resources is, from Equation 2.60,

$$J_{\min} = C(\hat{\mathbf{x}}) = \mathbf{w}^{\mathrm{T}}\hat{\mathbf{x}} = -\mathbf{w}^{\mathrm{T}}\mathbf{B}^{-1}\mathbf{a} + (2\hat{q} + \mathbf{a}^{\mathrm{T}}\mathbf{B}^{-1}\mathbf{a})^{1/2}(\mathbf{w}^{\mathrm{T}}\mathbf{B}^{-1}\mathbf{w})^{1/2} \tag{2.71}$$

This results in the value of the minimum production cost, which is a function of the quantity output q. We will use this result in a later example in this section to maximize the profit of the firm expressed as a function of the production output q.

We can show[5] that the problem of minimizing the input factor costs subject to a fixed production output is equivalent to that of maximizing production q subject to the constraint of a fixed budget, C_f. To resolve this equivalent problem we adjoin the constraint to the cost function and obtain

$$J'' = q + \lambda[C_f - C(\mathbf{x})] = \mathbf{a}^{\mathrm{T}}\mathbf{x} + 0.5\mathbf{x}^{\mathrm{T}}\mathbf{B}\mathbf{x} + \lambda(C_f - \mathbf{w}^{\mathrm{T}}\mathbf{x}) \tag{2.72}$$

[5]See Problem 18 at the end of this chapter. These two problems are mathematical equivalents of one another.

To obtain the minimum input factor costs, we set

$$\frac{\partial J''}{\partial \mathbf{x}} = \mathbf{0} = \mathbf{a} + \mathbf{B}\mathbf{x} + \lambda\mathbf{w} \tag{2.73}$$

$$\frac{\partial J''}{\partial \lambda} = 0 = C_f - \mathbf{w}^{\mathrm{T}}\mathbf{x} \tag{2.74}$$

Then we solve for the resulting optimum Lagrange multiplier and factor inputs. From Equation 2.73 we obtain

$$\hat{\mathbf{x}} = -\mathbf{B}^{-1}(\mathbf{a} + \hat{\lambda}\mathbf{w}) \tag{2.75}$$

and by premultiplying this by $\mathbf{w}^{\mathrm{T}}$ and using Equation 2.74, we see that

$$\hat{\lambda} = -\frac{\mathbf{w}^{\mathrm{T}}\mathbf{B}^{-1}\mathbf{a} + C_f}{\mathbf{w}^{\mathrm{T}}\mathbf{B}^{-1}\mathbf{w}} \tag{2.76}$$

$$\hat{\mathbf{x}} = -\mathbf{B}^{-1}\mathbf{a} + \mathbf{B}^{-1}\mathbf{w}(\mathbf{w}^{\mathrm{T}}\mathbf{B}^{-1}\mathbf{w})^{-1}(\mathbf{w}^{\mathrm{T}}\mathbf{B}^{-1}\mathbf{a} + C_f) \tag{2.77}$$

For this particular example the optimum factor input increases linearly with increases in the budget for factor inputs. However, $\hat{\mathbf{x}}$ is not linear with respect to changes in $\mathbf{w}$.

We should be careful to ensure that the results obtained and the models used for an optimization procedure are physically meaningful. In this particular example, for instance, the unconstrained maximum production output occurs at

$$\frac{\partial q}{\partial \mathbf{x}} = \mathbf{0} = \frac{\partial f(\mathbf{x})}{\partial \mathbf{x}} = \mathbf{a} + \mathbf{B}\mathbf{x}$$

and yields

$$\hat{\mathbf{x}}_{\mathrm{m}} = -\mathbf{B}^{-1}\mathbf{a} \tag{2.78}$$

The cost of this input factor is

$$C_m(\mathbf{x}) = \mathbf{w}^{\mathrm{T}}\hat{\mathbf{x}}_m = -\mathbf{w}^{\mathrm{T}}\mathbf{B}^{-1}\mathbf{a} \tag{2.79}$$

In this example, we should never pay any more for the factor inputs than this. If we do, we are forcing a wasteful production system to accommodate the input factor cost greater than $C_m(\mathbf{x})$. The cost of the factor input vector of Equation 2.77 is C_f. If $C_f < C_m(\mathbf{x})$, we should use $\hat{\mathbf{x}}$ of Equation 2.77 as the optimum factor input. If $C_f > C_m(\mathbf{x})$, we should use $\hat{\mathbf{x}}_m$ of Equation 2.78 and accept the fact that we do not need all of the budget for the factor inputs to the production process.

A reason for our dilemma here is that we really should have maximized output production q subject to the inequality constraint that the factor input cost is less than or equal to some budget amount, $\mathbf{w}^{\mathrm{T}}\mathbf{x} \leq C_f$. We will now extend our treatment of the theory of the firm to include inequality constraints. This will allow us to consider problems, such as this, in a more realistic manner. ■

The nonbudget-constrained problem of the firm, which is the classic long-term problem of the firm, consists of maximizing profits as given by

$$\Pi(\mathbf{x}) = pf(\mathbf{x}) - \mathbf{w}^{\mathrm{T}}\mathbf{x} - \mathrm{FC} \tag{2.80}$$

subject to the constraint that we can have only nonnegative input factors, or

$$\mathbf{x} \geq \mathbf{0} \tag{2.81}$$

We accomplish this maximization by choosing the input factor vector $\mathbf{x}$. This is a classic problem in nonlinear programming and there are a number of nonlinear programming algorithms that could resolve this problem for a specified production function.

One approach is the basic one of expanding the function to be extremized in a Taylor's series. We assume that there is a local maximum of Equation 2.80 at $\mathbf{x} = \hat{\mathbf{x}}$. Then for any small change $\boldsymbol{\Delta}\mathbf{x}$ we must have

$$\Pi(\hat{\mathbf{x}}) \geq \Pi(\hat{\mathbf{x}} + \boldsymbol{\Delta}\mathbf{x}) \tag{2.82}$$

If we assume that Π is differentiable in $\mathbf{x}$, we can expand the right-hand side of Equation 2.45 in a Taylor's series about $\boldsymbol{\Delta}\mathbf{x}=\mathbf{0}$ to obtain the relation

$$\Pi(\hat{\mathbf{x}} + \Delta\mathbf{x}) \cong \Pi(\hat{\mathbf{x}}) + \Delta\mathbf{x}^{\mathrm{T}} \frac{\partial\Pi(\hat{\mathbf{x}})}{\partial\hat{\mathbf{x}}} + \frac{1}{2}\Delta\mathbf{x}^{\mathrm{T}} \frac{\partial^2\Pi(\hat{\mathbf{x}})}{\partial\hat{\mathbf{x}}^2}\Delta\mathbf{x} + \cdots \tag{2.83}$$

We drop all terms here of order higher than the first in $\Delta\mathbf{x}$. When we compare the foregoing two equations, we see that we must have

$$\Delta\mathbf{x}^{\mathrm{T}} \frac{\partial\Pi(\hat{\mathbf{x}})}{\partial\hat{\mathbf{x}}} \leq 0 \tag{2.84}$$

Suppose that we are on the boundary where $\hat{\mathbf{x}} = \mathbf{0}$. On the boundary, we must have $\Delta\mathbf{x} > \mathbf{0}$, since $\mathbf{x} \geq \mathbf{0}$ is a constraint (Eq. 2.81). The quantity $\Delta\mathbf{x}$ is otherwise arbitrary and so the foregoing is equivalent to

$$\frac{\partial\Pi(\hat{\mathbf{x}})}{\partial\hat{\mathbf{x}}} \leq \mathbf{0} \quad \text{if } \hat{\mathbf{x}} = \mathbf{0} \tag{2.85}$$

If we assume that we are not at the inequality boundary where some $x_i=0$, then we must require that $\partial\Pi(\hat{\mathbf{x}})/\partial x_i$ be zero for an arbitrary Δx_i. Thus for all x_i solutions not on the boundary, where $x_i=0$, we require, to ensure that Equation 2.84 holds, that

$$\frac{\partial\Pi(\hat{\mathbf{x}})}{\partial\hat{x}_i} = 0, \quad x_i \neq 0, i = 1, 2, \ldots, n \tag{2.86}$$

TABLE 2.3. Basic Constrained Nonlinear Optimization Problem

Objective function to be maximized	$J = \Pi(\mathbf{x})$
Constraint	$\mathbf{x} \geq \mathbf{0}$
Necessary conditions for optimum	$\hat{\mathbf{x}} \geq \mathbf{0}$
1.	$\dfrac{\partial \Pi(\hat{\mathbf{x}})}{\partial \hat{\mathbf{x}}} \leq \mathbf{0}$
2.	$\hat{\mathbf{x}}^{\mathrm{T}} \dfrac{\partial \Pi(\hat{\mathbf{x}})}{\partial \hat{\mathbf{x}}} = 0$
Alternate necessary conditions for optimum (scalar representation)	$\dfrac{\partial \Pi(\hat{\mathbf{x}})}{\partial \hat{x}_i} = 0, \ \hat{x}_i > 0$
3.	$\dfrac{\partial \Pi(\hat{x})}{\partial \hat{x}_i} < 0, \ \hat{x}_i = 0$

For solutions on the boundary we cannot have arbitrary $\Delta\mathbf{x}$, since we require that $\hat{x}_i = 0$. We can combine this with the foregoing requirement and obtain, as an equivalent requirement to Equation 2.84,

$$\frac{\partial \Pi(\hat{\mathbf{x}})}{\partial \hat{x}_i} \hat{x}_i = 0 \tag{2.87}$$

We can sum this requirement over all i and obtain the necessary condition, sometimes called the complementary slackness conditions,

$$\hat{\mathbf{x}}^{\mathrm{T}} \frac{\partial \Pi(\hat{\mathbf{x}})}{\partial \hat{\mathbf{x}}} = 0 \tag{2.88}$$

Table 2.3 represents the basic nonlinear programming problem and the necessary conditions for a solution, which we have obtained as Equations 2.81, 2.85, and 2.88. We may replace these three conditions, for $\hat{x}_i > 0$, by the requirement that

$$\frac{\partial \Pi(\hat{\mathbf{x}})}{\partial \hat{x}_i} = 0, \quad \hat{x}_i > 0, i = 1, 2, \ldots, n \tag{2.89}$$

For $\hat{x}_i = 0$, we have the requirement

$$\frac{\partial \Pi(\hat{\mathbf{x}})}{\partial \hat{x}_i} \leq 0, \quad \hat{x}_i = 0, i = 1, 2, \ldots, n \tag{2.90}$$

Equations 2.89 and 2.90 clearly indicate the complementary slackness condition in that either x_i or $\partial \Pi(\mathbf{x})/\partial x_i$ is zero.

Example 2.10: As a simple example of profit optimization, we assume that the production function for a firm is of the form

$$f(\mathbf{x}) = f(x_1, x_2) = 16x_1x_2 - x_1^3x_2^3$$

and that the total costs are given by

$$C = \mathbf{w}^{\mathrm{T}}\mathbf{x} + \mathrm{FC} = x_1 + 2x_2 + \mathrm{FC}$$

Here, we see that the profit is, for $p=1$, given by the expression

$$\Pi(x_1, x_2) = 16x_1x_2 - x_1^3x_2^3 - x_1 - 2x_2 - \mathrm{FC}$$

If we assume that the optimum is not on the boundary, then solution of Equation 2.89 for this example yields

$$16\hat{x}_2 - 3\hat{x}_1^2\hat{x}_2^3 - 1 = 0$$

$$16\hat{x}_1 - 3\hat{x}_1^3\hat{x}_2^2 - 2 = 0$$

which can be combined to give $\hat{x}_1 = 2\hat{x}_2$. We therefore have to solve the non-linear algebraic equation $12\hat{x}_2^5 - 16\hat{x}_2 + 1 = 0$. The solution of this equation represents our solution for this example and is $\hat{x}_1 = 2.11668$ and $\hat{x}_2 = 1.05834$. We obtain a reasonable solution for this example, since the example is one in which the marginal productivity becomes negative for sufficiently large $\mathbf{x}$. Thus, there is a relevant economic region and the optimum solution lies in this region. Often the production function is one, such as the linear or Cobb–Douglas production function, in which the marginal productivity is always positive and this unconstrained optimization is not meaningful.

To verify whether or not a solution obtained from the necessary conditions of our optimization procedures is actually a correct and meaningful one, we need to examine the sufficiency conditions that we can obtain from the second variation term in Equation 2.83. We obtain

$$\frac{\partial^2\Pi}{\partial\mathbf{x}} = \begin{bmatrix} \dfrac{\partial^2\Pi}{\partial x_1^2} & \dfrac{\partial^2\Pi}{\partial x_1\partial x_2} \\ \dfrac{\partial^2\Pi}{\partial x_2\partial x_1} & \dfrac{\partial^2\Pi}{\partial x_2^2} \end{bmatrix} = \begin{bmatrix} -3x_1x_2^3 & 16 - 9x_1^2x_2^2 \\ 16 - 9x_1^2x_2^2 & -6x_1^3x_2 \end{bmatrix}$$

This is negative definite at the obtained values for $\hat{x}_1$ and $\hat{x}_2$, and so we do have a meaningful problem and solution. ■

We can draw some useful general conclusions from the problem of unconstrained profit maximization in which we desire to maximize[6]

$$\Pi(\mathbf{x}) = pf(\mathbf{x}) - \mathbf{w}^{\mathrm{T}}\mathbf{x} - \mathrm{FC}, \quad \mathbf{x} \geq 0 \tag{2.91}$$

[6]The conditions for optimality do not depend on FC, so we can let FC=0 without loss of generality, as we have noted previously.

Use of the necessary conditions of Table 2.2 leads to the requirements that for all i

$$\frac{\partial \Pi(\mathbf{x})}{\partial \mathbf{x}} = p\frac{\partial f(\mathbf{x})}{\partial \mathbf{x}} - \mathbf{w} \leq 0 \tag{2.92}$$

and

$$\mathbf{x}^{\mathrm{T}}\frac{\partial \Pi(\mathbf{x})}{\partial \mathbf{x}} = \mathbf{x}^{\mathrm{T}}\left(p\frac{\partial f(\mathbf{x})}{\partial \mathbf{x}} - \mathbf{w}\right) = 0 \tag{2.93}$$

with

$$\mathbf{x} \geq \mathbf{0} \tag{2.94}$$

Usually we require presence of some of all the input factor components x_i, and so Equation 2.94 is routinely satisfied. Equation 2.93 will then be satisfied by the requirement of Equation 2.92 that

$$\frac{\partial f(\mathbf{x})}{\partial \mathbf{x}} = \mathbf{MP}(\mathbf{x}) = \frac{\mathbf{w}}{p} \tag{2.95}$$

We have just obtained the important result that the price multiplied by the marginal productivity, $p\mathbf{MP}(\mathbf{x})$, is the wage or value for each input. In scalar component form, this relation states that

$$p\mathrm{MP}_i(\mathbf{x}) = p\frac{\partial f(\mathbf{x})}{\partial x_i} = w_i$$

An important consequence of this result is that the optimum combination of input factors is such that the ratio of their marginal products is the ratio of their wages or values. We have from the foregoing discussion

$$\frac{\mathrm{MP}_i(\mathbf{x})}{\mathrm{MP}_j(\mathbf{x})} = \frac{w_i}{w_j} \tag{2.96}$$

This is the important result that we refer to in Equation 2.59 of Example 2.8.

Example 2.11: We consider the quadratic production function where $\mathbf{a} > \mathbf{0}$ and $\mathbf{B}$ is a negative definite symmetric matrix,

$$Q = f(\mathbf{x}) = \mathbf{a}^{\mathrm{T}}\mathbf{x} + 0.5\mathbf{x}^{\mathrm{T}}\mathbf{B}\mathbf{x} \tag{2.97}$$

and find the optimum factor input vector to maximize the profit of the firm (in the long run). This profit is, for the production function of Equation 2.97,

$$\Pi(\mathbf{x}) = qp - \mathbf{w}^{\mathrm{T}}\mathbf{x} = \mathbf{a}^{\mathrm{T}}\mathbf{x}p + 0.5\mathbf{x}^{\mathrm{T}}\mathbf{B}\mathbf{x}p - \mathbf{w}^{\mathrm{T}}\mathbf{x} \tag{2.98}$$

We set

$$\frac{\partial \Pi(\mathbf{x})}{\partial \mathbf{x}} = \mathbf{0} = \mathbf{a}p + \mathbf{B}\mathbf{x}p - \mathbf{w} \tag{2.99}$$

and obtain

$$\hat{\mathbf{x}} = -\mathbf{B}^{-1}\left(\mathbf{a} - \frac{\mathbf{w}}{p}\right) \tag{2.100}$$

as the optimum input factor vector. For this example we will assume that $\mathbf{a}p \geq \mathbf{w}$, that is, $a_j p > w_j \forall i$. See Example 2.13 for a discussion of the case where not all $a_i p > w_i$. We can draw several important conclusions from this result. If the price $\mathbf{w}$ of the input factors is zero or if the price of the output product is infinite, then we will produce at the maximum value given by $\mathbf{x} = -\mathbf{B}^{-1}\mathbf{a}$. For any finite p and nonzero $\mathbf{w}$ the factor input is less than the value that yields the maximum output.

The optimum output $\hat{q}$ to maximize profit is, from Equations 2.97 and 2.100,

$$\hat{q} = -0.5\left(\mathbf{a} + \frac{\mathbf{w}}{p}\right)^{\mathrm{T}}\mathbf{B}^{-1}\left(\mathbf{a} - \frac{\mathbf{w}}{p}\right) \tag{2.101}$$

The optimum value of the factor input becomes zero for $\mathbf{w} = \mathbf{a}p$, and for any $\mathbf{w} > \mathbf{a}p$ the optimum factor input is zero. This simply illustrates the fact that the marginal return, or marginal revenue, $\mathbf{MR}(\mathbf{x}) = p\mathbf{MP}(\mathbf{x}) = p\mathbf{a} + p\mathbf{B}\mathbf{x}$, reaches its maximum value at $\mathbf{x} = \mathbf{0}$, and if this is less than the wage $\mathbf{w}$, we should not produce anything at all.

These results may be obtained in another way that yields considerable insight into the optimization involved. If we could find the minimum (production) cost $C(q)$ for a productive firm, we could maximize the profit as a function of output product. Since revenue R is equal to pq and since profit is revenue minus cost, we have

$$\Pi(q) = pq - C(q) \tag{2.102}$$

and obtain

$$\frac{\mathrm{d}\Pi(q)}{\mathrm{d}q} = 0 = p - \frac{\mathrm{d}C(q)}{\mathrm{d}q} \tag{2.103}$$

and so we see that, for maximum profit, the price of the output product is equal to the marginal cost of the input factors

$$p = \frac{\mathrm{d}C(q)}{\mathrm{d}q} \tag{2.104}$$

In Example 2.9 we found the minimum input factor cost for a productive firm as Equation 2.71, which is

$$C(q) = -\mathbf{w}^{\mathrm{T}}\mathbf{B}^{-1}\mathbf{a} + (2q + \mathbf{a}^{\mathrm{T}}\mathbf{B}^{-1}\mathbf{a})^{1/2}(\mathbf{w}^{\mathrm{T}}\mathbf{B}^{-1}\mathbf{w})^{1/2} \tag{2.105}$$

Use of Equation 2.104 leads to a relationship for the price

$$p = (2q + \mathbf{a}^{\mathrm{T}}\mathbf{B}^{-1}\mathbf{a})^{-1/2}(\mathbf{w}^{\mathrm{T}}\mathbf{B}^{-1}\mathbf{w})^{1/2}$$

which we may use to obtain for the production function

$$q = -0.5\mathbf{a}^{\mathrm{T}}\mathbf{B}^{-1}\mathbf{a} + \frac{0.5\mathbf{w}^{\mathrm{T}}\mathbf{B}^{-1}\mathbf{w}}{p^2} \tag{2.106}$$

This is identical to Equation 2.101, as it should be. Equations 2.101 and 2.106 are very important results here: they indicate the impact of the relationship between wage and price on production and form the supply curve for this example. ■

The classic economic problem of the firm in the short run is to maximize profit

$$\mathrm{J} = \Pi(\mathbf{x}) \tag{2.107}$$

subject to a constraint on input factor availability,

$$\mathbf{C}(\mathbf{x}) \le \mathbf{b}, \;\; \mathbf{x} \ge 0 \tag{2.108}$$

Let us develop a set of conditions with which to accomplish this optimization. We convert the inequality constraint to an equality constraint by introducing a nonnegative *slack variable* $\mathbf{s}$ of appropriate dimension:

$$\mathbf{s} = \mathbf{b} - \mathbf{C}(\mathbf{x}) \tag{2.109}$$

Our problem now becomes one of maximizing $J = \Pi(\mathbf{x})$ subject to the equality constraint

$$\mathbf{b} - \mathbf{C}(\mathbf{x}) - \mathbf{s} = \mathbf{0} \tag{2.110}$$

and the inequality constraints

$$\mathbf{x} \ge \mathbf{0} \tag{2.111}$$

$$\mathbf{s} \ge \mathbf{0} \tag{2.112}$$

The basic optimization procedure we have just derived is applicable. We add a special form of nothing, Equation 2.110 multiplied by a Lagrange multiplier vector $\boldsymbol{\lambda}$ such that the resultant product is a scalar, to the cost function of Equation 2.107 to obtain the Lagrangian

$$L' = \Pi(\mathbf{x}) + \boldsymbol{\lambda}\mathrm{T}[\mathbf{b} - \mathbf{C}(\mathbf{x}) - \mathbf{s}] \tag{2.113}$$

Now when Equation 2.110 is satisfied, the cost function of Equation 2.113 is precisely the same as that of Equation 2.107. We will maximize Equation 2.113 and then adjust the Lagrange multiplier such that Equation 2.110 is valid. This is a standard approach in optimization theory, generally called the calculus of variations, and can easily be shown to be a valid one.

Table 2.4 illustrates the necessary conditions for an optimum easily obtained by this suggested procedure. It is obtained by using Table 2.3 with the cost function of Equation 2.113 and the inequality constraints of Equations 2.111 and 2.112. We obtain, considering **x**, **s**, and $\boldsymbol{\lambda}$ as variables,

$$\mathbf{x} \geq \mathbf{0}$$

$$\mathbf{s} \geq \mathbf{0}$$

$$\frac{\partial L'}{\partial \mathbf{x}} = \frac{\partial \Pi(\mathbf{x})}{\partial \mathbf{x}} - \frac{\partial \mathbf{C}^{\mathrm{T}}(\mathbf{x})}{\partial \mathbf{x}} \boldsymbol{\lambda} \leq \mathbf{0}$$

$$\frac{\partial L'}{\partial \mathbf{s}} = -\boldsymbol{\lambda} \leq \mathbf{0}$$

$$\mathbf{x}^{\mathrm{T}} \frac{\partial L'}{\partial \mathbf{x}} = \mathbf{x}^{\mathrm{T}} \left(\frac{\partial \Pi(\mathbf{x})}{\partial \mathbf{x}} - \frac{\partial \mathbf{C}^{\mathrm{T}}(\mathbf{x})}{\partial \mathbf{x}} \boldsymbol{\lambda} \right) = 0$$

$$\mathbf{s}^{\mathrm{T}} \frac{\partial L'}{\partial \mathbf{s}} = -\mathbf{s}^{\mathrm{T}} \boldsymbol{\lambda} = 0$$

$$\frac{\partial L'}{\partial \boldsymbol{\lambda}} = \mathbf{0} = \mathbf{b} - \mathbf{C}(\mathbf{x}) - \mathbf{s}$$

Using $\mathbf{b} - \mathbf{C}(\mathbf{x})$ to replace **s** leads immediately to Table 2.4. In Table 2.4 the equivalent Lagrangian, which does not contain **s**, has been defined. We could show that these conditions are both necessary and sufficient if the objective function is concave and the constraint function is convex, but to do this would take us far afield of our principle goal here, the study and engineering of economic systems. Unless we can demonstrate this concavity–convexity property or otherwise show that a locally optimal solution is also a globally optimal solution, we cannot claim sufficiency. The algorithms describing the conditions for optimality in Table 2.4 are known as the Kuhn–Tucker conditions and are discussed in most operations research texts; they are very useful theoretical constructs.

Although the Kuhn–Tucker conditions of Table 2.4 may provide necessary and sufficient conditions for optimality, they provide little if any clues concerning how to seek the optimum. There are a number of methods, usually involving various gradient-type algorithms, discussed in nonlinear programming texts in operations research that can be used to determine numerical solutions to the Kuhn–Tucker conditions or the basic nonlinear programming problem from which these conditions result.

TABLE 2.4. Nonlinear Optimization with Constraints: The Kuhn–Tucker Conditions for Optimality

Objective function to be maximized	$J = \Pi(\mathbf{x})$
Constraints	$\mathbf{C}(\mathbf{x}) \leq \mathbf{b}$
	$\mathbf{x} \geq \mathbf{0}$
Lagrangian	$L = \Pi(\mathbf{x}) + \boldsymbol{\lambda}^{\mathrm{T}}[\mathbf{b} - \mathbf{C}(\mathbf{x})]$
Necessary conditions: Kuhn–Tucker conditions	$\frac{\partial L}{\partial \mathbf{x}} = \frac{\partial \Pi(\mathbf{x})}{\partial \mathbf{x}} - \frac{\partial \mathbf{C}^{\mathrm{T}}(\mathbf{x})}{\partial \mathbf{x}} \boldsymbol{\lambda} \leq \mathbf{0}$
	$\mathbf{x}^{\mathrm{T}} \frac{\partial L}{\partial \mathbf{x}} = \mathbf{x}^{\mathrm{T}} \left(\frac{\partial \Pi(\mathbf{x})}{\partial \mathbf{x}} - \frac{\partial \mathbf{C}^{\mathrm{T}}(\mathbf{x})}{\partial \mathbf{x}} \boldsymbol{\lambda} \right) = 0$
	$\frac{\partial L}{\partial \boldsymbol{\lambda}} = \mathbf{b} - \mathbf{C}(\mathbf{x}) \geq \mathbf{0}$
	$\boldsymbol{\lambda}^{\mathrm{T}} \frac{\partial L}{\partial \boldsymbol{\lambda}} = \boldsymbol{\lambda}^{\mathrm{T}}[\mathbf{b} - \mathbf{C}(\mathbf{x})] = 0$
	$\mathbf{x} \geq \mathbf{0}$
	$\boldsymbol{\lambda} \geq \mathbf{0}$

Example 2.12: We assume a specific production function of the Cobb–Douglas type,

$$q = f(x) = x_1^2 x_2^4 \tag{2.114}$$

that has increasing return to scale, as we can easily show. The factor prices are w_1 and w_2 and the output product price is p. We desire to maximize profit and this is given by

$$\Pi(\mathbf{x}) = p x_1^2 x_2^4 - w_1 x_1 - w_2 x_2 - \mathrm{FC} \tag{2.115}$$

subject to a budget constraint given by

$$C(\mathbf{x}) = w_1 x_1 + w_2 x_2 + \mathrm{FC} = b \tag{2.116}$$

We adjoin the equality constraint of Equation 2.116 to the cost function of Equation 2.115 by means of a Lagrange multiplier to obtain the Lagrangian

$$L = \Pi(\mathbf{x}) + \lambda[b - C(\mathbf{x})] = p x_1^2 x_2^4 - w_1 x_1 - w_2 x_2 - \mathrm{FC} + \lambda(b - w_1 x_1 - w_2 x_2 - \mathrm{FC}) \tag{2.117}$$

We set

$$\frac{\partial L}{\partial x_1} = \frac{\partial L}{\partial x_2} = \frac{\partial L}{\partial \lambda} = 0$$

and obtain

$$2 p x_1 x_2^4 - w_1(1 + \lambda) = 0 \tag{2.118}$$

$$4px_1^2x_2^3 - w_2(1+\lambda) = 0 \tag{2.119}$$

$$w_1x_1 + w_2x_2 + \mathrm{FC} = b \tag{2.120}$$

By dividing Equation 2.118 by Equation 2.119, we obtain

$$\frac{2x_1}{x_2} = \frac{w_2}{w_1} \tag{2.121}$$

This relation is just a special case of Equation 2.96, which states that the ratio of marginal products is just the ratio of their input values.

When we combine Equations 2.120 and 2.121, we obtain a solution for the optimum input factors and the solution to the example is complete. It is a straightforward matter for us to show that the solution to the specific problem posed here is equivalent, in that the resource inputs for a given output are equal, to that of maximizing the output production q subject to the budget equality constraint. ■

Example 2.13: We again consider the quadratic production function, where $\mathbf{a} > \mathbf{0}$ and $\mathbf{B}$ is symmetric negative definite, as given by the expression

$$Q = f(\mathbf{x}) = \mathbf{a}^{\mathrm{T}}\mathbf{x} + 0.5\mathbf{x}^{\mathrm{T}}\mathbf{B}\mathbf{x} \tag{2.122}$$

For this production function the profit, excluding fixed costs, is given by

$$\Pi(\mathbf{x}) = pq - \mathbf{w}^{\mathrm{T}}\mathbf{x} = \mathbf{a}^{\mathrm{T}}\mathbf{x}p + 0.5\mathbf{x}^{\mathrm{T}}\mathbf{B}\mathbf{x}p - \mathbf{w}^{\mathrm{T}}\mathbf{x} \tag{2.123}$$

As we have seen, the problem of the firm in the long run is solved if we set the marginal profit equal to zero to obtain

$$\mathbf{MP}(\mathbf{x}) = \frac{\partial\Pi(\mathbf{x})}{\partial\mathbf{x}} = \mathbf{a}p + \mathbf{B}\mathbf{x}p - \mathbf{w} = \mathbf{0} \tag{2.124}$$

We should purchase the optimum input factor vector, which is determined from this relation as

$$\hat{\mathbf{x}} = -\mathbf{B}^{-1}(\mathbf{a} - \mathbf{w}/p) \tag{2.125}$$

as long as $\mathbf{x} \geq \mathbf{0}$. If some components of $\hat{\mathbf{x}}$ obtained from Equation 2.125 are negative, we set those values of $\hat{\mathbf{x}}_i$ equal to zero and modify the profit equation (Eq. 2.123) and the optimum factor input of Equation 2.125 accordingly. To illustrate this we assume that the initial solution of Equation. 2.125 leads to

$$\hat{\mathbf{x}} = \begin{bmatrix} \hat{\mathbf{x}}_1 \\ \hat{\mathbf{x}}_2 \end{bmatrix} \tag{2.126}$$

where $\hat{\mathbf{x}}_1 \geq \mathbf{0}$ and $\hat{\mathbf{x}}_2 \leq \mathbf{0}$. We can rewrite Equation 2.123 in terms of $\mathbf{x}_1$ and $\mathbf{x}_2$ as

$$\Pi(\mathbf{x}) = \mathrm{a}_1^{\mathrm{T}}\mathbf{x}_1 p + \mathrm{a}_2^{\mathrm{T}}\mathbf{x}_2 p + 0.5\begin{bmatrix}\mathbf{x}_1 \\ \mathbf{x}_2\end{bmatrix}^{\mathrm{T}}\begin{bmatrix}\mathbf{B}_{11} & \mathbf{B}_{12} \\ \mathbf{B}_{12}^{\mathrm{T}} & \mathbf{B}_{22}\end{bmatrix}\begin{bmatrix}\mathbf{x}_1 \\ \mathbf{x}_2\end{bmatrix} p - \mathbf{w}_1^{\mathrm{T}}\mathbf{x}_1 - \mathbf{w}_2^{\mathrm{T}}\mathbf{x}_2 \quad (2.127)$$

We set purchases of factor $\mathbf{x}_2$ equal to zero and Equation 2.127 becomes, with $\mathbf{x}_2$=0,

$$\Pi(\mathbf{x}_1) = \mathbf{a}_1^{\mathrm{T}}\mathbf{x}_1 p + 0.5\mathbf{x}_1^{\mathrm{T}}\mathbf{B}_{11}\mathbf{x}_1 p - \mathbf{w}_1^{\mathrm{T}}\mathbf{x}_1 \quad (2.128)$$

and we see that the optimum input factor vector is

$$\hat{\mathbf{x}}_1 = -\mathbf{B}_{11}^{-1}(\mathbf{a}_1 - \mathbf{w}_1/p), \ \hat{\mathbf{x}}_2 = \mathbf{0} \quad (2.129)$$

If we are concerned with the problem of the firm in the short run, we should maximize profits subject to the constraint of limited input resources, perhaps constraints that are due to budget limits or resource shortages. We suppose here that this constraint is, for $\mathbf{d} > \mathbf{0}$ and $\mathbf{C}$ symmetric positive semidefinite,

$$\mathbf{Cx} \leq \mathbf{d} \quad (2.130)$$

If some of the components of $\hat{\mathbf{x}}$ obtained by unconstrained minimization are negative, they will be set equal to zero for both the constrained and the unconstrained optima. Thus we will minimize

$$\Pi(\mathbf{x}_1) = \mathbf{a}_1^{\mathrm{T}}\mathbf{x}_1 p + 0.5\mathbf{x}_1^{\mathrm{T}}\mathbf{B}_{11}\mathbf{x}_1 p - \mathbf{w}_1^{\mathrm{T}}\mathbf{x}_1 \quad (2.131)$$

subject to a constraint of the form

$$\mathbf{C}_{11}\mathbf{x}_1 \leq \mathbf{d}_1 \quad (2.132)$$

Use of Table 2.3 leads to the conditions for optimality of $\mathbf{x}_1$, where we assume that we have already satisfied $\mathbf{x}_1 \geq \mathbf{0}$. These conditions are

$$\mathbf{a}_1 p + \mathbf{B}_{11}\mathbf{x}_1 p - \mathbf{w}_1 - \mathbf{C}_{11}^{\mathrm{T}}\boldsymbol{\lambda} = \mathbf{0} \quad (2.133)$$

$$\mathbf{C}_{11}\mathbf{x}_1 \leq \mathbf{d}_1 \quad (2.134)$$

$$\boldsymbol{\lambda} \geq \mathbf{0} \quad (2.135)$$

$$\boldsymbol{\lambda}^{\mathrm{T}}(\mathbf{d}_1 - \mathbf{C}_{11}\mathbf{x}_1) = 0 \quad (2.136)$$

There will generally be a number of rows of Equation 2.136 that will be satisfied owing to the inequality of Equation 2.134, and a number that will be satisfied owing to the equality of Equation 2.133. In other words, some of the inequality constraints in Equation 2.132 will become equality constraints and the rest of the inequality constraints in Equation 2.132 are really not needed, since satisfaction of other inequality constraints routinely ensures their satisfaction. For example, the inequality constraints $x_1 < 2$ and $x_1 + x_2 < 3$ are ineffective if we also have $x_1 < 1$ and $x_1 + x_2 < 2$. Unfortunately the Kuhn–Tucker conditions do not generally give any clues as to which is the set of truly needed constraints from the total set of inequality constraints of Equation 2.132. As we have mentioned, there are numerical algorithms from nonlinear programming that accomplish this. The result of using these algorithms is that we replace Equation 2.132 by a reduced set of equality constraints

$$\mathbf{C}_{11}\mathbf{x}_1 = \boldsymbol{\delta}_1 \tag{2.137}$$

In terms of these, the necessary conditions for optimality are given by

$$\mathbf{a}_1 p + \mathbf{B}_{11}\mathbf{x}_1 p - \mathbf{w}_1 - \mathbf{C}_{11}^{\mathrm{T}}\boldsymbol{\lambda} = \mathbf{0} \tag{2.138}$$

$$\mathbf{b}_1 - \mathbf{C}_{11}\mathbf{x}_1 = \mathbf{0} \tag{2.139}$$

where $\boldsymbol{\lambda}$ is the Lagrange multiplier for the minimally sized effective constraints. From Equations 2.138 and 2.139 we obtain

$$\hat{\boldsymbol{\lambda}} = (\mathbf{C}_{11}\mathbf{B}_{11}^{-1}\mathbf{C}_{11}^{\mathrm{T}})^{-1} p[\boldsymbol{\delta}_1 + \mathbf{C}_{11}\mathbf{B}_{11}^{-1}(\mathbf{a}_1 - \mathbf{w}_1/p)] \tag{2.140}$$

$$\hat{\mathbf{x}}_1 = -\mathbf{B}_{11}^{-1}(\mathbf{a}_1 - \mathbf{w}_1/p - \mathbf{C}_{11}^{\mathrm{T}}\hat{\gamma}/p) \tag{2.141}$$

as the optimum factor input.

To illustrate this further, let us assume a specific case with three input factors and

$$\mathbf{a} = \begin{bmatrix} 1 \\ 3 \\ 2 \end{bmatrix}, \quad \mathbf{B} = \begin{bmatrix} -1 & 0 & 0.5 \\ 0 & -2 & 1 \\ 0.5 & 1 & -1 \end{bmatrix}, \quad \mathbf{w} = \begin{bmatrix} 5 \\ 10 \\ 30 \end{bmatrix}, \quad p = 10$$

First we determine the unconstrained optimum input factors. Since

$$\mathbf{B}^{-1} = -\begin{bmatrix} 2 & 1 & 2 \\ 1 & 1.5 & 2 \\ 2 & 2 & 4 \end{bmatrix}$$

we have from Equation 2.125

$$\hat{\mathbf{x}} = \begin{bmatrix} 2 & 1 & 2 \\ 1 & 1.5 & 2 \\ 2 & 2 & 4 \end{bmatrix} \begin{bmatrix} 0.5 \\ 2 \\ -1 \end{bmatrix} = \begin{bmatrix} 1 \\ 1.5 \\ 1 \end{bmatrix}$$

which results in q=6.25 and $\Pi(\mathbf{x})$=12.5. Even though $w_3 > pa_3$, we do not obtain a negative x_3 factor in our solution. This is because of the enhancing value x_3 contributes to the output product q owing to the cross-product terms $0.5x_1x_3 + x_2x_3$ that appear in the quadratic production equation. If the **B** matrix were, say,

$$\mathbf{B} = \begin{bmatrix} -1 & 0 & 0 \\ 0 & -2 & 0.5 \\ 0 & 0.5 & -1 \end{bmatrix}$$

then these enhancing features disappear and x_3 is not nearly so desirable in influencing production. With this new **B** matrix and the same **a**, **w**, and p, we obtain from Equation 2.125

$$\hat{\mathbf{x}} = \begin{bmatrix} 1 & 0 & 0 \\ 0 & 0.571 & 0.286 \\ 0 & 0.286 & 1.143 \end{bmatrix} \begin{bmatrix} 0.5 \\ 2 \\ -1 \end{bmatrix} = \begin{bmatrix} 0.500 \\ 0.856 \\ -0.571 \end{bmatrix}$$

We also obtain q=0.905 and $\Pi(\mathbf{x})$=15.12. We see that we desire a negative input factor 3 with this new **B**. This is because of this new **B** matrix and the fact that **a** is not greater than $\mathbf{w}/p$. Component x_3 is negative in that a_3=2 and w_3/p=3. Thus we find it desirable to sell x_3 from the point of view of profit, and this is not allowed in our problem formulation. To allow this, would be, in effect, to allow the use of an invalid production model.

So we set x_3=0 and then reformulate the problem as a two-input factor problem where

$$\mathbf{a}_1 = \begin{bmatrix} 1 \\ 3 \end{bmatrix}, \quad \mathbf{B}_{11} = \begin{bmatrix} -1 & 0 \\ 0 & -2 \end{bmatrix}, \quad \mathbf{w}_1 = \begin{bmatrix} 5 \\ 10 \end{bmatrix}, \quad p = 10 \tag{2.142}$$

The optimum unconstrained factor input is now $\hat{x}_3 = 0$ and

$$\hat{\mathbf{x}}_1 = -\mathbf{B}_{11}^{-1}\left(\mathbf{a}_1 - \frac{\mathbf{w}_1}{p}\right) = \begin{bmatrix} 0.5 \\ 1 \end{bmatrix} \tag{2.143}$$

This results in an output q=2.375 and a profit $\Pi(\hat{\mathbf{x}}) = 11.25$. As we should expect, our profit is reduced by forcing $\hat{\mathbf{x}}_3 = 0$. But the "profit" obtained with a negative x_3 is not realistic owing to the use of an invalid model.

Now suppose that we introduce the budget constraint

$$w_1x_1 + w_2x_2 + w_3x_3 \leq 20 \tag{2.144}$$

and a constraint, imposed, perhaps, by rationing or some other resource availability constraint,

$$x_1 + x_2 + x_3 \leq 1 \tag{2.145}$$

Examination of the constraint of Equation 2.144 indicates that it is satisfied by the optimum solution of Equation 2.143. However, the constraint of Equation 2.145 is not satisfied by the solution of Equation 2.143. Thus we use Equation 2.140, where

$$\mathbf{C}_{11} = [1 \quad 1], \quad \delta_1 = 1$$

to obtain

$$\hat{\gamma} = 3.33, \ \hat{\mathbf{x}}_1 = \begin{bmatrix} \hat{x}_1 \\ \hat{x}_2 \end{bmatrix} = \begin{bmatrix} 0.167 \\ 0.833 \end{bmatrix}$$

Because of the input resource constraint, we cannot use a budget that is anywhere near that of Equation 2.144 and the productivity and profit of the firm are reduced somewhat. With this constraint we spend $\mathbf{w}^T\hat{\mathbf{x}} = 9.168$ of the 20 units available. We have $q=1.958$ and $\Pi(\mathbf{x})=10.417$. Thus we see that we are able to determine the factor inputs, subject to an inequality constraint, such that the profit is maximized. ■

2.5 IMPERFECT COMPETITION

Technological and market constraints affect the firm. All production schemes are not possible and, as we have seen, one of the problems of the firm is to determine production schemes that are technically feasible. Market conditions preclude decisions by firms only concerning their production quantity. Other agents and mechanisms influence the determination of production quantities. However, firms will certainly have some control over the prices to be charged for goods. If a firm is "large enough," it can unilaterally set the prices it charges for products, or the wages that it will pay for factor inputs to production. We will examine some characteristics associated with monopoly and monopsony, the names given to these activities, in this section.

A purely monopolistic firm is one that is able to set the price of its product in accordance with an estimate of the demand for the product. The monopolist, in effect, has information that a firm in "perfect competition" does not have, and thus the monopolist should always be able to obtain greater profits than firms that accept prices as givens, if we assume that the monopolist's demand estimate is correct. We will present some precise discussions of the implications of perfect competition in Chapter 4 and, especially, Chapter 6.

We will assume that efficient factor inputs are known[7] so that we can write the cost of production in terms of the production level as $C(q)$. We can write an expression for the revenue to the firm as

[7]These are obtained by minimizing the cost of production, given by $C=\mathbf{w}^T\mathbf{x}$, for a prescribed output production level $q=f(\mathbf{x})$ and then finding the minimum cost $C(q)$ in terms of this production level.

$$R(q) = Þ(q)q \tag{2.146}$$

such that the profit of the firm is given by

$$\Pi = R(q) - C(q) \tag{2.147}$$

The first-order necessary conditions for profit maximization are obtained as

$$\frac{d\Pi}{dq} = \frac{dR(q)}{dq} - \frac{dC(q)}{dq} = \text{MR}(q) - \text{MC}(q) = 0 \tag{2.148}$$

This general relation, that a profit-maximizing firm will set marginal revenue equal to marginal cost, is a very important one. Ideally, it would be "nice" to obtain a marginal revenue that is much higher than marginal cost, but this is completely unrealistic. By using the revenue relation (Eq. 2.146), we obtain

$$\frac{dR(q)}{dq} = Þ(q) + q\frac{dÞ(q)}{dq} = Þ(q)\left(1 + \frac{1}{\varepsilon(q)}\right) \tag{2.149}$$

where $\varepsilon(q)$ is the price elasticity of demand, or factor elasticity,

$$\varepsilon(q) = \frac{Þ(q)}{q}\frac{dq}{dÞ(q)} \tag{2.150}$$

which is easily shown to be negative. By using Equation 2.149, we obtain from Equation 2.148

$$Þ(q)\left(1 + \frac{1}{\varepsilon(q)}\right) = \frac{dC(q)}{dq} \tag{2.151}$$

From this relation we see that the price a monopolist will charge for a product is greater than that which would be justified by just setting the price equal to the marginal cost. Generally this will be higher than the price charged by a price-taking firm with the same minimum production cost function. Generally also, the quantity produced by a monopolistic firm will be less than that produced by a price-taking firm with the same minimum production cost function.

Example 2.14: As a simple illustration of the computations for a monopolist firm, and for comparison with an equivalent price-taking firm, we consider the production cost function

$$C(q) = 1 + q^2 \tag{2.152}$$

With this production cost, we have for average and marginal costs

$$\text{AC}(q) = \frac{C(q)}{q} = \frac{1}{q} + q \tag{2.153}$$

$$\mathrm{MC}(q) = \frac{\mathrm{d}C(q)}{\mathrm{d}q} = 2q \tag{2.154}$$

For a price-taking firm we obtain the optimum production level by setting price equal to marginal cost and we have as a result

$$p = 2q \tag{2.155}$$

This is the supply function for the pure price-taking firm. There is no equivalent of a supply curve for a monopolistic firm. This is obviously so, since the monopolist will adjust production as a function of the available information base, which includes knowledge of the consumer demand curve. We assume that the monopolist's estimate of demand is

$$\hat{p}(q) = a - bq \tag{2.156}$$

From Equation 2.5.5, we have for the price elasticity of demand

$$\varepsilon(q) = 1 - \frac{a}{bq} \tag{2.157}$$

We see that this is negative, since we must have $a > bq$ for the demand to be positive. The marginal revenue is obtained as

$$\mathrm{MR}(q) = \frac{\mathrm{d}R(q)}{\mathrm{d}q} = a - 2bq \tag{2.158}$$

Thus, we have for the operating condition for maximum profit where marginal revenue equals marginal cost

$$\mathrm{MR}(q) = \mathrm{MC}(q) = a - 2bq = 2q \tag{2.159}$$

$$q_{\mathrm{M}} = \frac{a}{2(1+b)} \tag{2.160}$$

We substitute this value into Equation 2.156 and obtain for the price charged by the monopolist

$$p_{\mathrm{M}} = \frac{a(2+b)}{2(1+b)} \tag{2.161}$$

If we assume that the pure price-taking firm faces the same demand curve as the monopolist, we have, from Equations 2.155 and 2.156, for the price and quantity produced, the relations

$$q_{\mathrm{PT}} = \frac{a}{(2+b)}, \quad p_{\mathrm{PT}} = \frac{2a}{(2+b)}$$

The profits of the two types of firms are obtained as

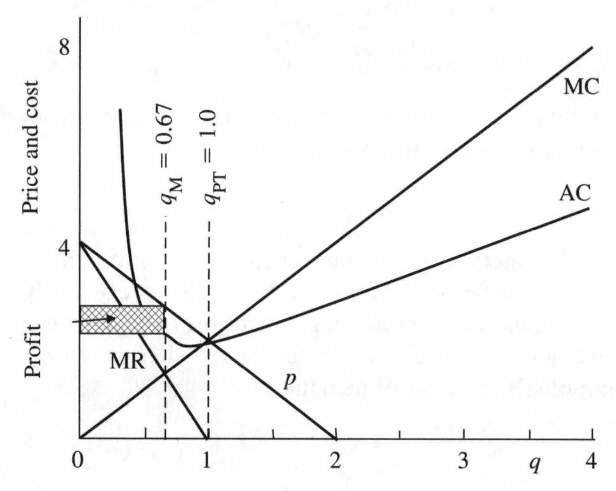

Figure 2.7. Graphic Solutions for Example 2.14.

$$\Pi_M = \frac{a^2}{4(1+b)} - 1, \ \Pi_{PT} = \frac{a^2}{(2+b)^2} - 1$$

As we see, the profit of the monopolistic firm is always at least as great as that of the price-taking firm, assuming only that $b > 0$.

Figure 2.7 illustrates the average cost, marginal cost, marginal revenue, and profit curves for this example for the particular case where a=4 and b=2. Of particular interest in Fig. 2.7 is the intersection of the various curves that determine the production level, price, and profit received. We easily obtain $q_M=2/3$, $p_M=8/3$, $\Pi_M=1/3$, $q_{PT}=1$, $p_{PT}=2$, $\Pi_{PT}=0$. ■

Example 2.15: We consider a monopolist with a unit return to scale Cobb–Douglas production function of the form

$$q = L^\alpha K^{1-\alpha} \tag{2.162}$$

The wages for labor and capital are w_L and w_K. We will first find the efficient cost of production relation. To do this, we maximize production subject to a constraint on production cost C. The Lagrangian associated with this maximization is given by

$$\Lambda = L^\alpha K^{1-\alpha} + \lambda(C - w_L L - w_K K)$$

We obtain from setting $\partial\Lambda/\partial L = \partial\Lambda/\partial K = \partial\Lambda/\partial\lambda = 0$, and then eliminating the Lagrange multiplier, the relations

$$\frac{K}{L} = \frac{1-\alpha}{\alpha}\frac{w_L}{w_K}, \ C = w_L L + w_K K$$

Using these results in Equation 2.162 leads to the efficient production relation

$$q = \frac{(1-\alpha)^{1-\alpha}\alpha^{\alpha}}{(w_L)^{\alpha}(w_K)^{1-\alpha}} C = \beta C$$

This indicates that the efficient quantity produced is a linear function of the production cost.

Let us assume that the monopolist knows that there exists a linear demand function $q = a - bp$. We are interested in determining the quantity that the monopolist will produce and the resulting profit. The monopolist's operating conditions are determined by equating marginal revenue to marginal cost. The revenue to the monopolist is given by

$$R(q) = p(q)q = \frac{a-q}{b}(q)$$

The operating costs are

$$C(q) = \frac{q}{\beta}$$

Here, we obtain for the marginal revenue

$$\frac{\mathrm{d}R(q)}{\mathrm{d}q} = \frac{a-2q}{b} = \frac{\mathrm{d}c(q)}{\mathrm{d}q} = \frac{1}{\beta}$$

This relation may be solved for the quantity produced as

$$q = \frac{a}{2} - \frac{b}{2\beta}$$

The difference between revenue and operating costs is the profit for the monopolist and this and the optimum sales price are obtained as

$$\Pi = \frac{(a\beta - b)(a\beta + b - 2b^3)}{4b^3\beta^2}, \quad p = \frac{a}{2b} + \frac{1}{2\beta}$$

It is of interest to examine conditions that could reduce the profit of the monopolist to the point where operation would become impossible. This would occur when the revenue equals the production cost, or when the average revenue equals the average production cost. We obtain for these results

$$R(q) = \frac{a-q}{b}(q) = C(q) = \frac{q}{\beta}, \quad q = a - \frac{b}{\beta}, \quad p = \frac{1}{\beta}, \quad \Pi = 0$$

This could be achieved by government action, such as through a price-freeze action to regulate the monopolist.

Some numerical results are of interest here. If we have $\alpha = 0.5$, $w_L = w_K = 0.1$, $a = 10$, and $b = 1$, then we obtain the following for the monopolist's operating conditions where there is no price fixing: $\beta = 5$, $q = 4.9$, $p = 5.1$, and $\Pi = 24.01$. With price fixing at $p = 0.20$, we get $q = 9.8$ and $\Pi = 0$. Generally, the monopolist operates by controlling prices. Price fixing by government action defeats a fundamental goal of a monopolist. ■

As we have seen in Example 2.14, under monopoly a productive unit will generally cut the output price of a product to sell more of the product. Thus, we have

$$\frac{\mathrm{d}p}{\mathrm{d}q} = \frac{\mathrm{d}\text{Þ}(q)}{\mathrm{d}q} < 0 \tag{2.163}$$

We must modify our previous price-taking relationships to incorporate this variable-price structure. For example, the revenue of the firm now becomes

$$R = pq = \text{Þ}(q)q \tag{2.164}$$

We now obtain *marginal revenue*, as we have just seen, as

$$\mathrm{MR}(q) = \frac{\mathrm{d}R}{\mathrm{d}q} = \text{Þ}(q) + \frac{\mathrm{d}\text{Þ}(q)}{\mathrm{d}q}q \tag{2.165}$$

We see that the marginal revenue now depends on the output quantity of the productive unit. The marginal revenue under monopoly is less than that obtained for the unit output price Þ(q) of the productive unit.

Monopsony is a term sometimes used in microeconomics. The monopsonist controls the price of input factors by adjusting input prices or wages as a function of the input factor purchases according to

$$w_i = \Omega_i(x_i) \tag{2.166}$$

Since a classic productive unit can generally only purchase more of a factor by paying a higher price for it, we require

$$\frac{dw_i}{dx_i} = \frac{d\Omega(x_i)}{dx_i} > 0 \tag{2.167}$$

The cost of the ith input factor is the factor production cost and is given by

$$PC_i = w_i x_i = \Omega_i(x_i)x_i \tag{2.168}$$

The *marginal cost of the ith factor input* is the derivative of the cost of this factor with respect to the factor input level. This is

$$\mathrm{MC}_i(x_i) = \frac{d\mathrm{PC}_i}{dx_i} = w_i(x_i) + \frac{d\Omega_i(x_i)}{dx_i}x_i \tag{2.169}$$

We see that this marginal cost exceeds the price of the input factor.

Under the imperfect competition of monopoly and monopsony, the unconstrained profit maximization problem is that of maximizing

$$\Pi = \text{Þ}(q)q - \sum_{i=1}^{n} \Omega_i(x_i) - \text{FC} \tag{2.170}$$

subject to the production function equality constraint

$$q = f(x) = f(x_1, x_2, \ldots, x_n) \tag{2.171}$$

We could insert q from Equation 2.171 into Equation 2.170 to eliminate the need for a Lagrange multiplier, but we will not do this here to give a special interpretation to the Lagrange multiplier for this problem.

To solve this problem, we adjoin Equation 2.171 to Equation 2.170 by means of a Lagrange multiplier and obtain the Lagrangian

$$L = \text{Þ}(q)q - \sum_{i=1}^{n} \Omega_i(x_i)x_i - \text{FC} - \lambda[q - f(x_1, x_2, \ldots, x_n)] \tag{2.172}$$

The necessary conditions for an extremum are given by the expressions

$$\frac{\partial L}{\partial q} = 0 = \frac{\partial \text{Þ}(q)}{\partial q} q + \text{Þ}(q) - \lambda \tag{2.173}$$

$$\frac{\partial L}{\partial x_i} = 0 = -\frac{\partial \Omega_i(x_i)}{\partial x_i} x_i - \Omega_i(x_i) + \lambda \frac{\partial f(\mathbf{x})}{\partial x_i} \tag{2.174}$$

$$\frac{\partial L}{\partial \lambda} = 0 = q - f(x_1, x_2, \ldots, x_n) \tag{2.175}$$

Equation 2.173 leads to the interesting and useful result that the optimum Lagrange multiplier is, for this particular problem formulation, just the marginal revenue

$$\lambda = \text{MR}(q) = \text{Þ}(q) + \frac{\partial \text{Þ}(q)}{\partial q} q \tag{2.176}$$

Equation 2.174 leads to the conclusion that the product of the marginal revenue (λ) and the marginal factor productivity $\partial f / \partial x_i$ is just the marginal cost of that factor. It is convenient to denote the *factor marginal revenue product* as this product, such that we have

$$\text{MRP}_i = [\text{MR}(q)]\text{MP}_i = \text{MC}_i(x_i) = \frac{\partial \Omega(x_i)}{\partial x_i} x_i + \Omega(x_i) \tag{2.177}$$

If there are a few productive units and if these few units mutually agree to cooperate such as to jointly price their outputs, we have what is known as an *oligopoly*. For two productive units (duopoly) we have the outputs $q^1 = f^1(x^1)$ and

$q^2 = f^2(x^2)$. The firms cooperate in setting prices such that $p^1 = \text{Þ}^1(q^1, q^2)$ and $p^2 = \text{Þ}^2(q^1, q^2)$. If two firms cooperate in such a way, as they could if they were the only buyers of a given factor input, to set prices and wages for a factor input according to their needs for the input factors $w_i = \Omega_i(x_i^1, x_i^2)$, then we have what is known as an *oligopsony*.

Problems involving monopoly, monopsony, oligopoly, and oligopsony can be formulated as optimization problems such that the economic optimization methods we have formulated in this section are applicable. We shall not do this here, since determination of the necessary conditions is a relatively straightforward task for optimization problems of the sort we have been discussing. Unfortunately closed-form solutions to the problems we have been discussing in this chapter are difficult or impossible to obtain except in very special circumstances for relatively simple production functions, and numerical methods are generally essential. Problems 23 to 34 at the end of this chapter expand considerably on the concepts presented here concerning monopoly, monopsony, oligopoly, and oligopsony and provide for some interesting comparisons with firms in perfect competition.

2.6 SUMMARY

In this chapter we have examined some elementary concepts from production theory or the theory of the firm. We have introduced the concept of a production function and a number of economic concepts concerning production that will be very useful to us later. We have briefly examined the fundamental or classic problem of the firm and used some elementary optimization concepts to illustrate how, in some very simple cases, profit maximization can be obtained.

There are many topics concerning production that we have not discussed. Perhaps the most important one is that we presently have no way of determining how much output a firm should produce to accommodate purchases of this output. This is related, of course, to the demand for the product of the firm, a subject known as the theory of the household or the theory of consumer. Also, we did not examine a number of factors, such as advertising, capital costs, taxes, and the effects of negative externalities such as pollution, or the effects of social and regulatory inputs to the production process. Chapter 3 will examine the theory of the household and later chapters will discuss some of the other important inputs to the production determination process.

PROBLEMS

1. Show that the marginal productivity and average productivity are equal at the point of maximum average productivity. Is there any physical significance to this point?
2. What do the curves of Fig. 2.3 become for

a. a linear production function,
b. a Cobb–Douglas production, and
c. the production function $q = 16x_1x_2 - x_1^3x_2^3$.

3. Repeat Examples 2.3 through 2.5 using the production function $q = 16x_1^2x_2^2 - x_1^3x_2^3$.

4. Reconstruct Table 2.1 for the production function of Problem 3.

5. Investigate the behavior and important properties of the following production functions, using Table 2.1 as a guide:
   a. $q = a[bx_1^{-c} + (1-b)x_2^{-c}]^{-1/c}$
   b. $q = ax_1^2x_2^2 - bx_1^4x_2^4$
   c. $q = ax_1^bx_2^{1-b} + cx_1dx_2$

6. Is it reasonable that a production function demonstrate diminishing average productivity? Is a law of diminishing average productivity implied by the law of diminishing marginal productivity? You may find it helpful to use the following production functions in support of your arguments. Examine the marginal and average productivities for input x_2.
   a. $q = \left[\dfrac{x_1^2 + 4x_2^2}{3x_1^2 + 4x_2^2}\right]x_2$
   b. $q = \left[\dfrac{x_1^2 + x_2^2}{3x_1^2 + 4x_2^2}\right]x_2$

7. Prove or disprove the statement that if marginal productivity is everywhere a monotonically decreasing function of an input factor, then average productivity is everywhere a monotonically decreasing function of that input factor. Consider a specific example to demonstrate your result.

8. Prove or disprove the statement that if average productivity is everywhere a monotonically decreasing function of an input factor, then marginal productivity is everywhere a monotonically decreasing function of that input factor. Consider a specific example to demonstrate your result.

9. Find the necessary conditions to maximize the output of a productive unit subject to an equality constraint in total cost.

10. What are the requirements on the Cobb–Douglas production function of Equation 2.23 such that it be (a) a homogeneous production function and (b) a linear homogeneous production function?

11. Suppose that the manufacture of a number of products, q, is specified by the Cobb–Douglas production function $q = 1.6x_1^{0.3}x_2^{0.2}x_3^{0.5}$ where x_1 represents the labor used in person-hours, x_2 the number of hours of machine time, and x_3 the aggregated amount of raw material in kilograms. Labor costs $200 per 40 h, machine use costs $300 per 40-h work week, and raw materials cost $50 per kilogram.
    a. What are the optimal proportions of labor, machines, and raw materials in the production process?
    b. What will be the effect of a 20% increase in labor costs on the production process?
    c. What will be the effect of a 100% increase in raw material costs on the production process?

12. Repeat Problem 11 for the production function $q = 1.6x_1 x_2^{0.2} x_3^{0.7}$.

13. Repeat Problem 11 for the linear production function $q=b_1x_1 + b_2x_2 + b_3x_3$.

14. Investigate the behavior of the production function $(1 + \kappa q) = \Pi_{i=1}^{n}(1 + \kappa k_i x_i^{b_i})$. Under what conditions will this become a linear production function? Under what conditions will this become a Cobb–Douglas production function?

15. A manufacturer produces a quantity q of a product using two inputs x_1 and x_2. The prices of the input and output quantities are fixed. Rationing is introduced and the manufacturer may purchase no more than $x_{2\ \max}$ units of x_2. Without rationing, the manufacturer would like to produce using more than $x_{2\ \max}$.

a. What are the conditions for profit maximization under rationing?

b. What are the marginal productivities of x_1 and x_2?

c. What is the relationship between x_1 and x_2?

d. The rate of technical substitution (RTS) or the marginal rate of substitution (MRS) or the marginal rate of technical substitution (MRTS) is defined as the ratio of marginal productivities at any point on the production curve. What is the RTS for this problem both with and without rationing?

16. We wish to allocate a scarce resource x_1 between two production processes such as to maximize their total returns. Thus we wish to maximize $q_1 + q_2$ where $q_1=f_1(x_{1,1})$ and $q_2=f_2(x_{1,2})$, and where we have the constraint $x_{1,1} + x_{1,2}=x_1=\text{fixed}=b$. Inputs to the production process other than x_1 may be ignored for simplicity.

a. What are the conditions for optimality?

b. If $f_1(x_{1,1})=25 - (1 - x_{1,1})^2$ and $f_2(x_{1,2})=50 - (2 - x_{1,2})^2$, what are the optimal inputs?

c. Under what conditions will the marginal revenue be greater than the marginal cost? What is the physical significance of this?

17. For a particular monopolistic situation, marginal revenue is a linear function and marginal cost a quadratic function of production output such that $\text{MR}=A + Bq$ and $\text{MC}=C + Dq + Eq^2$. We will assume that the fixed cost of the product, FC, is known, as are A, B, C, D, and E.

a. What are reasonable signs for A, B, C, D, and E?

b. Find the revenue R, the production cost PC, the total cost C, and the profit Π of the firm.

c. What are optimality conditions for profit maximization?

d. Suppose that the government introduces a sales tax ST that is a linear function of the number of units sold and given by $\text{ST}=Tq$. What is the optimum value of T that produces maximum tax revenue? The tax is paid by the manufacturer.

18. Show that the problems of maximizing output production with a constraint on input costs and minimizing input factor costs subject to a constraint on total output are equivalent problems in the sense that the Lagrangian and the optimum solutions are the same for both problems.

19. The output of a firm is given by $q=\mathbf{b}^T x - 0.5\mathbf{x}^T\mathbf{A}\mathbf{x}$, where $\mathbf{A}$ is a positive definite matrix. Find the values of $\mathbf{x}$ that result in the following:

a. maximum output subject to the budget constraint $\mathbf{x}^T\mathbf{w} \leq B$;

b. minimum input factor cost $C(\mathbf{x})=\mathbf{w}^T\mathbf{x}$ for a fixed output $q=q_f$;

c. maximum profit;

d. maximum profit subject to a budget constraint of the form $\mathbf{x}^T\mathbf{w} \leq B$; and

e. maximum profit subject to the budget constraint of d and rationing of the form $\mathbf{x}^T\mathbf{1} \leq k$, where $\mathbf{1}$ is an appropriately dimensioned vector of all ones.

20. Obtain numerical results for Problem 19 for the case and illustrate your results as a function of B and k where

$$\mathbf{A} = \begin{bmatrix} 1 & 1 & 0.5 \\ 1 & 3 & -2 \\ 0.5 & -2 & 2 \end{bmatrix}, \quad \mathbf{w} = \begin{bmatrix} 3 \\ 4 \\ 5 \end{bmatrix}, \quad p = 10, \quad \mathbf{b} = \begin{bmatrix} 1 \\ 1 \\ 1 \end{bmatrix}.$$

21. Repeat Problem 19 for the case where $\mathbf{A}$ is negative definite. Show that you are now minimizing output and profit!

22. Repeat Example 2.9 for the case where $\mathbf{a}=\mathbf{0}$ and $\mathbf{B}$ is positive definite.

23. The price equation of a monopolist is $\text{Þ}(q)=100 - 5q$. The optimum production cost of the monopolist is given by $\text{PC}=25q$.

a. What is the profit equation for the monopolist?

b. What is the optimum production and price that maximize profit?

c. What is the profit that would result if the monopolist were to determine production by setting the price $\text{Þ}(q)$ equal to the marginal production cost? This is the result obtained under perfect competition conditions. How do these two profits compare?

24. Suppose that a monopsonist sells products at a price $p=4$. The production function for the monopsonist is $q=10x - 0.5x^2$. The wage the monopsonist pays for the input factors x is $\Omega(x)=20 + 3x$.

a. What is the profit equation for the monopsonist?

b. What is the optimum production and factor input that maximize profit?

c. What is the profit that would result if the monopsonist were to determine production by setting marginal revenue $\partial pf(x)/\partial x$ equal to the wage for the factor input w? This is the result obtained under perfect competition conditions. How do these two profits compare?

25. Suppose that the production function for a monopolist–monopsonist firm is $q=\mathbf{a}^T\mathbf{x}$. The price of the product sold decreases as the quantity increases according to $\text{Þ}(q)=c - dq$ and the wages increase as the input factor increases according to the relation $\boldsymbol{\Omega}(\mathbf{x})=\mathbf{e} + \mathbf{F}\mathbf{x}$.

a. What is the profit equation for the monopolist–monopsonist firm?

b. What is the optimum factor input $\mathbf{x}$, wage $\boldsymbol{\Omega}$, price Þ, and quantity q?

c. What is the profit that would result if the firm were to act as a firm in perfect competition and set the product of the price $\text{Þ}(\mathbf{x})$ and the marginal productivity equal to the wages $\boldsymbol{\Omega}(x)$? How does this profit compare to the profit under imperfect competition obtained in part b?

26. A *discriminating monopolist* will sell the product of the firm in two or more markets at different prices. Purchasers are assumed to be unable to purchase the product of the firm in one commodity market and sell it in another. The revenue to the firm for the two-market case is $R=R_1(q_1) + R_2(q_2)$ where $R_1(q_1)=\text{Þ}_1(q_1)\, q_1$ and $R_2(q_2)=\text{Þ}_2(q_2)\, q_2$ represent the revenues in markets 1 and 2, and $\text{Þ}_1(q_1)$ and $\text{Þ}_2(q_2)$

the prices in these markets, respectively. The total production of the firm is $q=q_1 + q_2$, and the cost of production is PC(q). What are the requirements for profit maximization for the discriminating monopolist? In which market will the price be lowest?

27. What are the results of Problem 26 for the case where $\tilde{P}_1(q_1)=100-6.33q_1$, $\tilde{P}_2(q_2)=100-20q_2$ and $PC(q_1+q_2)=25(q_1+q_2)$. What is the profit of the firm? How do the results of this problem compare with those in Problem 23? Show that the aggregation of the two markets in this example results in Problem 23.

28. For an oligopolist production with two firms (generally called a duopoly) the total quantity of production is a function of price $p=g(q)=g(q_1+q_2)$ where q_1 and q_2 are the production levels of the two firms with production costs $PC_1(q_1)$ and $PC_2(q_2)$.

a. What is the expression for the profit of each firm?

b. What is the requirement for each profit to be maximized with respect to the output of the firm?

c. What is the total profit of the two firms?

d. What is the requirement that this total profit be maximized?

29. What do the results of Problem 28 become if $g(q_1+q_2)=100-5(q_1+q_2)$ and $PC_1(q_1)=20q_1$, $PC_2(q_2)=3q_2$?

30. An oligopsonist purchases factor inputs to production according to $W(\mathbf{x})=W(\mathbf{x}_1+\mathbf{x}_2)=g(\mathbf{x}_1+\mathbf{x}_2)$, where $\mathbf{x}_1$ and $\mathbf{x}_2$ are the factor inputs for the two productive units whose production functions are $q_1=f_1(x_1)$ and $q_2=f_2(x_2)$.

a. What is the equation for the profit of each firm?

b. What is the requirement that the profit of each firm be maximized by choice of the factor input for each firm?

c. What is the total profit of both firms?

d. What is the requirement that this profit be maximized with respect to the factor inputs $\mathbf{x}_1$ and $\mathbf{x}_2$?

31. What are the results of Problem 30 for the case where

$$w(x)=5+x_1+x_2$$
$$q_1(x_1)=20x_1-0.5x_1^2$$
$$q_2(x_2)=10x_2-0.5x_2^2$$
$$p_1=8$$
$$p_2=5$$

32. It is sometimes stated that a firm, rather than maximizing profits, will seek to maximize revenue subject to the constraint that profits be above some minimum level. Formulate this problem and determine optimality conditions for

a. a perfectly competitive firm,

b. a firm that practices monopoly and monopsony, and

c. the monopolistic firm of Problem 23.

33. A potential manufacturer is contemplating the purchase of a factory. Two options are available: (a) $q=\alpha KL/(0.6L+0.4K)$ and (b) $q=\beta K^{0.25}L^{0.75}$, where K is capital and L is labor. Determine the appropriate decision rules to enable the selection of a production process in terms of the wages for capital and labor, and α and β.

34. Suppose that the demand curve for a product is $q=10-p$. The cost of making the product is $c(q)=5q$ and the firm is a monopoly.

a. What will be the equilibrium price that maximizes profit, the profit, and the quantity produced?

b. Suppose that price ceilings $p=4.5$ and 6 are set by the government. What will be the production level and profit under these conditions?

c. Sketch these results graphically.

d. What would the response of a firm in perfect competition be to the price ceilings of b? Compare the resulting profit with that for the monopolist.

35. Advertising can generally act to increase the price at which a given quantity of a product will sell. Only firms with monopoly capability will find it desirable to advertise. Suppose that the demand function for a monopolist is $q=25 + A^{0.5}-p$ where A is the cost of advertising. The production and advertising cost of the product is $C=5q+q^2+A$.

a. What are the optimum values of p, q, and A to maximize profit?

b. What is the effect of the advertising?

c. What would be the results for parts a and b for the firm if it operates under perfect competition conditions?

36. Consider a monopolist who produces a product using labor only, with $q=L+0.1L^2-0.0001L^3$. The wage rate for labor is given by $w=20+0.5L$. The price for the product q is $p=3$. Determine the relations for revenue and profit. Determine the conditions that yield maximum profit.

37. Suppose that there are a large number of possible oil firms in the world and each is equally efficient, such that each has the production cost function $\mathrm{PC}(q)=1+q+q^2$, where the cost is measured in millions of dollars and q is oil production in millions of barrels. Suppose that the world demand for oil is $p=10-q/10$.

a. What is the perfectly competitive equilibrium for the oil industry? What is the resulting number of firms in competitive equilibrium?

b. This number of firms now agrees to form a cartel. What price do they now set? At what production level? What is the resulting profit per firm? The form of control for the cartel is the quota. Members are given a maximum production level.

c. What happens if the form of control used to restrict output is the price that is achieved by pooling revenue? What is the best output and pooled revenue per firm?

d. Suppose that a new firm decides not to join the cartel; rather, it decides to sell as much as desired at the price at which the cartel is selling. What are the conditions under which the new firm operates?

e. The government makes the activities in d illegal. It gives the cartel the right to force all oil firms into the cartel. What is the new equilibrium number of firms in the cartel under these conditions?

f. The cartel falls out of favor with the government, which then forbids the restrictive setting of output quotas. The same number of firms is in the cartel as in part f. What are the new equilibrium conditions?

g. Generalize the results of this example to the extent that you find meaningful. Write a few sentences concerning the general lessons learned from this example.

BIBLIOGRAPHY AND REFERENCES

This chapter is a very basic one and there are many economic texts that discuss the theory of the firm. Some of the more classic recent discussions appear in

Baumol WJ, Blinder AS. Microeconomics: principles and policy. 9th ed. Mason Ohio: South-Western; 2009.

Besanko D, Braeutigam RR. Microeconomics: an integrated approach. Hoboken New Jersey: Wiley; 2001.

Jehle GA, Reny PJ. Advanced microeconomic theory. 2nd ed. Reading Massachusetts: Addison Wesley; 2000.

Pindyck RS, Rubinfeld DL. Microeconomics. 5th ed. Englewood Cliffs, New Jersey: Prentice Hall; 2000.

Samuelson PA. Microeconomics. New York: McGraw Hill; 1995.

Varian HR. Microeconomic analysis. New York: W.W. Norton & Company; 1992.

Varian HR. Intermediate microeconomics: a modern approach. 8th ed. New York: W.W. Norton & Company; 2009.

CHAPTER 3

THE THEORY OF THE CONSUMER

3.1 INTRODUCTION

We will define a *consumer* as an individual or group of individuals who each possess income and values and who purchase goods and services, and who generally perform labor for this income. The fundamental problem of the consumer is one of determining the quantity of goods and services to purchase by the given prices for goods and services and the income of the consumer. The consumer will be assumed to make this choice such that the value, or satisfaction or utility, derived from consuming the goods and services is the greatest possible. We will assume, in this chapter, a single consumer, and we will also assume that the consumer is "rational" by being aware of alternative goods and services that may be purchased and capable of evaluating their worth in terms of their value and cost.

To do this we will postulate the existence of a utility function that measures the satisfaction or value derived from alternative goods and services. We will examine some concepts of utility maximization and determine demand functions that measure the quantity of goods and services a consumer will purchase by their price. Then we will examine various properties of this demand function. This will enable us to combine the theory of the firm and the theory of the household and determine various supply–demand or market equilibrium conditions, which is the subject of Chapter 4.

3.2 ECONOMIC UTILITY THEORY AND ITS AXIOMS

Suppose that we wish to purchase a quantity of meat or tofu for preparing a meal for a fixed number of people. It seems almost intuitive that prudent consumers would reduce the quantity of meat or tofu purchased as the price of meat or tofu rises and substitute an increased quantity of other items of lesser cost. Alternatively, suppose that we have the choice of spending n dollars for a trip to Europe or on a new automobile. We decide on the trip to Europe. This is because, in this specific instance, a trip to Europe has greater utility to us as a

Economic Systems Analysis and Assessment,
by Andrew P. Sage and William B. Rouse

consumer than does an automobile. We assume that the consumer is able to make choices "as if" there existed a utility function that could be used to determine them.

We would like to determine consumer demand for various goods and services and we adopt certain, generally reasonable, axioms of the theory of rational choice to accomplish this. First, we define the consumer's vector or bundle of goods and services, which we will call the *commodity bundle*

$$\mathbf{x} = \begin{bmatrix} x_1 \\ x_2 \\ \vdots \\ x_n \end{bmatrix}$$

of n available commodities, or goods and services, for the consumer to purchase. We will not worry about fractional numbers of commodities, such as one and one-half automobiles, but we must assume that only nonnegative commodities are purchased, such that $x_i \geq 0$ for all I, or $\mathbf{x} \geq \mathbf{0}$. The commodity space is the set of all possible commodity bundles $C = \{\mathbf{x} | \mathbf{x} \geq 0\}$, and this commodity space is a closed convex set in the nonnegative part of the n-dimensional Euclidean space.

The "numbers" that we assign to various commodity bundles are irrelevant as long as any bundle x_i that is preferred over any other bundle x_j is assigned a higher utility number, such that we have $U(x_i) > U(x_j)$ if and only if (iff or IFF) $x_i > x_j$. In the foregoing expression, we require only ordinal utility functions. For many purposes, however, we need to trade off commodities. Then we will typically need to know whether the differences between utility levels are comparable. Thus, as a specific comparison, we will be able to determine whether $U(x_i) - U(x_j) = U(x_k) - U(x_i)$, and this will require a cardinal utility function. An ordinal utility function will provide information only about preference orderings among commodity bundles. A cardinal utility function will provide information about differences between utility functions in terms of their value; for example, we would be able to say that one values the difference between a 10-day vacation in Miami and one in Atlanta as equivalent to the difference between a 10-day vacation in Miami and one in Las Vegas through the use of a cardinal utility function.

The requirement that the relative magnitudes of the differences between utilities of individual commodities have explicit meaning is essential in dealing with normative or prescriptive issues that involve economic systems. It requires interpersonal comparisons of values, something that is not easily accomplished. However, this is needed if we are to decompose a utility function into various attributes and aggregate attribute scores to obtain the overall utility.

A utility function $U(\mathbf{x})$ is effectively ordinal if it can be replaced by any monotonically increasing transformation of itself and still preserve all the properties of the utility function. A cardinal utility function $U(\mathbf{x})$ can be replaced only by an affine function of itself and the properties of the utility function as an indicator of preference are preserved. Although many of our developments in utility theory in this chapter apply to ordinal utilities, we find it convenient to view utilities as if they were cardinal utilities. Our initial discussions here apply to ordinal utilities and, since cardinal utilities are certainly ordinal, they apply also to cardinal utilities. We state four axioms applicable to ordinal or cardinal utilities. The first two are the most important and they are as follows:

Axiom 1. The Preference and Indifference Axiom

The choice between two commodity bundles depends on the consumer's basic primitive notions of preference or taste. We use the symbol $>$ to denote *is preferred to*, the symbol $\sim$ to denote *is indifferent to*, and the symbol $\geq$ to denote *is preferred to or indifferent to*. Thus, when we write $\mathbf{x}_1 > \mathbf{x}_2$ we mean that commodity bundle 1 is preferred to commodity bundle 2; when we write $\mathbf{x}_1 \sim \mathbf{x}_2$ we mean that we are indifferent between commodity bundles 1 and 2; and when we write $\mathbf{x}_1 \geq \mathbf{x}_2$ we mean that we prefer or are indifferent to commodity bundles $\mathbf{x}_1$ and $\mathbf{x}_2$.

This axiom assumes that preferences can be expressed, that every pair of commodities gives rise to a choice. Also, we assume that the weak preference relation $\geq$ is complete so that there are no gaps in commodity space. Finally, we assume that weak preference relations are continuous. This *continuity condition* ensures that there will be no rapid jumps in preference for very small changes in bundle composition.

Axiom 2. The Transitivity Axiom

We assume that preferences and indifferences are transitive. This requires the following:

a. If we prefer bundle $\mathbf{x}_1$ to bundle $\mathbf{x}_2$ and bundle $\mathbf{x}_2$ to bundle $\mathbf{x}_3$, then we must prefer bundle $\mathbf{x}_1$ to bundle $\mathbf{x}_3$; that is, if $\mathbf{x}_1 > \mathbf{x}_2$ and $\mathbf{x}_2 > \mathbf{x}_3$, then, by transitivity, $\mathbf{x}_1 > \mathbf{x}_3$.
b. If we are indifferent between bundles $\mathbf{x}_1$ and $\mathbf{x}_2$ and between bundles $\mathbf{x}_2$ and $\mathbf{x}_3$, then we must be indifferent between bundles $\mathbf{x}_1$ and $\mathbf{x}_3$; that is, if $\mathbf{x}_1 \sim \mathbf{x}_2$ and $\mathbf{x}_2 \sim \mathbf{x}_3$, then $\mathbf{x}_1 \sim \mathbf{x}_3$.
c. If $\mathbf{x}_1$ is preferred to or indifferent to $\mathbf{x}_2$ and $\mathbf{x}_2$ is preferred or indifferent to $\mathbf{x}_3$, then $\mathbf{x}_1$ is either preferred or indifferent to $\mathbf{x}_3$; that is, if $\mathbf{x}_1 \geq \mathbf{x}_2$ and $\mathbf{x}_2 \geq \mathbf{x}_3$, then $\mathbf{x}_1 \geq \mathbf{x}_3$.

Axioms 1 and 2 represent rationality suppositions for a consumer.[1] If these are not assumed, various maladies result, the most common of which is that the consumer may be converted into a money pump.

Example 3.1: To illustrate a consequence of intransitivity, suppose that we can find a person who will say, "I prefer beer to milk, I prefer milk to water, and I

[1]We should remark that these are normative assumptions or assumptions for a prescriptive theory of consumer behavior. In a descriptive sense, consumers are often unintentionally nontransitive or intransitive. There is a wealth of literature on this subject and, sadly, we will be unable to cover it here in any detail due to lack of space. We will devote some more attention to this subject in Chapters 5 and 10.

prefer water to beer." If we can get the consumer to agree to set a price differential converting the preference to an indifference such that we have the cardinal relations beer = milk + 25¢, milk = water + 15¢, and water = beer + 10¢, then we can convert the consumer into a money pump. We start by *giving* the consumer a glass of water; then we offer to trade the consumer a glass of milk for the glass of water and 15¢. The consumer should agree, since they are indifferent between the two choices. Now we offer to trade the consumer a glass of beer for the milk if they will give us 25¢. Again, the consumer is indifferent between these, and so a trade is made. Finally we offer to trade the consumer a glass of water in exchange for the beer and 10¢ and the consumer agrees that this is reasonable, since indifference exists. In this process the consumer starts with a glass of water and winds up with a glass of water. We make 50¢ and can continue this pumping operation if we so desire. The consumer is happy to accommodate this, assuming that they truly have this preference intransitivity!. ■

The two axioms of transitive preference and transitive indifference and the fact that we have assumed that preference relations are complete and continuous allow us to define a *utility function*, a real continuous function defined in the commodity space C. The primary properties of an ordinal utility function are: $U(\mathbf{x}_1) > U(\mathbf{x}_2)$ iff $\mathbf{x}_1 > \mathbf{x}_2$, $U(\mathbf{x}_1) = U(\mathbf{x}_2)$ iff $\mathbf{x}_1 \sim \mathbf{x}_2$, and $U(\mathbf{x}_1) \geq U(\mathbf{x}_2)$ iff $\mathbf{x}_1 \geq \mathbf{x}_2$. This utility function is defined with respect to various attributes of the bundle purchased. Satisfaction values depend on this time interval and the attributes of the commodity bundle.

Example 3.2: Considerable care must be exercised in combining preferences based on ordinal utilities. Even though individual preferences may well not be intransitive, there does not generally exist any reasonable basis on which to combine ordinal preferences for aspects of various alternative commodity bundles; intransitive preferences may therefore easily result.

To illustrate this, suppose that we wish to establish a preference ordering among three automobiles: A, B, and C. We determine that cost, performance, and style are the attributes of importance to us. Suppose that the preference ordering with respect to cost is such that A is preferred to B, which is preferred to C; that is, $A >_c B >_c C$. Suppose further that B is preferred to C, which is preferred to A with respect to performance, or $B >_p C >_p A$. Finally, suppose that C is preferred to A, which is preferred to B with respect to style, or $C >_s A >_s B$. It might not seem unreasonable to prefer car i over car j if it is better on a majority of the attributes of importance. So we prefer car A to B since it is better on two of the three attributes. Similarly, we prefer car B to C since it is better on two of the three attributes. A common method of judgment consists of making pairwise comparisons and discarding the inferior alternative from further consideration. Usually transitivity is assumed and the failure to examine potentially disconfirming comparisons may lead us to select alternatives that we believe are "best" but which may be in fact inferior. Here, for example, we have determined that $B > C$; so we discard C from further comparison. Then we could determine that $A > B$ and infer transitivity such that we believe that $A > C$. But this is not correct, since C is preferred to A on two out of the three attribute scores! The problem is that we are using deficient judgment heuristics[2] and that these are not recognized.

The use of cardinal utilities, which allows expressions of preference weights across attributes and the expression of alternative preferences within attributes, is potentially capable of resolving difficulties associated with the poor judgment heuristics that are often associated with attempts to combine ordinal preferences, often by making binary preference comparisons across two alternatives and rejecting the one judged inferior. Here, for example, we may determine the following utility scores of the alternative commodity bundles on the three attributes:

	Attributes		
Car	Cost	Performance	Style
A	1.0	0.0	0.5
B	0.5	1.0	0.0
C	0.0	0.5	1.0

These scores are fully consistent with the ordinal preferences given earlier. If we can assume a linear aggregation rule $U = w_c U_c + w_p U_p + w_s U_s$ with which to obtain a single scalar cardinal utility function for the three cars, then we will have for the utilities of the three automobiles: $U(A) = w_c + 0.5w_s$, $U(B) = 0.5w_c + w_p$, and $U(C) = 0.5w_p + w_s$. For convenience we assume that the weights are in the interval 0 to 1 and that they sum to 1, $w_c + w_p + w_s = 1$, such that the foregoing utilities become $U(A) = 0.5 + 0.5w_c - 0.5w_p$, $U(B) = 0.5w_c + w_p$, and $U(C) = 1.0 - w_c - 0.5w_p$. With these utilities, we can obtain any preference ordering that we wish by appropriate selection of the weights. However, all preferences will be transitive. If we are to obtain, for example, $A > B > C$, this requires $U(A) > U(B) > U(C)$, or $w_p < 1/3$ and $w_c + w_p > 2/3$. These two inequalities also infer that $w_c > 1/3$ and $w_s < 1/3$. ■

As we have noted, utility functions are not unique. For example, if an ordinal utility function for a given bundle is $U_1(x_1)$, then any monotonically increasing function of the utility U_1 is also a valid utility function. An affine, or linear, transformation of cardinal utility functions $U_1(x) = a + bU(x)$, for $b > 0$, is often used to scale utilities to the range 0–1 or 0–100. Cardinal utilities are

[2]Discussion on this topic in detail would carry us far away from our principal objectives. The reader is referred to Sage's 'Behavioral and organizational' considerations in the design of information systems and processes for planning and decision support. IEEE Trans Syst Man Cybern 1981;SMC-11(9):640–678, for a somewhat dated discussion of these issues, as well as references to much related literature. Also see Kahneman D, Tversky A, editors. Choices, values and frames. Cambridge: Cambridge University Press; 2000. We examine these issues further in Chapter 5.

unique only up to affine transformations. In many cases the anchor on the cardinal utility scale is such that the constant term must be zero and we can only have a linear transformation $U_1(x) = bU(x)$. We are now in a position to state the final two formal axioms of economic utility theory.

Axiom 3. The Nonsatiation Axiom

If bundle $\mathbf{x}_1$ contains at least as much of any constituent commodity as bundle $\mathbf{x}_2$, then $\mathbf{x}_1$ must be preferred or indifferent to $\mathbf{x}_2$. Thus, the utility of $\mathbf{x}_1$ must be greater than the utility of $\mathbf{x}_2$. So $\mathbf{x}_1 > \mathbf{x}_2$ implies $\mathbf{x}_1 \succ \mathbf{x}_2$ and $\mathbf{x}_1 \succ \mathbf{x}_2$ implies $U(\mathbf{x}_1) > U(\mathbf{x}_2)$. Also, $\mathbf{x}_1 \geq \mathbf{x}_2$ implies $\mathbf{x}_1 \succ \mathbf{x}_2$ and $\mathbf{x}_1 \succeq \mathbf{x}_2$ implies $U(\mathbf{x}_1) \geq U(\mathbf{x}_2)$.

Axiom 4. The Strict Convexity Axiom

The axiom of continuity ensures that there will be at least one commodity bundle with preference equivalent to $\mathbf{x}_3$ that can be made up of a combination of bundles $\mathbf{x}_1$ and $\mathbf{x}_2$, where $\mathbf{x}_1 \succ \mathbf{x}_3 \succ \mathbf{x}_2$. This axiom[3] can also be stated in a slightly different form as $\mathbf{x}_1 \succ a\mathbf{x}_1 + (1 - a)\mathbf{x}_2 \succ \mathbf{x}_2$ for all $0 < a < l$ if $\mathbf{x}_1 \succ \mathbf{x}_2$.

3.3 PROPERTIES OF UTILITY FUNCTIONS

Economic utility functions have a number of important properties. In this section we will investigate some of them, based on the assumption that we are dealing with cardinal utility functions. We assume that utility functions are twice differentiable with respect to the independent variable $\mathbf{x}$, the commodity bundle. In Chapter 2 the symbol $\mathbf{x}$ is used to represent the inputs to the production process, which will generally consist of commodities (goods) and factors (services). Here we use it to represent the commodity bundle, which generally could also consist of goods and services. The first derivative of the utility function with respect to $\mathbf{x}$ is known as the marginal utility $\mathbf{MU}(\mathbf{x})$:

$$\mathbf{MU}(\mathbf{x}) = \frac{\partial U(\mathbf{x})}{\partial \mathbf{x}} = \begin{bmatrix} \dfrac{\partial U(\mathbf{x})}{\partial x_1} \\ \dfrac{\partial U(\mathbf{x})}{\partial x_2} \\ \vdots \\ \dfrac{\partial U(\mathbf{x})}{\partial x_n} \end{bmatrix} \tag{3.1}$$

From the nonsatiation axiom, we see that marginal utilities are always positive, or $\mathbf{MU}(\mathbf{x}) > \mathbf{0}$. At each and every point in the commodity space C, the marginal

[3]The two statements of the axiom are equivalent as we easily see by letting $\mathbf{x}_3 = a\mathbf{x}_1 + (1 - a)\mathbf{x}_2$ such that the ith component of $\mathbf{x}_3$ contains $ax_{1,i} + (1 - a)x_{2,i}$ units of commodity i.

utility must increase for changes in any single component of the commodity bundle, in that

$$\mathrm{MU}_i(\mathbf{x}) = \frac{\partial U(\mathbf{x})}{\partial x_i} > 0, \qquad i = 1, 2, \ldots, n \tag{3.2}$$

We examine the strictly convex *indifference curve* defined by

$$U(\mathbf{x}) = b \tag{3.3}$$

Along this indifference curve the derivative or gradient with respect to the commodity bundle must be zero such that we have the requirement that $\frac{\mathrm{d}U(\mathbf{x})}{\mathrm{d}\mathbf{x}} = \mathbf{0}$. More importantly, the differential must be zero on a surface of constant utility, and so $\left(\frac{\mathrm{dU}(\mathbf{x})}{\mathrm{d}\mathbf{x}}\right)^{\mathrm{T}} \mathrm{d}\mathbf{x} = 0$. We can rewrite this expression, since we recognize the definition of the marginal utility in the expression, as $\mathbf{MU}^{\mathrm{T}}(\mathbf{x})\,\mathrm{d}\mathbf{x} = 0$ or

$$\sum_{i=1}^{n} \mathrm{MU}_i(\mathbf{x}) \mathrm{d}x_i = 0 \tag{3.4}$$

This simply says that the sum of the changes in utility produced by a unit change in x_i times the change $\mathrm{d}x_i$ will be zero. Thus for the *two-dimensional case*, if we decrease x_1 such that $\mathrm{d}x_1$ is negative, then $\mathrm{d}x_2$ must be positive as the marginal utilities are always positive and Equation 3.4 will hold. The *marginal rate of commodity substitution* MRCS is an important quantity that can be obtained from this relationship. For the two-commodity case we have, from Equation 3.4,

$$\mathrm{MRCS}_{1,2} = -\frac{dx_2}{dx_1} = \frac{\mathrm{MU}_1(\mathbf{x})}{\mathrm{MU}_2(\mathbf{x})} \tag{3.5}$$

and for the general case, where we keep all commodity bundle components constant except for x_i and x_j and trade an increase in consumption of x_i for a decrease in consumption of x_j (or vice versa), we have

$$\mathrm{MRCS}_{i,j} = -\frac{dx_j}{dx_i} = \frac{\mathrm{MU}_i(\mathbf{x})}{\mathrm{MU}_j(\mathbf{x})} \tag{3.6}$$

We will obtain some important implications of these properties of utility functions in Section 3.4. A number of special utility functions can be examined. The case of *strictly complementary* commodities exists when for two commodities, use of x_1 requires the use of ax_1 units of x_2 for a fixed a and excess purchases of x_2 do not increase utility. The utility function for strictly complementary commodities is $U(x_1, x_2) = f[\min(ax_1, x_2)]$. If the utility function is additive in individual utilities, we have

$$U(\mathbf{x}) = \sum_{i=1}^{n} a_i U_i(x_i) \tag{3.7}$$

and we say that the commodities are utility independent. A more general form of utility independence is represented by the multiplicative form

$$1 + KU(\mathbf{x}) = \prod_{i=1}^{n} [1 + Kk_i U_i(x_i)] \tag{3.8}$$

This expression reduces to the additive form of Equation 3.7 when $K = 0$, and to the multiplicative form

$$U(\mathbf{x}) = \prod_{i=1}^{n} U_i(x_i) \tag{3.9}$$

when $K = \infty$, $k_i = 0$, $k_i K = 1$.

3.4 THE FUNDAMENTAL PROBLEM OF THE CONSUMER

In general the consumer will have a given budget or income I to spend on a commodity bundle of goods and services $\mathbf{x}$. The fundamental problem for the consumer is to pick a particular commodity bundle that maximizes satisfaction, value, or utility for the consumer. We note that a realistic consumer problem would be more difficult than this, as there would be questions of savings, taxes, and the likes involved. Extensions to include these broader aspects of economic behavior will be considered later; here our objective is to obtain a reasonably simple, although incomplete, picture of the economic behavior of the consumer.

We assume that a constant known *price vector*

$$\mathbf{p} = \begin{bmatrix} p_1 \\ p_2 \\ \vdots \\ p_n \end{bmatrix}$$

of prices per unit item for the items in the commodity bundle has been established. Both the price vector $\mathbf{p}$ and the *budget constraint* I are nonnegative. The consumer's utility function has been established and so the fundamental problem of the consumer is to maximize consumer utility

$$J = U(\mathbf{x}) \tag{3.10}$$

subject to the budget constraint

$$\mathbf{p}^T \mathbf{x} \leq I \tag{3.11}$$

We assume a positive budget and positive prices and require that the consumer not accept any negative goods, and so we have the constraint

$$\mathbf{x} \geq \mathbf{0} \tag{3.12}$$

We recognize this as a simple problem in nonlinear optimization for which the Kuhn–Tucker conditions of Table 2.4 are directly applicable. We form the

Lagrangian by adjoining the inequality constraint of Equation 3.11 to the cost function of Equation 3.10 and obtain

$$J' = U(\mathbf{x}) + \lambda(I - \mathbf{p}^{\mathrm{T}}\mathbf{x}) \tag{3.13}$$

Use of the Kuhn–Tucker conditions obtained in Chapter 2,

$$\frac{\partial J'}{\partial \mathbf{x}} \leq \mathbf{0}, \qquad \mathbf{x}^{\mathrm{T}}\frac{\partial J'}{\partial \mathbf{x}} = 0 \tag{3.14}$$

$$\frac{\partial J'}{\partial \lambda} \geq 0, \qquad \lambda\frac{\partial J'}{\partial \lambda} = 0 \tag{3.15}$$

$$\mathbf{x} \geq \mathbf{0}, \qquad \lambda \geq 0 \tag{3.16}$$

or direct use of Table 2.4 leads to the necessary (and sufficient) conditions for optimality:

$$\frac{\partial U(\mathbf{x})}{\partial \mathbf{x}} - \lambda\mathbf{p} \leq 0 \tag{3.17}$$

$$\mathbf{x}^{\mathrm{T}}\left(\frac{\partial U(\mathbf{x})}{\partial \mathbf{x}} - \lambda\mathbf{p}\right) = 0 \tag{3.18}$$

$$I - \mathbf{p}^{\mathrm{T}}\mathbf{x} \geq 0 \tag{3.19}$$

$$\lambda(\mathbf{I} - \mathbf{p}^{\mathrm{T}}\mathbf{x}) = 0 \tag{3.20}$$

$$\mathbf{x} \geq \mathbf{0} \tag{3.21}$$

$$\lambda \geq 0 \tag{3.22}$$

where the variables $\mathbf{x}$ and λ are evaluated at the optimum values $\hat{\mathbf{x}}$ and $\hat{\lambda}$.

We can draw some very important general conclusions from this result. We note that we must have $\mathbf{x} \geq \mathbf{0}$ from Equation 3.21 and our initial constraint on the commodity bundle. For $\mathbf{x} = \mathbf{0}$ we see that Equation 3.18 is satisfied and so we must require, from Equation 3.17, that

$$\mathbf{MU}(\mathbf{x}) = \frac{\partial U(\mathbf{x})}{\partial \mathbf{x}} \leq \lambda\mathbf{p} \tag{3.23}$$

Here, we need the partial derivative signs to define marginal utility, since utility is an implicit function of I. When $\mathbf{x} > \mathbf{0}$, we see that we must have, from Equation 3.18,

$$\mathbf{MU}(\mathbf{x}) = \lambda\mathbf{p} \tag{3.24}$$

This relation will generally be the applicable relation rather than Equation 3.23, since the consumer will normally purchase a nonzero bundle. The important conclusion to be gained from this is that the ratio of marginal utility to price is the same constant for all components in the commodity bundle vector such that, from Equation 3.24,

$$\frac{MU_i(\mathbf{x})}{p_i} = \lambda, \qquad i=1,2,\ldots,n \tag{3.25}$$

It is not reasonable that all the marginal utilities be zero (unless all the prices are infinite), and so we see that the Lagrange multiplier will not, in general, be zero. From Equation 3.22 we see that λ must then be positive. Thus we must have, from Equation 3.20,

$$\mathbf{p}^{\mathrm{T}}\mathbf{x} = I \tag{3.26}$$

which simply states that the consumer will spend their entire budget or income I. This is a most reasonable conclusion, in that it is, under the mildest of conditions, guaranteed by the nonsatiation axiom. There is no penalty, in our problem description thus far, for spending the entire budget. We can increase the quantity of the bundle $\mathbf{x}$ by spending more. The nonsatiation axiom insists that this will necessarily increase utility. Of course, we could introduce savings as a term in the utility function and we would then devote a portion of the budget to savings.

Example 3.3: The results of this optimization effort have particular significance for the case of a two-dimensional bundle. From Equations 3.25 and 3.26 we have

$$\mathrm{MU}_1(x) = p_1\lambda \tag{3.27}$$

$$\mathrm{MU}_2(x) = p_2\lambda \tag{3.28}$$

and

$$p_1x_2 + p_2x_2 = 1 \tag{3.29}$$

If we consider the isoquants, or indifference curves of constant utility where $U(\mathbf{x}) = \text{constant}$, we can obtain the slope of these lines, as in Equation 3.6,

$$\mathrm{MU}_1(\mathbf{x})\mathrm{d}x_1 + \mathrm{MU}_2(\mathbf{x})\mathrm{d}x_2 = 0 \tag{3.30}$$

By substituting the values of the marginal utilities from Equations 3.27 and 3.28 into Equation 3.30, we obtain, after canceling the Lagrange multiplier,

$$p_1\mathrm{d}x_1 + p_2\mathrm{d}x_2 = 0 \tag{3.31}$$

or

$$\frac{\mathrm{d}x_2}{\mathrm{d}x_1} = -\frac{p_1}{p_2} \tag{3.32}$$

We see that the slope of the utility indifference curve at the point where it touches the budget line, Equation 3.29, is $-p_1/p_2$. This is to be expected. Figure 3.1 illustrates the tangency solution for this classic optimization problem of the consumer. ■

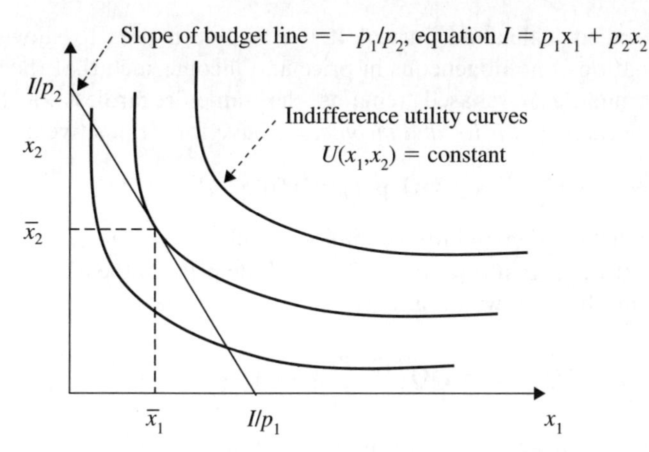

Figure 3.1. Two-Dimensional Graphic Solution for the Problem of the Consumer.

An interesting and useful interpretation can be given to the Lagrange multiplier λ. We combine Equation 3.25, which can be written as

$$\frac{\partial U_i(\mathbf{x})}{\partial x_i} = \lambda p_i$$

and the partial derivative with respect to I of Equation 3.26,

$$\sum_{i=1}^{n} p_i \frac{\partial x_i}{\partial I} = 1$$

so as to eliminate p_i. We then obtain the expression

$$\sum_{i=1}^{n} \frac{\partial U(\mathbf{x})}{\partial x_i} \frac{\partial x_i}{\partial I} = \frac{\partial U(\mathbf{x})}{\partial I} = \lambda$$

This very important result indicates that the Lagrange multiplier for this problem is the *marginal utility of income* MUI,

$$\lambda = \mathrm{MUI} = \frac{\partial U(\mathbf{x})}{\partial I} \tag{3.33}$$

where we have dropped the optimality symbol for the sake of simplicity.

The nonlinear algebraic equations 3.17 to 3.22 will not, in general, be easy to solve analytically, and there are a number of nonlinear programming algorithms we can use to obtain numerical solutions. The result of this is that we obtain a solution for the optimum commodity bundle as a function of price and income,

$$\hat{\mathbf{x}} = \mathbf{DQ}(\mathbf{p}, I) = \mathbf{f}_d(\mathbf{p}, I) \tag{3.34}$$

and this is called a demand curve.[4] It is a simple matter to show that the demand equation is homogeneous in price and income such that the optimum commodity bundle purchased remains the same, regardless of the same *proportionate change in price and income*; we have for all positive a,

$$\hat{\mathbf{x}} = \mathbf{DQ}(\mathbf{p}, I) = \mathbf{DQ}(a\mathbf{p}, aI) \tag{3.35}$$

Thus we see that the demand for a commodity will depend on *relative prices* and *real income*. If we let a in Equation 3.35 equal the price of the ith component in the commodity bundle, we obtain

$$\hat{\mathbf{x}} = \mathbf{DQ}(\mathbf{p}, I) = \mathbf{DQ}\left(\frac{p_1}{p_i}, \frac{p_2}{p_i}, \cdots, 1, \frac{p_i+1}{p_i}, \cdots, \frac{p_n}{p_i}, \frac{I}{p_i}\right) \tag{3.36}$$

Alternately, we can set $a = I^{-1}$ to obtain for Equation 3.35

$$\hat{\mathbf{x}} = \mathbf{DQ}\left(\frac{p_1}{I}, \frac{p_2}{I}, \cdots, \frac{p_n}{I}, 1\right) \tag{3.37}$$

This is a sometimes used normalized expression for a consumer with unit budget.

Example 3.4: As our second example of consumer optimization, let us suppose that the consumer's utility function is linear in the commodity vector and given by

$$U(\mathbf{x}) = \mathbf{a}^{\mathrm{T}}\mathbf{x} = \sum_{i=1}^{n} a_i x_i$$

We also assume that we have a budget constraint given by

$$\mathbf{p}^{\mathrm{T}}\mathbf{x} = \sum_{i=1}^{n} p_i x_i \leq I$$

The nonsatiation axiom ensures that the consumer will spend the entire budget. We are able to solve this linear programming problem by simple intuitive means. We should determine the component of the commodity bundle that has the maximum marginal utility–cost ratio and spend the entire budget on this item. Thus we find

$$\frac{\mathrm{MU}_i(\mathbf{x})}{p_i} = \frac{a_i}{p_i}$$

Also we find the i that yields the largest ratio. We will denote this amount as $\hat{i}$. Thus, we purchase an amount given by

[4] It is clear that the resulting demand curve is a useful prescriptive or normative result in that it describes how a consumer should behave. Often these results are essentially correct for descriptive behavior. However, in many cases, people do not behave as they "should".

$$\hat{x}_i = \begin{cases} 0, & i \neq \hat{i} \\ \dfrac{I}{p_i}, & i = \hat{i} \end{cases}$$

The resulting maximum utility is given by

$$U(\hat{\mathbf{x}}) = \frac{a_{\hat{i}} I}{p_{\hat{i}}}$$

The utility function we have assumed here is not strictly concave and does not satisfy the principle of diminishing marginal utility. Thus we do not obtain for this example the result that the ratio of marginal utility to price is constant for all consumer bundle components. ■

Example 3.5: As a fairly complete illustrative example that has an analytical solution, let us consider a quadratic consumer utility function

$$U(\mathbf{x}) = \mathbf{a}^{\mathrm{T}}\mathbf{x} + 0.5\mathbf{x}^{\mathrm{T}}\mathbf{B}\mathbf{x} \tag{3.38}$$

in which $\mathbf{a} > \mathbf{0}$, the $\mathbf{B}$ matrix is symmetric negative definite, and

$$\mathbf{a} + \mathbf{B}\mathbf{x} > \mathbf{0} \tag{3.39}$$

The consumer's budget constraint is

$$I \geq \mathbf{p}^{\mathrm{T}}\mathbf{x} \tag{3.40}$$

We have indicated earlier in this chapter that the nonsatiation principle guarantees that the consumer spends the entire budget. Thus we need only consider the equality constraint in Equation 3.40. But we must be very careful to ensure that all the conditions on which our previous results are established are valid. One of these conditions is that the marginal utility always be greater than zero, or in equation form that

$$\mathbf{MU}(\mathbf{x}) = \mathbf{a} + \mathbf{B}\mathbf{x} > \mathbf{0} \tag{3.41}$$

Unfortunately, this will not always be true for sufficiently large $\mathbf{x}$ since $\mathbf{B}$ is negative definite. Thus we cannot necessarily assume that the nonsatiation axiom holds for $U(\mathbf{x})$ of Equation 3.38, and it then will not be always true that all of the budget I is spent. When the budget constraint does not apply, we have $\lambda = 0$ in Equation 3.22. Then Equation 3.18 becomes, since $\mathbf{x} \neq 0$,

$$\frac{\partial U(\mathbf{x})}{\partial \mathbf{x}} = \mathbf{0} \tag{3.42}$$

and we see that this is just the result for unconstrained maximization of Equation 3.38. This unconstrained maximization using Equation 3.38 yields

$$\hat{\mathbf{x}}_{\mathrm{u}} = -\mathbf{B}^{-1}\mathbf{a} \tag{3.43}$$

as the optimum commodity bundle. The total unconstrained purchases using this $\hat{\mathbf{x}}_u$ are the unconstrained purchases $\mathbf{p}^T\mathbf{x} = -\mathbf{p}^T\mathbf{B}^{-1}\mathbf{a}$. If this is not greater than the budget, unconstrained purchases are going to be less than or equal to I, and then $\hat{\mathbf{x}}_u$ of Equation 3.42 is indeed the optimum solution.

There is likely going to be a modeling problem inherent in this example if these results hold. If the budget I is sufficiently large such that we do not need to use the entire budget, we are in the utility satiation region. A utility function of this sort is unrealistic. Thus, if our assumed utility function is to be a reasonable model for the range of commodity bundles that provide feasible solutions, we must restrict the budget to a value less than that spent on the unconstrained purchases. So we will assume that $I < \mathbf{p}^T\hat{\mathbf{x}}_u$, or from Equation 3.43,

$$I < -\mathbf{p}^T\mathbf{B}^{-1}\mathbf{a} \tag{3.44}$$

for this example. We will then, under this restriction, use the entire budget I.

The Kuhn–Tucker conditions are directly applicable here, but we will use a somewhat simpler approach, since we assume a budget equality constraint, to obtain the commodity bundle that maximizes utility. We adjoin the equality constraint of Equation 3.40 to the cost function of Equation 3.38 using a Lagrange multiplier and obtain

$$J = U(\mathbf{x}) + \lambda(I - \mathbf{p}^T\mathbf{x}) \tag{3.45}$$

The necessary conditions of optimality are, at $\mathbf{x} = \hat{\mathbf{x}}$ and $\lambda = \hat{\lambda}$,

$$\frac{\partial J}{\partial \mathbf{x}} = \mathbf{0} = \frac{\partial U(\mathbf{x})}{\partial \mathbf{x}} - \mathbf{p}\lambda \tag{3.46}$$

$$\frac{\partial J}{\partial \lambda} = 0 = I - \mathbf{p}^T\mathbf{x} \tag{3.47}$$

From this, we obtain the relationship that, under optimality conditions, the ratio of the marginal utility for the ith bundle component to the ith bundle component price is a constant for all i. For the specific problem considered here, we obtain for Equation 3.46

$$\frac{\partial J}{\partial \mathbf{x}} = \mathbf{0} = \mathbf{a} + \mathbf{B}\mathbf{x} - \mathbf{p}\lambda$$

Thus, we have

$$\hat{\mathbf{x}} = -\mathbf{B}^{-1}(\mathbf{a} - \mathbf{p}\hat{\lambda}) \tag{3.48}$$

We must adjust $\hat{\lambda}$ such that the equality constraint of Equation 3.47

$$\mathbf{p}^T\mathbf{x} = I \tag{3.49}$$

is satisfied. Premultiplying Equation 3.48 by $\mathbf{p}^T$ results in

$$\mathbf{p}^T\hat{\mathbf{x}} = I = -\mathbf{p}^T\mathbf{B}^{-1}\mathbf{a} + \mathbf{p}^T\mathbf{B}^{-1}\mathbf{p}\hat{\lambda}$$

Thus, we have on solving for $\hat{\lambda}$,

$$\hat{\lambda} = (\mathbf{p}^T\mathbf{B}^{-1}\mathbf{p})^{-1}(I + \mathbf{p}^T\mathbf{B}^{-1}\mathbf{a}) \tag{3.50}$$

as the value of the Lagrange multiplier. The optimum commodity bundle is obtained by substituting Equation 3.50 into Equation 3.48 to obtain

$$\hat{\mathbf{x}} = -\mathbf{B}^{-1}\mathbf{a} + \mathbf{B}^{-1}\mathbf{p}(\mathbf{p}^T\mathbf{B}^{-1}\mathbf{p})^{-1}(I + \mathbf{p}^T\mathbf{B}^{-1}\mathbf{a}) \tag{3.51}$$

For this particular example the commodity bundle increases linearly with respect to increases in budget or income I. It is interesting to note that the optimum commodity bundle is not a linear function of price, however. We note from Equation 3.51 that it is possible to obtain negative components of $\hat{\mathbf{x}}$ in the foregoing for sufficiently small I. These are not reasonable values, as they violate Equation 3.21. We do not allow the consumer to sell back some items to get money to pay for other items. We shall not explore problem solution for this case in any depth. Generally, it is sufficient to set these computed negative elements of $\mathbf{x}$ equal to zero, delete the corresponding coefficients for the zero $\mathbf{x}$ elements, and use the Kuhn–Tucker conditions to find a solution to the problem. ■

Example 3.6: As a special case of Example 3.5, let us consider the simplest case where the commodity bundle is a one vector. For this scalar case the results simplify considerably and we have

$$U(x) = ax + 0.5bx^2, \qquad a > 0, \qquad b < 0$$

$$I \geq px$$

$$\hat{x}_u = -a/b$$

$$U(\hat{x}_u) = -a^2/2b$$

$$\hat{x} = -I/p$$

where the last equation follows from the general optimal result of Equation 3.51. We note that x for this example is just the value that satisfies the equality budget constraint. However, when $\mathrm{MU}(x) = a + bx$ becomes zero (or less), then we should use the inequality constraint budget and actually obtain a larger utility by using less than the full budget. Of course, this is because of the (now) unrealistic utility function. Thus we have, from Equation 3.44, that when $I > -pa/b$, we should use the unconstrained optimum $\hat{x}_u$. Figures 3.2 and 3.3 illustrate salient features of this example for the particular case where $a = 1$ and $b = -0.1$. ■

Example 3.7: We may also obtain some interesting graphical results for the two-commodity case. For the case where there are two commodities, we may rewrite Equation 3.51 as

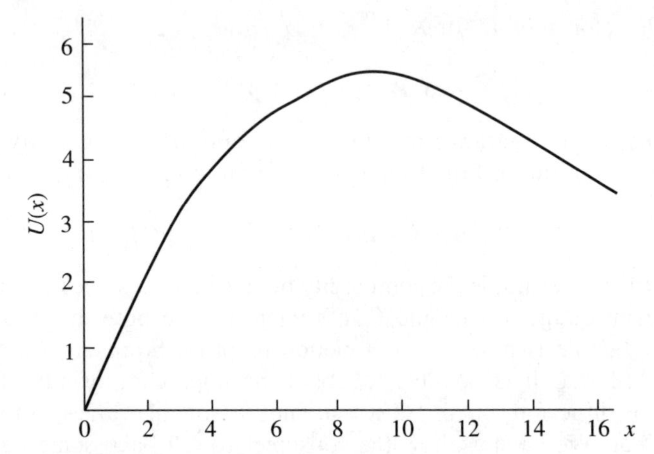

Figure 3.2. Utility Function for Example 3.5.

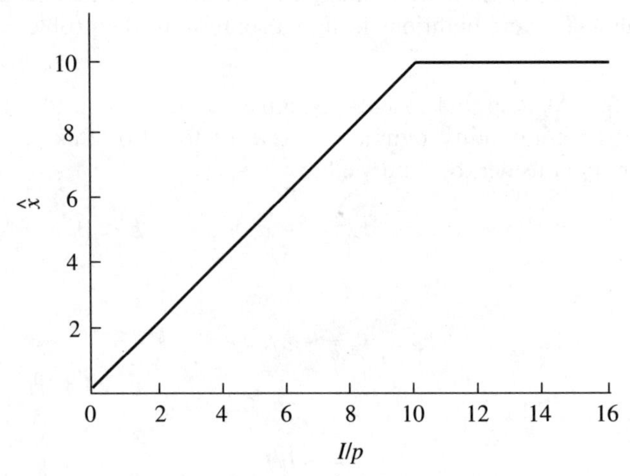

Figure 3.3. Optimal $\hat{x}$ versus I/p for Example 3.5.

$$\hat{x}_1 = \Xi[a_2 p_1 p_2 - a_1 p_2^2 + (b_{22} p_1 - b_{12} p_2) I] \tag{3.52}$$

$$\hat{x}_1 = \Xi[a_1 p_1 p_2 - a_2 p_1^2 + (b_{11} p_2 - b_{12} p_1) I] \tag{3.53}$$

where

$$\Xi = (b_{22}^2 p_1^2 - 2 b_{12} p_1 p_2 + b_{11} p_2^2)^{-1} \tag{3.54}$$

We consider the specific case where

$$\mathbf{a} = \begin{bmatrix} 1 \\ 3 \end{bmatrix}, \qquad \mathbf{B} = \begin{bmatrix} -0.2 & 0.1 \\ 0.1 & -0.2 \end{bmatrix} = -\begin{bmatrix} 6.67 & 3.33 \\ 3.33 & 6.67 \end{bmatrix}^{-1}$$

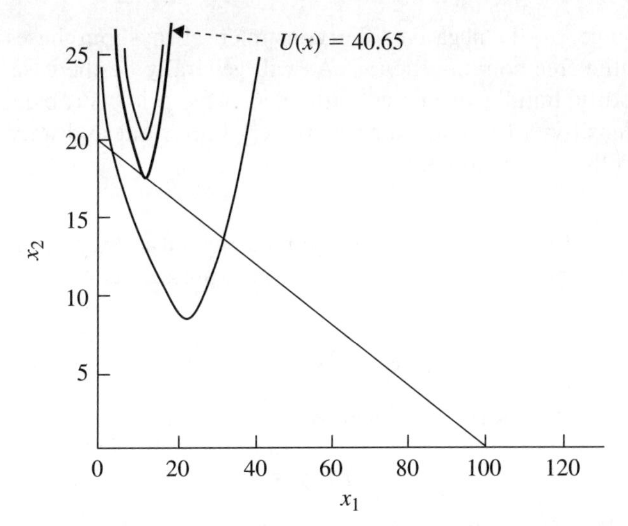

Figure 3.4. Budget Lines and Utility Indifference Curves for Example 3.6.

such that

$$U(\mathbf{x}) = x_1 + 3x_2 - 0.1x_1^2 - 0.1x_2^2 + 0.1x_1x_2 \tag{3.55}$$

The unconstrained optimum solution is

$$\hat{\mathbf{x}}_u = -\mathbf{B}^{-1}\mathbf{a} = \begin{bmatrix} 16.67 \\ 23.34 \end{bmatrix} \tag{3.56}$$

Figure 3.4 shows contours of constant utility, the utility indifference curves; also shown is a budget line for $I = \$100$, $p_1 = \$1$, $p_2 = \$5$. The optimum $\hat{\mathbf{x}}$ can be determined either from Equations 3.52 to 3.54 or by careful construction of the curves in Fig. 3.4 as

$$\hat{x}_1 = 12.90, \qquad \hat{x}_2 = 17.42$$

and we see that both are less than the components of $\hat{\mathbf{x}}_u$. The amount of money required to purchase $\hat{\mathbf{x}}_u$ is $\mathbf{p}^T\mathbf{x} = \$133.33$ and the consumer has only \$100.

It is of great interest to determine the behavior of the consumer if the prices p_1 and p_2 change. If, for example, $p_1 = 0$, we obtain from Equations 3.52 through 3.54

$$\hat{x}_1 = -\frac{a_1}{b_{11}} - \frac{b_{12}I}{b_{11}p_2}$$

$$\hat{x}_2 = \frac{I}{p_2}$$

and we see that the full consumer budget is now being spent on commodity x_2 and so we use whatever value of x_1 results in unconstrained maximization of $U(\mathbf{x})$. Since b_{11} is negative, the consumer always purchases a positive number of the free commodities x_1. As will generally be the case, the size of the commodity bundle decreases with increasing price. Problems involving consumer sensitivity to price changes are very important and we will turn our attention to this in Section 3.5. ■

Example 3.8: In this example, we examine the sufficiency requirements for consumer maximization of a utility function expressed as a general quadratic of the form

$$U(\mathbf{x}) = \mathbf{a}^{\mathrm{T}}\mathbf{x} + 0.5\mathbf{x}^{\mathrm{T}}\mathbf{B}\mathbf{x} \tag{3.57}$$

where the budget constraint is given by

$$I \geq \mathbf{p}^{\mathrm{T}}\mathbf{x} \tag{3.58}$$

and where $\mathbf{B}$ is symmetric and possesses an inverse. To proceed, we define the Lagrangian, which is a function of the variables $\mathbf{x}$ and λ:

$$L(\mathbf{x}, \lambda) = \mathbf{a}^{\mathrm{T}}\mathbf{x} + 0.5\mathbf{x}^{\mathrm{T}}\mathbf{B}\mathbf{x} + \lambda(I - \mathbf{p}^{\mathrm{T}}\mathbf{x}) \tag{3.59}$$

We replace $\mathbf{x}$ and λ with $\hat{\mathbf{x}} + \boldsymbol{\delta}\mathbf{x}$ and $\hat{\lambda} + \delta\lambda$ and assume that $\boldsymbol{\delta}\mathbf{x}$ and $\delta\lambda$ are sufficiently small such that the three-term Taylor series

$$L(\hat{\mathbf{x}} + \boldsymbol{\delta}\mathbf{x}, \hat{\lambda} + \delta\lambda) \cong L(\hat{\mathbf{x}}, \hat{\lambda}) + \boldsymbol{\delta}\mathbf{x}^{\mathrm{T}} \frac{\partial L(\hat{\mathbf{x}}, \hat{\lambda})}{\partial \hat{\mathbf{x}}} + \delta\lambda \frac{\partial L(\hat{\mathbf{x}}, \hat{\lambda})}{\partial \lambda}$$

$$+ \frac{1}{2}[\delta\mathbf{x}^{\mathrm{T}} \quad \delta\lambda] \begin{bmatrix} \dfrac{\partial^2 L}{\partial \hat{\mathbf{x}}^2} & \dfrac{\partial}{\partial \hat{\lambda}} \dfrac{\partial L}{\partial \hat{\mathbf{x}}} \\ \left(\dfrac{\partial}{\partial \hat{\lambda}} \dfrac{\partial L}{\partial \hat{\mathbf{x}}} \right)^{\mathrm{T}} & 0 \end{bmatrix} \begin{bmatrix} \delta\mathbf{x} \\ \delta\lambda \end{bmatrix} \tag{3.60}$$

is a valid approximation. For the case where $\boldsymbol{\delta}\mathbf{x}$ and $\delta\lambda$ are completely arbitrary, the necessary conditions for an extremum (either maximum or minimum) of L are that

$$\frac{\partial L}{\partial \mathbf{x}} = \mathbf{0} \tag{3.61}$$

$$\frac{\partial L}{\partial \lambda} = 0 \tag{3.62}$$

For $L(\mathbf{x}, \lambda)$, or $U(\mathbf{x})$ subject to the constraint, to be maximum, we require that the quadratic expression

$$[\boldsymbol{\delta}\mathbf{x}^{\mathrm{T}} \quad \delta\lambda]\begin{bmatrix} \dfrac{\partial^2 L}{\partial \mathbf{x}^2} & \dfrac{\partial}{\partial \lambda}\dfrac{\partial L}{\partial \mathbf{x}} \\ \left(\dfrac{\partial}{\partial \lambda}\dfrac{\partial L}{\partial \mathbf{x}}\right)^{\mathrm{T}} & 0 \end{bmatrix}\begin{bmatrix} \boldsymbol{\delta}\mathbf{x} \\ \delta\lambda \end{bmatrix} \tag{3.63}$$

be negative, or at least nonpositive, for arbitrary $\boldsymbol{\delta}\mathbf{x}$ and $\delta\lambda$ with $\mathbf{x} \geq \mathbf{0}$. This is equivalent to the requirement that $\partial^2 L/\partial \mathbf{x}^2$ be negative definite subject to the inequality constraint of Equation 3.58. We could use this requirement as the sufficiency requirement associated with the use of Table 2.2.

Equations 3.61 and 3.62 yield, assuming that the full budget is spent,

$$\mathbf{A} + \mathbf{B}\mathbf{x} - \lambda\mathbf{p} = \mathbf{0} \tag{3.64}$$

and

$$I = \mathbf{p}^{\mathrm{T}}\mathbf{x} \tag{3.65}$$

from which we obtain, as in Example 3.5, the optimum commodity bundle

$$\hat{\mathbf{x}} = -\mathbf{B}^{-1}\mathbf{a} + \mathbf{B}^{-1}\mathbf{p}(\mathbf{p}^{\mathrm{T}}\mathbf{B}^{-1}\mathbf{p})^{-1}(I + \mathbf{p}^{\mathrm{T}}\mathbf{B}^{-1}\mathbf{a}) \tag{3.66}$$

Here we assume that the solution to Equation 3.66 yields $\hat{\mathbf{x}} \geq \mathbf{0}$. This result clearly shows us that $\mathbf{B}$ must have an inverse, but it does not require that $\mathbf{B}$ be either positive or negative definite matrix; in fact, it can be indefinite.

Either the full budget I or less than this amount will be spent. For the case where less than the full budget is spent, there is really no need for the constraint of Equation 3.58. The necessary conditions for optimality are obtained directly from Equation 3.64 as the unconstrained optimum

$$\mathbf{x}_{\mathrm{u}} = -\mathbf{B}^{-1}\mathbf{a}$$

and the sufficiency conditions become

$$\boldsymbol{\delta}\mathbf{x}^{\mathrm{T}}\frac{\partial^2 U(\mathbf{x})}{\partial \mathbf{x}^2}\boldsymbol{\delta}\mathbf{x} = \boldsymbol{\delta}\mathbf{x}^{\mathrm{T}}\mathbf{B}\boldsymbol{\delta}\mathbf{x} < 0$$

and we see that we must require $\mathbf{B}$ to be negative definite.

For the case where the full budget is spent, the necessary conditions Equations 3.64 and 3.65 prevail. From Equation 3.57 we see that if $\mathbf{B}$ is negative definite, then surely $\dfrac{\partial^2 U(\mathbf{x})}{\partial \mathbf{x}^2}$ or $\dfrac{\partial^2 \mathbf{L}}{\partial \mathbf{x}^2}$ is negative definite and

$$\boldsymbol{\delta}\mathbf{x}^{\mathrm{T}}\frac{\partial^2 L(\mathbf{x})}{\partial \mathbf{x}^2}\boldsymbol{\delta}\mathbf{x} = \boldsymbol{\delta}\mathbf{x}^{\mathrm{T}}\mathbf{B}\boldsymbol{\delta}\mathbf{x} < \mathbf{0}$$

Thus we see that negative definiteness of $\mathbf{B}$ is sufficient to establish a maximum of $U(\mathbf{x})$ under full budget utilization conditions, regardless of the budget path.

Other conditions can also be obtained from the general requirement that

$$\begin{bmatrix} \dfrac{\partial^2 L}{\partial \mathbf{x}^2} & \dfrac{\partial}{\partial \lambda}\dfrac{\partial L}{\partial \mathbf{x}} \\ \left(\dfrac{\partial}{\partial \lambda}\dfrac{\partial L}{\partial \mathbf{x}}\right)^{\mathrm{T}} & \mathbf{0} \end{bmatrix}$$

be negative definite. One of these is that the first $n-1$ trailing principal minors of this matrix alternate in sign, with the sign of the first principal minor being $(-1)^n$. Thus if $U(\mathbf{x}) = x_1 x_2$ such that

$$\mathbf{B} = \begin{bmatrix} 0 & 1 \\ 1 & 0 \end{bmatrix}$$

we need to examine the first trailing principal minor of the matrix

$$\begin{bmatrix} \mathbf{B} & -\mathbf{p} \\ -\mathbf{p}^{\mathrm{T}} & 0 \end{bmatrix} = \begin{bmatrix} 0 & 1 & -p_1 \\ 1 & 0 & -p_2 \\ -p_1 & -p_2 & 0 \end{bmatrix}$$

which is just the determinant of the entire matrix. This is $2p_1p_2$. The sign of this is positive and so we see that the **B** matrix does not necessarily have to be negative definite. The indefinite matrix used here satisfies the sufficiency condition quite nicely.

We are really just determining the conditions under which $\boldsymbol{\delta}\mathbf{x}^{\mathrm{T}} \dfrac{\partial^2 L(\mathbf{x})}{\partial \mathbf{x}^2} \boldsymbol{\delta}\mathbf{x} < \mathbf{0}$ along the budget income line. For $U = x_1 x_2$ we need to examine

$$\boldsymbol{\delta}\mathbf{x}^{\mathrm{T}} \mathbf{B} \boldsymbol{\delta}\mathbf{x} = [\,\delta x_1 \quad \delta x_2\,] \begin{bmatrix} 0 & 1 \\ 1 & 0 \end{bmatrix} \begin{bmatrix} \delta x_1 \\ \delta x_2 \end{bmatrix} = 2\delta x_1 \delta x_2$$

In general, this expression is not definite in sign. However, since $I = \mathbf{p}^{\mathrm{T}}\mathbf{x}$, we have $0 = \mathbf{p}^{\mathrm{T}}\boldsymbol{\delta}\mathbf{x}$ or $p_1\delta x_1 + p_2\delta x_2 = 0$ for the example we are considering. Thus $\delta x_2 = -p_1\delta x_1/p_2$ and

$$\boldsymbol{\delta}\mathbf{x}^{\mathrm{T}} \frac{\partial^2 L(\mathbf{x})}{\partial \mathbf{x}^2} \boldsymbol{\delta}\mathbf{x} = \boldsymbol{\delta}\mathbf{x}^{\mathrm{T}} \mathbf{B} \boldsymbol{\delta}\mathbf{x} = -\frac{2p_1}{p_2}(\delta x_1)^2$$

This expression is surely negative for all positive prices, and so we see that $U(\mathbf{x}) = 2x_1x_2$ is a valid utility function for utility maximization. This utility function has a convex utility indifference curve, as we may easily verify. ■

Example 3.9: We now consider a logarithmic utility function of the form

$$U(\mathbf{x}) = \ln\left(\prod_{i=1}^{n} x_i^{a_i}\right) = \sum_{i=1}^{n} a_i \ln x_i \tag{3.67}$$

which is strictly quasi-convex for all positive a_i (and x_i). We wish to maximize this utility function subject to, because of the applicable nonsatiation axiom, the equality budget constraint given by

$$\sum_{i=1}^{n} p_i x_i = I \tag{3.68}$$

We adjoin the equality constraint of Equation 3.68 to the cost function of Equation 3.67 and write the Lagrangian as

$$\mathbf{J} = U(\mathbf{x}) + \lambda(\mathbf{I} - \mathbf{p}^{\mathrm{T}}\mathbf{x}) = \sum_{i=1}^{n} a_i \ln x_i + \lambda\left(\mathbf{I} - \sum_{i=1}^{n} p_i x_i\right)$$

We obtain necessary (and sufficient) conditions for an optimum from $\partial\mathbf{J}/\partial\mathbf{x} = 0$ and $\partial\mathbf{J}/\partial\boldsymbol{\lambda} = 0$ as

$$\frac{a_i}{\hat{x}_i} - \hat{\lambda} p_i = 0, \qquad I = 1, 2, \ldots, n \tag{3.69}$$

$$\sum_{i=1}^{n} p_i \hat{x}_i = I \tag{3.70}$$

Multiplying both sides of Equation 3.69 by $\hat{x}_i$ and summing over all i from $i = 1$ to n yields

$$\sum_{i=1}^{n} a_i = \hat{\lambda} I \tag{3.71}$$

Thus, we have, on substituting the value of $\hat{\lambda}$ obtained in Equation 3.71 into Equation 3.69,

$$\hat{x}_i = \frac{I a_i}{p_i \sum_{i=1}^{n} a_i}, \qquad i = 1, 2, \ldots, n \tag{3.72}$$

The a_i are assumed to be fixed by the consumer's utility function. Here, the amount of a commodity that a consumer purchases will vary inversely with the price of the commodity. A very interesting property of this result is that the budget allocation to each commodity is

$$p_i \hat{x}_i = \frac{a_i}{\sum_{i=1}^{n} a_i} I = \frac{a_i}{\hat{\lambda}} \tag{3.73}$$

In many ways this is a very attractive utility function, because this result makes calibration of the utility function a relatively simple matter once we invoke its form. For example, suppose that we observe purchase data, averaged perhaps over a year, for a consumer, in the table shown next:

Commodity number	Number of commodities purchased	Total price paid for commodities
1	250	$1000
2	300	450
3	1000	300
4	650	250
5	30	500
Total	2230	$2500

We have, on invoking the utility function of Equation 3.67 and using Equation 3.73, the results $1000 = a_1/\hat{\lambda}$, $450 = a_2/\hat{\lambda}$, $300 = a_3/\hat{\lambda}$, $250 = a_4/\hat{\lambda}$, and $500 = a_5/\hat{\lambda}$. Because of the nonuniqueness of utility functions we are free to pick $\hat{\lambda}$ at any convenient value and so we choose $\hat{\lambda} = 2500$. Thus for this consumer that we have modeled, we obtain the vectors

$$\mathbf{p} = \begin{bmatrix} 4.00 \\ 1.50 \\ 0.30 \\ 0.38 \\ 16.67 \end{bmatrix}, \quad \mathbf{a} = \begin{bmatrix} 0.40 \\ 0.18 \\ 0.12 \\ 0.10 \\ 0.20 \end{bmatrix}$$

Product substitutability is a general function of economic life. If the a_i in this example are truly constants, then the utility function chosen is somewhat unrealistic in that the consumer always denotes a fixed percentage of their total budget to the purchase of each commodity. We can make the a_i a function of the price of the commodities in the bundle and obtain product substitutability. It would not seem unreasonable to let $a_i = \dfrac{b_i p_{in}}{p}$, where p_{in} is the fixed nominal or normal price of commodity i, p_i the actual price of item i, and b_i a constant. It is also not at all unreasonable that consumer utility for a commodity depend on the price of the commodity. This is the substance, in effect, of the theory of *revealed preference*, from which a consumer's utility indifference curve can be obtained. According to this theory, the fact that a consumer purchases a bundle x_1 rather than x_2 only indicates a consumer preference for x_1 over x_2 if the total price paid for x_1 is greater than that for x_2. In other words, where we use $>$ for revealed preference, we have $\mathbf{x}_1 > \mathbf{x}_2$ iff $\mathbf{p}_1^T \mathbf{x}_1 \geq \mathbf{p}_1^T \mathbf{x}_2$ for all $\mathbf{p}_1$.

There are two axioms of revealed preference: a weak axiom that states that revealed preferences are asymmetric, in that if $\mathbf{x}_1 > \mathbf{x}_2$, it is not possible for $\mathbf{x}_2 > \mathbf{x}_1$, and a strong axiom that states the transitivity of revealed preferences, in that if $\mathbf{x}_1 > \mathbf{x}_2$ and $\mathbf{x}_2 > \mathbf{x}_3$, then it must be true that $\mathbf{x}_1 > \mathbf{x}_3$. ■

A number of extensions of the basic theory of the consumer are possible. For example, consumers not only desire to purchase commodities but also desire leisure. Thus we might postulate a utility function for a commodity bundle $\mathbf{x}$ and leisure time l and denote it by $U(\mathbf{x}, l)$. Income and leisure are both desirable and the substitution of income for leisure to maintain constant utility is given by the expression

$$\mathrm{d}\mathbf{x}^{\mathrm{T}} \frac{\partial U(\mathbf{x}, l)}{\partial \mathbf{x}} + \frac{\partial U(\mathbf{x}, l)}{\partial l} \mathrm{d}l = 0 \tag{3.74}$$

We may alternately write this as

$$\frac{[\partial U(\mathbf{x}, l)/\partial \mathbf{x}]^{\mathrm{T}}}{\partial U(\mathbf{x}, l)/\partial l} \frac{\mathrm{d}\mathbf{x}}{\mathrm{d}l} = -1$$

We may maximize consumer utility in much the same way as before. If we let t represent the total amount of available time, W the time devoted to work, and w the wage rate, then we see that we wish to maximize

$$J = U(\mathbf{x}, l) \tag{3.75}$$

subject to the constraints that the total time t is fixed, and given by the sum or work time W and leisure time l:

$$t = l + W \tag{3.76}$$

Also, the total income for commodity purchase is determined by the amount of time worked and is given by

$$Ww = \mathbf{p}^{\mathrm{T}}\mathbf{x} \tag{3.77}$$

We can eliminate Equation 3.76 by substituting W from this equation into Equation 3.77 to obtain

$$(t - l)w - \mathbf{p}^{\mathrm{T}}\mathbf{x} = 0 \tag{3.78}$$

We have thus obtained the result that maximization of Equation 3.75 subject to the constraint of Equation 3.78 is the problem of the consumer with leisure. To obtain a solution, we adjoin Equation 3.78 to Equation 3.75 and obtain

$$J = U(\mathbf{x}, l) + \lambda(tw - lw - \mathbf{p}^{\mathrm{T}}\mathbf{x}) \tag{3.79}$$

The necessary conditions for optimality

$$\frac{\partial J}{\partial \mathbf{x}} = \mathbf{0} = \frac{\partial U(\mathbf{x}, l)}{\partial \mathbf{x}} - \mathbf{p}\lambda \tag{3.80}$$

$$\frac{\partial J}{\partial l} = 0 = \frac{\partial U(\mathbf{x}, l)}{\partial l} - \lambda w \tag{3.81}$$

$$\frac{\partial J}{\partial \lambda} = 0 = tw - lw - \mathbf{p}^{\mathrm{T}}\mathbf{x} \tag{3.82}$$

result. The first two of these equations may be combined to eliminate the Lagrange multiplier λ such that we have

$$\frac{\partial l}{\partial \mathbf{x}} = \frac{\mathbf{p}}{w} \tag{3.83}$$

This is equivalent to the expression

$$\frac{\partial \mathbf{p}^{\mathrm{T}}\mathbf{x}}{\partial l} = w \tag{3.84}$$

Thus, we see that the rate of substitution of budget income for leisure is the wage rate. This is a potentially useful result, although obtained under simplified assumptions, in that it tells us something about how much leisure a consumer would like. This is the sort of result possible from the theory of the consumer and we will examine issues like this further in Chapter 5.

We may modify the basic problem of the consumer in any number of ways that are basically similar to our modification to include the utility for leisure. Some of these are suggested for exploration in the problems at the end of this chapter. Others will be considered in Chapter 5 and the subsequent chapters.

Example 3.10: For our next example in this section we consider the very general translog (transcendental logarithmic) utility function as defined by the log quadratic function

$$\ln U(\mathbf{x}, l) = a + \mathbf{b}^{\mathrm{T}}\mathbf{z} + 0.5\mathbf{z}^{\mathrm{T}}\mathbf{C}\mathbf{z} \tag{3.85}$$

where a is a scalar, $\mathbf{b}$ a vector, and $\mathbf{C}$ a matrix. All three are assumed constant. The vector $\mathbf{z}$ is comprised of the logarithm of all commodities and leisure:

$$\mathbf{z} = \begin{bmatrix} \ln x_1 \\ \ln x_2 \\ \vdots \\ \ln x_n \\ \ln l \end{bmatrix} \tag{3.86}$$

We formulate precisely the same optimization of consumer utility problem that we have just discussed. Here, Equations 3.80 to 3.82 are the necessary conditions for consumer optimality. We obtain from Equation 3.85, for $i = 1, 2, \ldots, n$,

$$\frac{\partial U(\mathbf{x}, l)}{\partial x_i} = \frac{U(\mathbf{x}, l)}{x_i} \frac{\partial[\ln U(\mathbf{x}, l)]}{\partial(\ln x_i)} = \frac{U(\mathbf{x}, l)}{x_i}(b_i + \mathbf{C}_i\mathbf{z}) \tag{3.87}$$

$$\frac{\partial U(\mathbf{x}, l)}{\partial l} = \frac{U(\mathbf{x}, l)}{l} \frac{\partial[\ln U(\mathbf{x}, l)]}{\partial(\ln l)} = \frac{U(\mathbf{x}, l)}{l}(b_{n+1} + \mathbf{C}_{n+1}\mathbf{z}) \tag{3.88}$$

where $\mathbf{C}_i$ is the ith row of the matrix $\mathbf{C}$. Using Equations 3.80 to 3.82 results in

$$\frac{U(\mathbf{x}, l)}{x_i}(b_i + \mathbf{C}_i\mathbf{z}) = p_i\lambda \tag{3.89}$$

$$\frac{U(\mathbf{x}, l)}{l}(b_{n+1} + \mathbf{C}_{n+1}\mathbf{z}) = w\lambda \tag{3.90}$$

$$tw - lw = \sum_{i=1}^{n} p_i x_i \tag{3.91}$$

We multiply both sides of Equation 3.89 by x_i and sum over i from 1 to n. Use of Equation 3.91 then gives us

$$U(\mathbf{x}, l)\left(\sum_{i=1}^{n}(b_i + \mathbf{C}_i\mathbf{z})\right) = \lambda(tw - lw) \tag{3.92}$$

Substituting for λ in Equation 3.89 with this result leads us to the relation

$$(t - l)w(b_i + \mathbf{C}_i\mathbf{z}) = p_i x_i\left(\sum_{j=1}^{n}(b_j + C_j\mathbf{z})\right), \qquad i=1,2,\ldots,n \tag{3.93}$$

Substitution of the value of λ obtained in Equation 3.92 into Equation 3.90 results in

$$(t - l)(b_{n+1} + \mathbf{C}_{n+1}\mathbf{z}) = l\left(\sum_{j=1}^{n}(b_j + \mathbf{C}_j\mathbf{z})\right) \tag{3.94}$$

Because of the linear way in which the parameters **b** and **C** enter into Equations 3.93 and 3.94, we can use linear regression techniques, such as those discussed in statistics, forecasting, or system identification texts, to identify the **b** and **C** parameters. In so doing we will have identified the parameters in the utility function of Equation 3.85.

Equations 3.93 and 3.94 are the central results of this example. From these equations we obtain the optimum commodity demand bundle **x** and the optimum amounts of leisure or work. An explicit solution of Equations 3.93 and 3.94 is not possible unless $\mathbf{C} = \mathbf{0}$. In this special case we obtain from Equations 3.93 and 3.94

$$p_i x_i = \frac{twb_i}{b_{n+1} + \sum_{j=1}^{n} b_j}, \qquad i = 1,2,\ldots,n \tag{3.95}$$

$$l = \frac{tb_{n+1}}{b_{n+1} + \sum_{j=1}^{n} b_j} \tag{3.96}$$

as the optimum consumer commodity demand bundle and leisure supply. ■

In this section, we have examined a relatively large number of examples. It would be instructive to consider some particular numerical cases. This is the purpose of some of the problems at the end of this chapter.

3.5 SENSITIVITY AND SUBSTITUTION EFFECTS

Often optimum economic consumer operating conditions have been determined and one or more parameters, most often income and prices, change. If these changes are small, we can make a linear perturbation of the original problem equations and then make a relatively simple determination of changes in the optimum consumer bundle as a function of the changed parameters. This linear perturbation about operating conditions is often termed a *sensitivity analysis*. To accomplish this sensitivity analysis we recall some of our results from Section 3.4 concerning the theory of the consumer.

The fundamental problem of the consumer is to maximize utility

$$J = U(\mathbf{x}) \tag{3.97}$$

subject to the budget constraint

$$\mathbf{p}^{\mathrm{T}}\mathbf{x} = I \tag{3.98}$$

where we assume that prices are nonnegative and that the full budget I will be spent. The Kuhn–Tucker optimality conditions for this problem are, where we recall that the optimum commodity bundle could and perhaps should be written as $\hat{\mathbf{x}}(I, \mathbf{p})$ to give explicit indication to the fact that the optimum bundle is a function of budget and prices, given by

$$\frac{\partial U(\mathbf{x})}{\partial \mathbf{x}} - \lambda\mathbf{p} = \mathbf{0} \tag{3.99}$$

$$I - \mathbf{p}^{\mathrm{T}}\mathbf{x} = 0 \tag{3.100}$$

A brief introduction to sensitivity-based analysis will provide useful results that we will use on the consumer demand problem. We obtain sensitivity analysis relations for an equation

$$\mathbf{f}(\mathbf{x}, \mathbf{z}) = \mathbf{0} \tag{3.101}$$

by letting the values $\mathbf{x}$ and $\mathbf{z}$ equal nominal values plus a perturbation,

$$\mathbf{x} = \mathbf{x}^n + \Delta\mathbf{x} \tag{3.102}$$

$$\mathbf{z} = \mathbf{z}^n + \Delta\mathbf{z} \tag{3.103}$$

and then expanding the resulting function in a Taylor series about $\Delta\mathbf{x}$ and $\Delta\mathbf{z}$ and then dropping terms higher than first order in $\Delta\mathbf{x}$ and $\Delta\mathbf{z}$ to yield[5] the result

[5]To simplify the notation, we define

$$\frac{\partial \mathbf{f}(\mathbf{x}^n, \mathbf{z}^n)}{\partial \mathbf{x}^n} = \left.\frac{\partial \mathbf{f}(\mathbf{x}, \mathbf{z})}{\partial \mathbf{x}}\right|_{\substack{\mathbf{x} = \mathbf{x}^n \\ \mathbf{z} = \mathbf{z}^n}}$$

We use a similar definition for the derivative with respect to $\mathbf{z}$.

$$\mathbf{f}(\mathbf{x},\mathbf{z}) = \mathbf{0} = \mathbf{f}(\mathbf{x}^n,\mathbf{z}^n) + \frac{\partial \mathbf{f}(\mathbf{x}^n,\mathbf{z}^n)}{\partial \mathbf{x}^n}\Delta\mathbf{x} + \frac{\partial \mathbf{f}(\mathbf{x}^n,\mathbf{z}^n)}{\partial \mathbf{z}^n}\Delta\mathbf{z} \tag{3.104}$$

The nominal solution $(\mathbf{x}^n, \mathbf{z}^n)$ should satisfy Equation 3.101, and so we have

$$\mathbf{0} = \frac{\partial \mathbf{f}(\mathbf{x}^n,\mathbf{z}^n)}{\partial \mathbf{x}^n}\Delta\mathbf{x} + \frac{\partial \mathbf{f}(\mathbf{x}^n,\mathbf{z}^n)}{\partial \mathbf{z}^n}\Delta\mathbf{z}$$

We have a linear relationship between $\Delta\mathbf{x}$ and $\Delta\mathbf{z}$ given, from the foregoing, by

$$\Delta\mathbf{x} = \left(-\frac{\partial \mathbf{f}(\mathbf{x}^n,\mathbf{z}^n)}{\partial \mathbf{x}^n}\right)^{-1}\frac{\partial \mathbf{f}(\mathbf{x}^n,\mathbf{z}^n)}{\partial \mathbf{z}^n}\Delta\mathbf{z} \tag{3.105}$$

assuming, of course, that the matrix inverse in this equation exists.

We will now apply these results to the necessary conditions for consumer optimality of Equations 3.99 and 3.100. We let

$$\mathbf{p} = \mathbf{p}^n + \Delta\mathbf{p} \tag{3.106}$$

$$I = I^n + \Delta I \tag{3.107}$$

We next note that, to terms of first order,

$$\mathbf{x} = \mathbf{x}(I,\mathbf{p}) \cong \hat{\mathbf{x}} + \Delta\mathbf{x} \tag{3.108}$$

$$\lambda = \lambda(I,\mathbf{p}) \cong \hat{\lambda} + \Delta\lambda \tag{3.109}$$

$$U(\mathbf{x}) = U(\hat{\mathbf{x}} + \Delta\mathbf{x}) \cong U(\hat{\mathbf{x}}) + \left(\frac{\partial U(\hat{\mathbf{x}})}{\partial \mathbf{x}}\right)^{\mathrm{T}}\Delta\mathbf{x} \tag{3.110}$$

Here we assume that the optimum consumer bundle is that which results when the budget income and prices are nominal. Inserting Equations 3.106 through 3.110 into Equations 3.99 and 3.100 results in the expressions

$$\frac{\partial U(\hat{\mathbf{x}})}{\partial \hat{\mathbf{x}}} + \frac{\partial^2 U(\hat{\mathbf{x}})}{\partial \hat{\mathbf{x}}^2}\Delta\mathbf{x} - (\hat{\lambda} + \Delta\lambda)(\mathbf{p}^n + \Delta\mathbf{p}) = \mathbf{0} \tag{3.111}$$

$$I^n + \Delta I - (\mathbf{p}^n + \Delta\mathbf{p})^{\mathrm{T}}(\hat{\mathbf{x}} + \Delta\mathbf{x}) = 0 \tag{3.112}$$

We subtract the identities of Equations 3.99 and 3.100 when they are evaluated at the nominal budget and prices I^n and $\mathbf{p}$, respectively, and at the resulting $\hat{\mathbf{x}}$ from the foregoing two equations. This yields, after neglecting the $\Delta\lambda\ \Delta\mathbf{p}$ and $\Delta\mathbf{p}^{\mathrm{T}}\ \Delta\mathbf{x}$ products of second-order terms, the expression

$$\mathbf{A}\begin{bmatrix}\Delta\mathbf{x}\\ \Delta\lambda\end{bmatrix} = \mathbf{B}\begin{bmatrix}\Delta I\\ \Delta\mathbf{p}\end{bmatrix} \tag{3.113}$$

On dropping the superscripts for convenience, we have the expressions

$$\mathbf{A} = \begin{bmatrix} \dfrac{\partial^2 U(\mathbf{x})}{\partial \hat{\mathbf{x}}^2} & -\mathbf{p} \\ -\mathbf{p}^{\mathrm{T}} & 0 \end{bmatrix} \tag{3.114}$$

$$\mathbf{B} = \begin{bmatrix} 0 & \lambda \\ -1 & \mathbf{x}^{\mathrm{T}} \end{bmatrix} \tag{3.115}$$

If the utility function obeys the restriction we imposed earlier, the $\partial^2 U(\mathbf{x})/\partial \mathbf{x}^2$ expression is negative definite. All the components of $-\mathbf{p}$ are negative since no price is negative. Thus the inverse of the $\mathbf{A}$ matrix exists. We have, as we may easily verify,[6]

$$\mathbf{A}^{-1} = \begin{bmatrix} \alpha \mathbf{K}^{-1}\mathbf{p}\mathbf{p}^{\mathrm{T}}\mathbf{K}^{-1} + \mathbf{K}^{-1} & \alpha \mathbf{K}^{-1}\mathbf{p} \\ \alpha \mathbf{p}^{\mathrm{T}}\mathbf{K}^{-1} & \alpha \end{bmatrix} \tag{3.116}$$

where

$$\mathbf{K} = \frac{\partial^2 U(\mathbf{x})}{\partial \mathbf{x}^2}, \qquad \alpha = \frac{-1}{\mathbf{p}^{\mathrm{T}}\mathbf{K}^{-1}\mathbf{p}}$$

We see from Equation 3.113 that we can write the relation between the commodity bundle change and changes in the budget and prices as

$$\Delta \mathbf{x} = \hat{\lambda}(\alpha \mathrm{K}^{-1}\mathbf{p}\mathbf{p}^{\mathrm{T}}\mathrm{K}^{-1} + \mathrm{K}^{-1})\Delta \mathbf{p} + \alpha \mathrm{K}^{-1}\mathbf{p}\mathbf{x}^{\mathrm{T}}\Delta \mathbf{p} - \alpha \mathrm{K}^{-1}\mathbf{p}\Delta \mathbf{I} \tag{3.117}$$

We can show[7] that the scalar coefficient α is, from Equation 3.33,

$$\alpha = -\frac{\partial \lambda}{\partial I} = -\frac{\partial}{\partial I}\frac{\partial U(\mathbf{x})}{\partial I} = -\frac{\partial^2 U(\mathbf{x})}{\partial I^2} \tag{3.118}$$

Thus, we see that this coefficient is the rate of decrease of the marginal utility of income. Also, we can easily obtain a set of partial differential equations equivalent to Equation 3.117. They are

$$\frac{\partial x}{\partial I} = -\alpha \mathrm{K}^{-1}\mathbf{p} \tag{3.119}$$

$$\frac{\partial \mathbf{x}}{\partial \mathbf{p}} = \alpha \mathrm{K}^{-1}\mathbf{p}\mathbf{x}^{\mathrm{T}} + \hat{\lambda}(\alpha \mathbf{K}^{-1}\mathbf{p}\mathbf{p}^{\mathrm{T}}\mathbf{K}^{-1} + \mathbf{K}^{-1}) \tag{3.120}$$

An interesting interpretation can be obtained from Equation 3.117 if we consider the effect of a compensated change in price where the income is adjusted to keep the consumer utility a constant. If $U(\mathbf{x})$ is a constant, then $\Delta U(\mathbf{x}) = 0$ and

$$\Delta U(\mathbf{x}) = \left(\frac{\partial U(\mathbf{x})}{\partial \mathbf{x}}\right)^{\mathrm{T}} \Delta \mathbf{x} = 0 \tag{3.121}$$

[6]We simply calculate $\mathbf{A}^{-1}\mathbf{A} = \mathbf{I}$, the identity matrix, and use this relation.

[7]See Problem 22.

From Equation 3.99 we see that the foregoing becomes, at the consumer optimal commodity bundle,

$$\Delta U(\mathbf{x}) = \lambda \mathbf{p}^{\mathrm{T}} \Delta \mathbf{x} = 0 \tag{3.122}$$

From Equation 3.100 we obtain the required change in income

$$\Delta I = \mathbf{x}^{\mathrm{T}} \Delta \mathbf{p} + \mathbf{p}^{\mathrm{T}} \Delta \mathbf{x} \tag{3.123}$$

which becomes, because of Equation 3.122,

$$\Delta I = \mathbf{x}^{\mathrm{T}} \Delta \mathbf{p} \tag{3.124}$$

This states the required relation between the budget income and prices to keep utility a constant. It states a compensated income change for preservation of constant utility. This is a very important relation in that it demonstrates a key property of the demand functions obtained from consumer optimization for maximum utility; *demand functions are homogeneous of degree zero* in budget I and prices $\mathbf{p}$. Thus we see that if all prices and consumer income are multiplied by the same positive constant a, then the optimum consumer bundle is unchanged and not a function of a.

We denote the solution for the optimum commodity bundle change under constant utility conditions as $(\Delta \mathbf{x})_{\mathrm{const}}$. Substituting Equation 3.124 into Equation 3.117, we see that this is given by

$$(\Delta \mathbf{x})_{\mathrm{const}} = \hat{\lambda}(\alpha \mathbf{K}^{-1} \mathbf{p} \mathbf{p}^{\mathrm{T}} \mathbf{K}^{-1} + \mathbf{K}^{-1}) \Delta \mathbf{p} \tag{3.125}$$

The $(\Delta \mathbf{x})_{\mathrm{const}}$ obtained from this relationship is often called the *substitution effect* of a compensated change in price on demand. The other terms in $(\Delta \mathbf{x})$ Equation 3.117 are called (1) the *income effect* $(\Delta \mathbf{x})_{income}$, which represents the effect of a change in income on demand, and (2) the change in demand quantity due to a change in income or budget from that required to purchase $\hat{\mathbf{x}}$. We have

$$(\Delta \mathbf{x})_{\mathrm{income}} = \alpha \mathbf{K}^{-1} \mathbf{p}(\mathbf{x}^{\mathrm{T}} \Delta \mathbf{p} - \Delta I) \tag{3.126}$$

and

$$\Delta \mathbf{x} = (\Delta \mathbf{x})_{\mathrm{const}} + (\Delta \mathbf{x})_{\mathrm{income}} \tag{3.127}$$

Or, in words, total effect equals substitution effect plus income effect.

We can obtain a partial differential equation corresponding to Equation 3.125 as

$$\left(\frac{\partial \mathbf{x}}{\partial \mathbf{p}}\right)_{\mathrm{const}} = \hat{\lambda}(\alpha \mathbf{K}^{-1} \mathbf{p} \mathbf{p}^{\mathrm{T}} \mathbf{K}^{-1} + \mathbf{K}^{-1}) \tag{3.128}$$

such that Equation 3.120 becomes, using the foregoing and Equation 3.119,

$$\frac{\partial \mathbf{x}}{\partial \mathbf{p}} = \left(\frac{\partial \mathbf{x}}{\partial \mathbf{p}}\right)_{\mathrm{const}} - \left(\frac{\partial \mathbf{x}}{\partial I}\right) \mathbf{x}^{\mathrm{T}} \tag{3.129}$$

From Equation 3.128 we see that the matrix $(\partial\mathbf{x}/\partial\mathbf{p})_{\text{const}}$ is symmetric and negative semidefinite. By equating the ijth coefficient of $(\partial\mathbf{x}/\partial\mathbf{p})_{\text{const}}$ in Equation 3.129 to the jith coefficient of $(\partial\mathbf{x}/\partial\mathbf{p})_{\text{const}}$, we obtain the interesting result[8] that, for any components in the commodity bundle,

$$\frac{\partial x_i}{\partial p_j} + x_j \frac{\partial x_i}{\partial I} = \frac{\partial x_j}{\partial p_i} + x_i \frac{\partial x_j}{\partial I} \tag{3.130}$$

This equation is known to economists as the *Slutsky equation* and can be used to account for the changes in the demand for one component in the commodity bundle due to changes in prices for another commodity bundle component.

Postmultiplying either Equation 3.125 or 3.128 by $\mathbf{p}$, we easily see that, because of the definition of α in Equation 3.116,

$$\left(\frac{\partial \mathbf{x}}{\partial \mathbf{p}}\right)_{\text{const}} \mathbf{p} = \mathbf{0} \tag{3.131}$$

This has a very interesting interpretation, since all prices $\mathbf{p}$ are positive. All elements of any row of $(\partial\mathbf{x}/\partial\mathbf{p})_{\text{const}}$ cannot have the same sign. The elements on the main diagonal must be negative.[9] Thus at least one of the off-diagonal elements must be a positive quantity. We say that commodities are *substitutes* if the off-diagonal element is positive, or

$$\left(\frac{\partial x_i}{\partial p_j}\right)_{\text{const}} > 0 \tag{3.132}$$

and that commodities are *complementary* if the off-diagonal element is negative, or

$$\left(\frac{\partial x_i}{\partial p_j}\right)_{\text{const}} < 0 \tag{3.133}$$

If commodities are substitutes, and utility is constant, an increase of price of one commodity will lead to an increased demand for the other commodity. All commodities must have at least one substitute, otherwise it will not be possible to keep the utility function invariant. For complementary commodities, an increase in price of one commodity will lead to a decreased demand for the other commodity. For example, coffee and tea are likely substitute commodities, whereas guns and bullets are likely complementary commodities.

The concept of substitute and complementary commodities is applicable only when utility is held invariant by the price–budget income relationship of Equation 3.124. When price and budget income are not so constrained, we say that a commodity is *normal* if the corresponding main diagonal component of $\partial\mathbf{x}/\partial\mathbf{p}$ is negative:

[8]See Problem 30. This result may also be obtained by using the necessary conditions, given by Equations 3.99 and 3.100, for combining optimum consumer behavior.

[9]The matrix $(\partial\mathbf{x}/\partial\mathbf{p})_{\text{const}}$ is negative semidefinite. Also, an element on the main diagonal of this matrix is its own substitution effect.

$$\frac{\partial x_i}{\partial p_i} < 0 \tag{3.134}$$

It is called a *Giffen commodity* if it is positive:

$$\frac{\partial x_i}{\partial p_i} > 0 \tag{3.135}$$

It is rare that a commodity is Giffen. A commodity is termed *superior* if

$$\frac{\partial x_i}{\partial I} > 0 \tag{3.136}$$

A commodity is called *inferior* if

$$\frac{\partial x_i}{\partial I} < 0 \tag{3.137}$$

For example, meat would likely be considered by many to be a superior commodity, and potatoes an inferior commodity.

Elasticities play an important role in the theory of the household, as they did in the theory of the firm. We define the (cross) *elasticity of the demand for commodity i with respect to the price of commodity j* as

$$\begin{aligned} \varepsilon_{i,j}^{p} &= \frac{\Delta x_i/x_i}{\Delta pj/pj} = \frac{\partial x_i/x_i}{\partial p_j/p_j} = \frac{p_j}{x_i}\frac{\partial x_i}{\partial p_j} \\ &= \frac{\text{price of marginal commodity purchased}}{\text{price of average commodity purchased}} \end{aligned} \tag{3.138}$$

The elasticity of a given commodity with respect to its own price will usually be negative (again, the Giffen commodity is rare). Such an elasticity is generally called the *price elasticity of demand* and is defined by

$$\varepsilon_i^p = \varepsilon_{i,i}^p = \frac{\partial x_i/x_i}{\partial p_i/p_i} = \frac{p_i}{x_i}\frac{\partial x_i}{\partial p_i}$$

The (price) elasticity of demand will depend on the shape of the demand curve and also on the equilibrium position on the demand curve. Let us examine three relevant examples here. More are present in the set of problems at the end of the chapter.

Example 3.11: For the particular case of a linear demand curve $x_i = a - bp_i$, we easily see that the price elasticity of demand is

$$\varepsilon_i^p = \frac{-bp_i}{a - bp_i}$$

and we see that the price elasticity of demand varies from 0 to $-\infty$ as the demand and price vary from a and 0 to 0 and a/b. For the case where $b = 0$, such that the demand is constant and independent of price, we have $\varepsilon_i^p = 0$. This zero elasticity is called a *completely inelastic* demand. The other extreme is where the price is extremely insensitive to demand and a constant price results. For this case $a = \infty$ and $b = \infty$ in such a way that $a/b = p_i$. Here the price elasticity of demand is $-\infty$ and we say that the demand is completely elastic. To avoid obtaining negative elasticities for normal commodities, many authors define elasticities as the *negative* of the ratio of the percentage of change in quantity to the change in price. Figure 3.5 illustrates typical demand curves and commonly used names for the price elasticity of demand. ■

Also, we can write for the elasticity of the demand for commodity i with respect to budget income

$$\varepsilon_i^I = \frac{\Delta x_i / x_i}{\Delta I / I} = \frac{I}{x_i}\frac{\partial x_i}{\partial I} = \frac{\text{income for marginal commodity purchased}}{\text{income for average commodity purchased}} \tag{3.139}$$

Either by using the necessary conditions for consumer optimality of Equations 3.99 and 3.100 or by postmultiplying Equation 3.129 by $\mathbf{p}$ and using Equation 3.131, we obtain

$$\frac{\partial \mathbf{x}}{\partial \mathbf{p}}\mathbf{p} + \frac{\partial \mathbf{x}}{\partial \mathbf{I}}\mathbf{I} = \mathbf{0} \tag{3.140}$$

In component form this is

$$\sum_{i=1}^{n} p_i \frac{\partial x_j}{\partial p_i} + I\frac{\partial x_j}{\partial I} = 0, \qquad j = 1, 2, \ldots, n \tag{3.141}$$

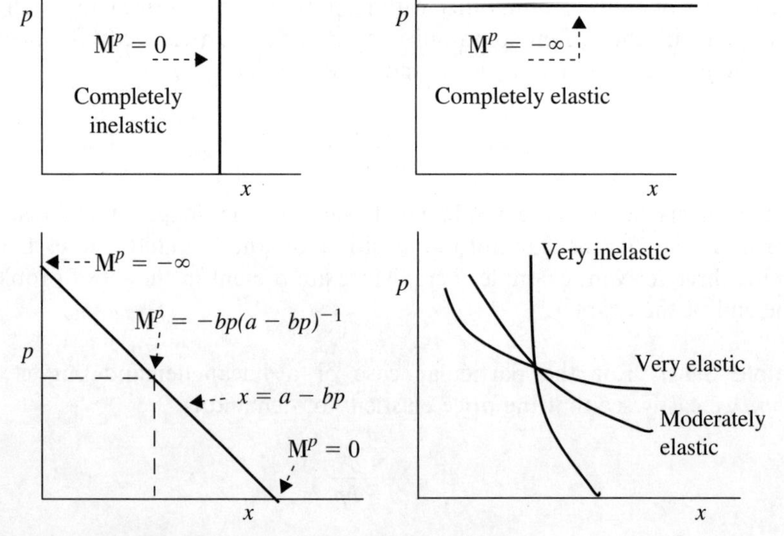

Figure 3.5. Price Elasticities for Various Demand Curves.

or

$$\sum_{i=1}^{n} \frac{p_i}{x_j} \frac{\partial x_j}{\partial p_i} + \frac{I}{x_j} \frac{\partial x_j}{\partial I} = 0, \qquad j = 1, 2, \ldots, n \tag{3.142}$$

and in terms of the price and income elasticities we have

$$\sum_{i=1}^{n} \varepsilon_{j,i}^{p} + \varepsilon_j^{I} = 0, \qquad j = 1, 2, \ldots, n \tag{3.143}$$

Thus we see that the sum of the income elasticity for commodity j and the summed demand cross-elasticity for commodity j with respect to the price of commodities i (for all i) is zero for all j.

Other interesting and useful results can also be obtained here. Premultiplying Equation 3.119 by $\mathbf{p}^{\mathrm{T}}$ we see that

$$\mathbf{p}^{\mathrm{T}} \frac{\partial \mathbf{x}}{\partial I} = 1 = \sum_{i=1}^{n} p^{i} \frac{\partial x_i}{\partial I} \tag{3.144}$$

which says that

$$1 = \sum_{i=1}^{n} \left(\frac{p_i x_i}{I} \right) \left(\frac{I}{x_i} \frac{\partial x_i}{\partial I} \right)$$

or

$$I = \sum_{i=1}^{n} p_i x_i \varepsilon_i^{I} \tag{3.145}$$

Thus the sum of total expenditures for each commodity times the income commodity sensitivity is the total budget income. This is called the *Engel aggregation condition* by economists.

If we premultiply Equation 3.129 by $\mathbf{p}^{\mathrm{T}}$ and use the results in Equations 3.131 and 3.144, we obtain

$$\mathbf{p}^{\mathrm{T}} \frac{\partial \mathbf{x}}{\partial \mathbf{p}} + \mathbf{x}^{\mathrm{T}} = \mathbf{0}$$

This is often known as the *Cournot aggregation* condition. In scalar form this says that

$$\sum_{i=1}^{n} p_i \frac{\partial x_i}{\partial p_j} + x_j = 0, \qquad j = 1, 2, \ldots, n$$

or

$$x_j = -\sum_{i=1}^{n} p_i \frac{\partial x_i}{\partial p_j} \tag{3.146}$$

So we see that the demanded commodity x_j is just the negative sum of the commodity price-weighted change of demanded quantities with respect to price.

Let us now suppose that budget income I is constant but that there is a change Δp_i in the price of the ith commodity. We might wish to obtain the ratio of the change in expenditure for commodity i to the original expenditure for commodity i. We have

$$\frac{\Delta(p_i x_i)}{p_i x_i} = \frac{x_i \Delta p_i + p_i \Delta x_i}{p_i x_i} = \frac{\Delta p_i}{p_i}\left(1 + \frac{p_i \Delta x_i}{x_i \Delta p_i}\right) = \frac{\Delta p_i}{p_i}(1 + \varepsilon_{i,i}^{p}) \tag{3.147}$$

Thus, we see that if $\varepsilon_{i,i}^{p} < -1$, we will spend less for commodity i if the price of commodity i rises. If $\varepsilon > -1$, a price rise for commodity i will cause an increase in expenditure for commodity i. Thus we see that if we have a highly elastic demand such that the magnitude of the price elasticity is large, $|\varepsilon_{i,i}^{p}| > 1$, we are dealing with either commodities for which substitutes can easily be found or optional commodities that we can easily do without. If the magnitude of the price elasticity is small, $|\varepsilon_{i,i}^{p}| < 1$, we say that we have a low demand or inelastic demand and we are dealing with a commodity that is either quite necessary or a very small part of the overall commodity bundle.

Sensitivity concepts are especially important in economic systems analysis and assessment. It is rare that we will have precise parameter information in realistic real-world situations. With sensitivity analysis we can determine which parameters are most influential in determining system behavior such that we can allocate parameter measurement resources appropriately. To accomplish sensitivity analysis on other than very simple problems requires the use of a digital computer for detailed computations, as the associated analysis efforts can become quite laborious, as we easily sense by the mathematics in this section.

Example 3.12: Suppose that the utility function for n goods is given by

$$U(\mathbf{x}) = \sum_{i=1}^{n} a_i \ln \mathbf{x}_i$$

and that the consumer income is I. We easily obtain the optimum commodity bundles from the Lagrangian expression

$$L = \sum_{i=1}^{n} a_i \ln x_i + \lambda\left(I - \sum_{i=1}^{n} p_i x_i\right)$$

as

$$x_j = \frac{a_j I}{p_j \sum_{i=1}^{n} a_i}, \qquad j = 1, 2, \ldots, n$$

This equation represents the demand equations for this example. Each commodity is normal and superior. The commodities are independent in that the demand function for one good does not depend on the price of another. Thus no good is a "substitute" or "complement" for another good *if we obtain this with respect to changes in price for a fixed income* or without regard for utility. We may remove this potential confusion over the meaning of "substitutability" by use of the terms *normal* and *superior*.

The elasticities are

$$\varepsilon_j^I = -\varepsilon_{j,j}^p = \frac{a_j I}{x_j p_j \sum_{i=1}^n a_i}$$

$$\varepsilon_{j,i}^p = 0, \qquad j \neq i$$

Thus, we see that the Engel aggregation condition of Equation 3.145 is satisfied. Equation 3.129 becomes

$$\left(\frac{\partial x_j}{\partial p_j}\right)_{\text{const}} = \frac{a_j I}{p_j^2 \sum_{i=1}^n a_i}\left(\frac{a_j}{\sum_{i=1}^n a_i} - 1\right)$$

$$\left(\frac{\partial x_j}{\partial p_k}\right)_{\text{const}} = \frac{a_j a_k I}{p_j p_k \left(\sum_{i=1}^n a_i\right)^2}$$

From application of the results of Equation 3.132 here, we see that the commodities are substitutes. Again we must emphasize that this concept of substitutability is applicable if and only if we keep utility invariant by appropriate adjustment of the price–income relationship. ■

Example 3.13: Consider now the utility function

$$U(x_1, x_2) = x_1^{0.5} + x_2^{0.5}$$

with the budget equation

$$I = p_1 x_1 + p_2 x_2$$

From the Lagrangian, which is given here by the expression

$$L = x_1^{0.5} + x_2^{0.5} + \lambda(I - p_1 x_1 - p_2 x_2)$$

we obtain the necessary conditions for utility maximization and the resulting optimal purchases as

$$x_1 = \frac{I p_2}{p_1(p_1 + p_2)}$$

$$x_2 = \frac{I p_1}{p_2(p_1 + p_2)}$$

We can easily obtain the results

$$\frac{\partial x_1}{\partial p_1} = -\frac{Ip_2(2p_1 + p_2)}{p_1^2(p_1 + p_2)^2}$$

$$\frac{\partial x_1}{\partial p_2} = \frac{I}{(p_1 + p_2)^2}$$

$$\frac{\partial x_2}{\partial p_2} = -\frac{Ip_1(2p_2 + p_1)}{p_2^2(p_2 + p_1)^2}$$

$$\frac{\partial x_2}{\partial p_1} = \frac{I}{(p_1 + p_2)^2}$$

$$\frac{\partial x_1}{\partial I} = \frac{p_2}{p_1(p_1 + p_2)}$$

$$\frac{\partial x_2}{\partial I} = \frac{p_1}{p_2(p_1 + p_2)}$$

From Equation 3.129 we obtain

$$\left(\frac{\partial x_1}{\partial p_1}\right)_{\text{const}} = -\frac{2Ip_2}{p_1(p_1 + p_2)^2}$$

$$\left(\frac{\partial x_1}{\partial p_2}\right)_{\text{const}} = -\frac{2I}{(p_1 + p_2)^2} = \left(\frac{\partial x_2}{\partial p_1}\right)_{\text{const}}$$

$$\left(\frac{\partial x_2}{\partial p_2}\right)_{\text{const}} = -\frac{2Ip_1}{p_2(p_2 + p_1)^2}$$

In this example, we see that the commodities are normal, superior, and substitutes for one another. ■

3.6 SUMMARY

In this chapter we have examined some elementary concepts from the microeconomic theory of the consumer. We introduced the concept of a utility function with which to express the satisfaction or value the consumer obtains from alternative goods and services. A number of axioms of utility theory were postulated to model a rational consumer. We examined the necessary conditions for a consumer to maximize utility subject to a constraint of a fixed budget income and with given prices for goods and services. The results of this optimization are important in that we obtain the consumer demand function for goods and services in terms of their price and the consumer budget or income for goods and services.

Finally, we established a number of important ancillary concepts using the necessary conditions for optimum consumer behavior. We obtained some sensitivity results concerning optimum consumer demand functions, as well as some linearized properties of the demand function. In Chapter 4 we will combine the results of this chapter concerning the consumer together with those of Chapter 2 concerning the firm and establish some important microeconomic supply–demand equilibrium properties.

We have not, in this chapter, considered problems of consumer demand where the outcomes involved with purchasing decisions are associated with risk and uncertainty. This is a very interesting and pertinent topic, but to go into it would greatly increase the size and scope of this book. Nor have we considered problems that involve preferences over time. Chapter 6 is devoted to a brief discussion of this topic where we examine cost–benefit and cost–effectiveness analyses and assessments.

PROBLEMS

1. Show pictorially the reasons for the following:

a. The indifference relation $\sim$ is symmetric and transitive.

b. The strict preference relation $>$ or $<$ is symmetric and transitive.

c. $U(\mathbf{x}_1) = U(\mathbf{x}_2)$ if and only if $\mathbf{x}_1 \sim \mathbf{x}_2$.

d. $U(\mathbf{x}_1) > U(\mathbf{x}_2)$ if and only if $\mathbf{x}_1 > \mathbf{x}_2$.

2. For the marginal rate of substitution defined by Equation 3.8,

$$\alpha = \mathrm{MRCS}_{ij} = \frac{\mathrm{MU}_i(\mathbf{x})}{\mathrm{MU}_j(\mathbf{x})} = -\frac{dx_j}{dx_i} = \frac{\partial U(\mathbf{x})/\partial x_i}{\partial U(\mathbf{x})/\partial x_j}$$

Show that the utility function is additive only if

$$\alpha \frac{\partial^2 \alpha}{\partial x_i \partial x_j} = \frac{\partial \alpha}{\partial x_i} \frac{\partial \alpha}{\partial x_j}$$

and that the demand for a commodity depends only on the price of the commodity and the total budget available for commodities.

3. Do any of the following utility functions have convex indifference curves? What is the marginal utility of each utility function?

a. $U(x_1,x_2) = x_1 x_2$

b. $U(x_1,x_2,x_3) = x_1^2 + x_2^2 + x_3^2$

c. $U(x_1,x_2) = \ln x_1 + \ln x_2$

d. $U(x_1,x_2) = 16x_1x_2 - x_1^3x_2^3$

e. $U(x_1,x_2) = x_1 + 4x_2 - x_1^2 - 0.5x_2^2$

4. Show that the utility functions

$$U_1(x_1, x_2) = x_1^{\alpha_1} x_2^{\alpha_2}, \qquad U_2(x_1, x_2) = A x_1^{\alpha_1/\alpha_2} x_2$$

can both describe the same preferences, in that one is a monotonic transformation of the other.

5. What is the optimum commodity bundle for a consumer with the following utility functions and budgets?

a. $U(\mathbf{x}) = x_1 x_2,\qquad x_1 + 4x_2 \leq 50$

b. $U(\mathbf{x}) = x_1 + 2x_2 + 3x_3,\qquad x_1 + 4x_2 + 2x_3 \leq 100$

c. $U(\mathbf{x}) = x_1^{\alpha_1} x_2^{\alpha_2},\qquad p_1 x_1 + p_2 x_2 \leq I$

6. A consumer purchases the commodity bundle $x_1 = 5$, $x_2 = 10$ at prices $p_1 = 3$, $p_2 = 1$. The same consumer also purchases the bundle $x_1 = 8$, $x_2 = 36$ at prices $p_1 = 10$, $p_2 = 6$. Is the consumer behavior consistent with the revealed preference concept?

7. Demonstrate that the necessary conditions of optimum consumer behavior are unchanged by any monotonically increasing transformation of the utility function.

8. For a two-commodity bundle, can both bundle components be superior or inferior goods?

9. Demonstrate for the case of additive independent utilities that

a. there are no complementary commodities,

b. there are no inferior commodities, and

c. there are no Giffen commodities.

10. We have defined substitute and complementary commodities in this chapter. Discuss the use of the *complementary utility of commodities i and j*

$$\frac{\partial^2 U(\mathbf{x})}{\partial x_i \partial x_j} > 0$$

and the *substitute utility of commodities i and j*

$$\frac{\partial^2 U(\mathbf{x})}{\partial x_i \partial x_j} < 0$$

as a replacement for our earlier definitions. Consider a simple quadratic utility function to illustrate your conclusions.

11. Contrast our definition of substitute and complementary commodities i and j to the alternate definitions $\partial x_i/\partial p_j > 0$ and $\partial x_i/\partial p_j < 0$. Consider a simple quadratic utility function to illustrate your result.

12. For the case where consumer utility is $U(\mathbf{x}) = 0.5\mathbf{x}^{\mathrm{T}}\mathbf{A}\mathbf{x}$ where $\mathbf{A}$ is positive semidefinite and where the purchase cost for a consumer bundle $\mathbf{x}$ is given by $b = \mathbf{x}^{\mathrm{T}}\mathbf{p}$ with $\mathbf{p}$ as the price vector for the commodity bundle, find the

a. optimum consumer bundle to maximize $U(\mathbf{x})$ where $b \leq b_{\max}$ and

b. optimum consumer bundle to minimize the purchase cost, $b = \mathbf{x}^{\mathrm{T}}\mathbf{p}$, subject to the constraint that $U(\mathbf{x}) \geq U_{\min}$. Show that these two problems are mathematical equivalents.

13. How will the revenue to the firm producing a commodity change as a function of price changes and the price elasticity? Equation 3.147 is an appropriate vehicle for your discussion.

14. What are the optimum commodity bundles for each of the utility functions of Problem 3 where there is a budget constraint that will limit consumer spending?

15. The following are the results of monotonically increasing transformations:

$$U_1 = U(x_1, x_2) = x_1 x_2$$

$$U_2 = U_1^{0.5} = x_1^{0.5} x_2^{0.5}$$

$$U_3 = \ln U_1 = \ln x_1 + \ln x_2$$

Suppose that the price vector is $\mathbf{p}^T = [0.2\ 0.6]$. What are the optimum commodity bundles for each utility function if we assume an income $I = 20$? Can these utility functions be interpreted as ordinal utility functions or must they be interpreted as cardinal utility functions? Do your conclusions change if we have the utility given by $U_4 = x_1^{0.5} x_2^{0.5}$?

16. For each of the utility functions of Problem 15, is commodity 1 superior or inferior? Are commodities 1 and 2 complements or substitutes?

17. Suppose that a consumer's utility function is

$$U(x_1, x_2) = 0.5x_1^2 + x_2 x_3$$

Determine the optimum commodity bundle as a function of the consumer income I and the price vector $\mathbf{p}$. Are any of these commodities inferior, superior, complements, or substitutes? Be sure to examine the conditions for optimality. Is there anything especially significant about $I \le 2p_1$?

18. Suppose that consumer utility is a function of the income of the consumer and the consumer's desire for leisure. Suppose further that the consumer has a constant marginal utility of income and a decreasing marginal utility of leisure and that the utility function is separable in leisure and commodities.

a. Show that a straight vertical or horizontal line will cross the utility indifference curves for income and leisure at points of constant slope.

b. Show that the effect of an income tax will result in less work and more leisure for the consumer.

c. Illustrate these results for the case where $U(I, l) = aI + bl - cl^2$ with a, b, and c constant and where I = income and l = leisure.

19. A consumer has a utility function for products and leisure $U(\mathbf{x}, l)$ and desires to maximize this utility by choice of the commodity bundle $\mathbf{x}$ and leisure l. This must be accomplished subject to the constraint $\mathbf{p}^T\mathbf{x} = \mathcal{I} + w\eta$, that the total dollar amount of products purchased must be equal to the nonwage income $\mathcal{I}$ plus the consumer's wages w times the number of hours worked η. The time available, 24 h per day, must be the sum of the hours worked per day η and the hours of leisure per day l: $24 = \eta + l$. Prices, wages, and nonwage income are fixed.

a. What are the necessary conditions for utility maximization?

b. What is the optimum number of work hours η for the consumer as a function of $\mathbf{p}$, w, and $\mathcal{I}$ if $U(\mathbf{x}, l) = x + 36l - l^2$?

20. A pure exchange economy is one in which there is no production and H households consume and exchange N commodities by bartering a number of items of a given commodity for a number of items of another commodity. Each household has an initial endowment of commodities $\mathbf{q}_0^g$ and seeks to increase its utility by exchanging a portion of this initial endowment for other commodities. Commodity prices $\mathbf{p}$ are fixed and so the consumer total initial worth is given by $I^h = \mathbf{p}^T\mathbf{q}_0^h$. $h = 1, 2, \ldots, H$. The worth of the commodities purchased and consumed must be equal to the initial worth of the consumer given by $\mathbf{I}^h = \mathbf{p}^T\mathbf{q}^h$, $h = 1, 2, \ldots, H$ where $\mathbf{q}^h$ may include some commodities in the initial endowment. Each household desires to maximize utility as given by $U^h = U^h(\mathbf{q}^h)$, $h = 1, 2, \ldots, H$, subject to the total worth constraint. Equilibrium requirements for the H households are $\sum_{h=1}^{H} \mathbf{q}_0^h = \sum_{h=1}^{H} \mathbf{q}^h$.

a. What are the necessary conditions for utility maximization for each consumer?

b. What are the equilibrium conditions for two households and two commodities where

$$\mathbf{q}_0^1 = \begin{bmatrix} 40 \\ 0 \end{bmatrix}, \qquad \mathbf{q}_0^2 = \begin{bmatrix} 0 \\ 200 \end{bmatrix}$$

$$U^1 = [2 \quad 7]\mathbf{q}^1 + (\mathbf{q}^1)^T \begin{bmatrix} 0 & 1 \\ 1 & 0 \end{bmatrix} \mathbf{q}^1, \qquad U^2 = [5 \quad 3]\mathbf{q}^2 + (\mathbf{q}^2)^T \begin{bmatrix} 0 & 1 \\ 1 & 0 \end{bmatrix} \mathbf{q}^2$$

c. What do these results become when you change the above two matrices to the identity matrix?

d. Determine (i) the optimum exchange price ratio p_1/p_2 and (ii) the optimum final commodity distributions $\mathbf{q}^1$ and $\mathbf{q}^2$ for the various conditions assumed here.

21. In times of scarcity, consumers must live not only within their individual budget or income constraints but also within constraints imposed by rationing. Develop the necessary conditions for maximization of the utility function $U(\mathbf{x})$ subject to a budget-rationing constraint $\mathbf{p}^T\mathbf{x} \leq I$ and a point-rationing constraint $\mathbf{r}^T\mathbf{x} \leq R$. How will the rationing of some commodities affect the consumption of those not rationed?

22. Demonstrate the validity of Equation 3.118. You may find the derivation leading to Equation 3.33 helpful in doing this.

23. Develop sensitivity and substitution effects for the theory of the firm that parallel our results in Section 3.5 concerning the theory of the household.

24. The effect of an *ad valorem* tax of $100v\%$, such that the total tax collected is $v\mathbf{p}^T\mathbf{x}$, can be considered by modifying the consumer budget constraint to $I > (1 + v)\mathbf{p}^T\mathbf{x}$. The effect of an income tax IT is to reduce the consumer disposable income such that the budget constraint is given by the expression $I - \mathrm{IT} = \mathbf{p}^T\mathbf{x}$. Investigate the effects of these two taxes on consumer utility. You may find it helpful to assume a quadratic utility $U(\mathbf{x}) = 0.5\mathbf{x}^T\mathbf{A}\mathbf{x}$. Be certain to indicate for the case of specified tax revenue whether consumer utility is greater with an income tax or an *ad valorem* tax. Can you explain your conclusion by substitution and income effects? Demonstrate your results graphically.

25. Suppose that two commodities x_1 and x_2 are perfect substitutes and that the price p_1 of x_1 is higher than the price p_2 of x_2. What will be the relative ratio of the two commodities in a consumer commodity bundle?

26. Will it necessarily be true that consumer utility will decrease with an increase in the price of a commodity?

27. For the quadratic utility function $U(\mathbf{x}) = \mathbf{a}^T\mathbf{x} + 0.5\mathbf{x}^T\mathbf{B}\mathbf{x}$ and the constraint $\mathbf{x}^T\mathbf{p} \le I$

a. What is the marginal rate of technical substitution?

b. What is the demand price elasticity?

c. What is the demand price cross-elasticity?

d. What is the condition(s) under which a commodity could be (i) normal, (ii) Giffen, (iii) superior, or (iv) inferior?

Assume that there are only components in the commodity bundle and obtain explicit results for this case.

28. Repeat Problem 27 for the case of a translog utility function such as that used in Example 3.10.

29. A person has the utility function given by the expression $U(x, l) = x^{0.5} + 2l^{0.5}$ where x is the quantity of goods produced and l is leisure. The person's income depends on the wage rate and time devoted to work. There is a maximum amount of time T available and so the income earned is given by $I = w(T - l)$. The price of the goods is p. Determine the labor supply and commodity demand functions. Suppose that there is an income tax rate t such that the actual income available for purchases is $I = (1 - t)w(T - l)$. How does this affect the labor supply and commodity demand functions?

30. The utility for food x_1 and clothing x_2 is given by $U(x_1, x_2) = x_1x_2$. The price of clothing is 1 and income is 250. What is the effect of a drop in the price of food from 2 to 1.5? How have consumers benefited? Illustrate your results graphically.

31. Is there an increase in the price of clothing that would leave consumer utility unchanged in Problem 30? Make these computations using both the fundamental theory of the consumer and the sensitivity results presented in this chapter.

32. The utility function of a consumer is given by $U(\mathbf{x}) = x_1x_2 + x_3x_4$. Prices are fixed and the consumer has a fixed income I. What is the optimum commodity bundle? Can there be any inferior goods? Can there be any complementary goods? Can there be any substitute goods?

33. Outline a numerical procedure to obtain the sensitivity-based results of Section 3.5 for cases where the consumer optimization problem cannot be explicitly solved. Use these results to find substitution effects for the case where $U(x_1, x_2, x_3) = x_1^{0.5} + x_2x_3$, $I = 100$, $p_1 = p_2 = p_3 = 1$.

34. A person's utility function is given by $U(\mathbf{x}) = \sum_{i=1}^{n} a_i u(x_i)$ where $U(x_i) = 1 - e^{-b_i x_i}$. The person is income constrained and prices are known constants.

a. If p_i increases, what are the possible behavior patterns for x_i?

b. If p_i increases, what are the possible behavior patterns for x_j?

c. Determine possible "demand curves" for cases **a** and **b** for the two-dimensional case.

d. Is it possible to have any inferior commodities in this example? Any Giffen commodities?

e. Repeat parts **a** to **d** for the two-dimensional case where the utility function is multiplicative and of the form $U(x_1, x_2) = k_1U_1(x_1) + k_2U_2(x_2) + k_1k_2KU_1(x_1)U_2(x_2)$.

BIBLIOGRAPHY AND REFERENCES

The texts referenced in Chapter 2 are basic references for this chapter, as well as for Chapter 2. Interesting discussions of the utility and preference concepts may be found in the following dated but seminal references.

Debreu G. Theory of value. New York: Wiley; 1959.

Edwards W. The theory of decision making. Psychol Bull 1954;5:380–417.

Edwards W. Behavioral decision theory. Annu Rev Psychol 1961;12:473–498.

Samuelson PA. Consumption theory in terms of revealed preferences. Econometrica 1948;15: 243–253.

CHAPTER 4

SUPPLY–DEMAND EQUILIBRIA AND MICROECONOMIC SYSTEMS ANALYSIS AND ASSESSMENT MODELS

4.1 INTRODUCTION

In Chapters 2 and 3, we studied the behavior of firms and consumers under the assumption that prices and wages are fixed. The producer was able to obtain a given price for a product and had to pay given prices and/or wages for the inputs required for the production process. The primary economic objective of a firm is to maximize profit. The consumer must pay a fixed (known) price for a commodity and purchase commodities to maximize a utility function. These assumptions are a bit artificial. In reality, prices are determined in a competitive marketplace where firms and consumers meet to exchange their given products. The consumer who has been paid wages for labor trades these wages for the final products or outputs of firms. Firms may also act as consumers, since the input to one firm may be the output of another. Similarly, consumers own a set of factors, primarily their labor, and obtain wages by trading these factors on a factor market. Figure 4.1 illustrates the general flow that results from the supply–demand interactions of firms and consumers. We will expand upon this diagram considerably as we move from the microeconomic analysis of individual firms and consumers to the economic analysis of the aggregate behavior of large groups of firms and consumers and government interactions with them.

We should state the conditions under which our equilibrium analysis is based in order to obtain a full understanding of the economic model we are using and its significance. We assume a *perfectly competitive economy* or a perfect market for labor and commodities with the following features:

Economic Systems Analysis and Assessment,
by Andrew P. Sage and William B. Rouse

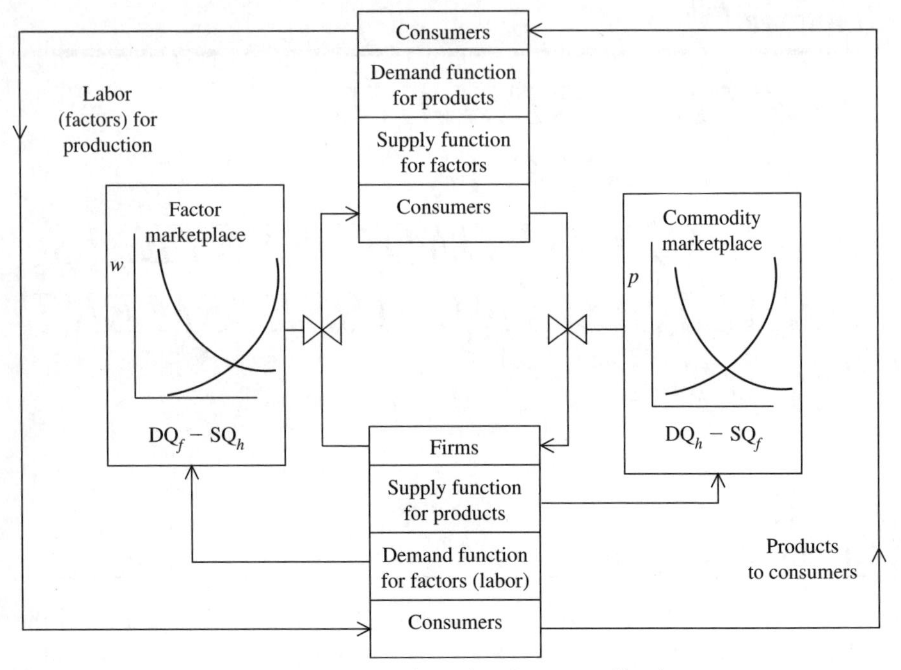

Figure 4.1. A Simple Supply–Demand Market Conceptualization.

1. The product of all firms is homogeneous in that all firms make exactly the same goods and all goods are of equal quality. Such things as trademarks, patents, and advertising do not exist, and consumers do not prefer the product of any one firm over that of any other.
2. There are a large number of firms and consumers. The individual transactions of buyers and sellers are very small compared with overall purchases and sales. This ensures that supply and demand are autonomous and individual consumers and firms have no influence on prices and wages.
3. Consumers and firms are *anonymous* in that there is no especially favored consumer or firm, with each consumer trying to buy products at the lowest price and each firm trying to find consumers that offer the highest price. Thus firms and consumers possess *perfect information* about prices. Each firm and consumer will *take* prices as given from the supply–demand results of market interactions.
4. Neither firms nor consumers make agreements among themselves or form coalitions to influence markets.
5. Firms and consumers are free to enter and leave the marketplace. Thus new firms will enter the marketplace if the profits of existing firms are positive, and unproductive firms, those that cannot earn a profit, leave the marketplace.

We may determine profit for a firm in the long run in which all factor inputs vary, or in the short run where either output production or input factors cannot vary. Each of these cases will have an optimum economic operation resulting in a production quantity that is a function of the price received for the product.

We have determined demand functions for consumers in which the consumer demand for a given commodity depends on the price for all commodities and consumer income. For supply–demand equilibrium, consumer budget income and all prices except that of the particular commodity being considered are assumed fixed. Thus the demand of a given consumer for a commodity will be a function of the price for the commodity. There are a number of consumers, and the total consumer demand for a given commodity will be just the sum of the individual consumer demands.

We will now proceed to establish conditions under which an economic supply–demand equilibrium exists; also, we will be concerned with the uniqueness of this equilibrium and whether the equilibrium is stable. Our efforts will first concentrate on the commodity marketplace. The factor marketplace is essentially the same as the commodity marketplace, with the roles of the firm and the consumer reversed. After a simplified look at the commodity marketplace only, and the establishment of conditions for existence, uniqueness, and stability of the commodity marketplace, we will turn our attention to a more general formulation that will allow us to consider both commodity and factor markets simultaneously. We will conclude this chapter with a discussion of linear input–output microeconomic behavior and other microeconomic models.

4.2 BASIC SUPPLY–DEMAND EQUILIBRIUM FOR A SINGLE GOOD

In this section, we assume that all prices, wages, and budget incomes are fixed except for the price of a single good, which we will denote by p. The total supply quantity SQ and demand quantity DQ will be a function of the price p of the good in question and is given by

$$\mathrm{DQ} = f_{\mathrm{d}}(p) = \mathrm{DQ}(p) \tag{4.1}$$

$$\mathrm{SQ} = f_{\mathrm{s}}(p) = \mathrm{SQ}(p) \tag{4.2}$$

We assume that these functions satisfy several propositions:

1. The demand for a good will increase as the price decreases, or, alternately, as a good becomes scarcer, its price will increase owing to the existing demand for the good.[1]

[1]This is not true for Giffen commodities, which are very rare.

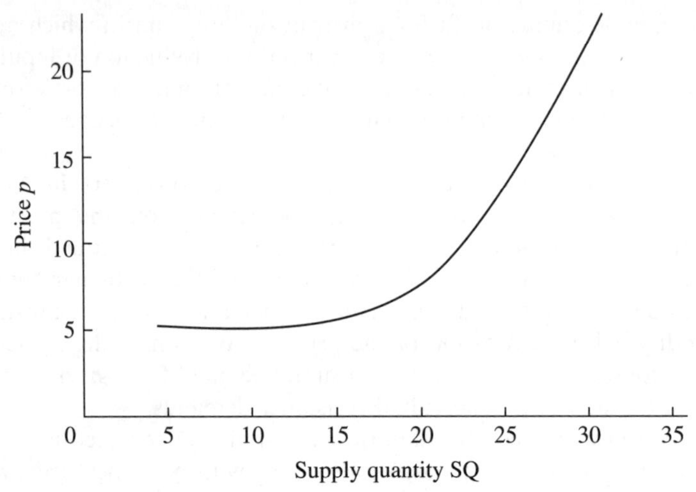

Figure 4.2. Typical Supply Curve.

2. The supply of a good will increase as its price increases, or, alternately, as the price of a good decreases, a decreased quantity of this good will be supplied by firms.
3. The supply and demand curves intersect at a point E, the equilibrium point.

With these assumptions, the first two of which are in full accord with our efforts in Chapters 2 and 3, we see that economic market equilibrium will exist at that price where supply and demand quantities are equal. If the price is raised above this equilibrium, the firm will supply more than the equilibrium value of the good, but the consumer will purchase less than the equilibrium value. The reduced demand and sales of the good will cause firms to lower prices and, hopefully, equilibrium will ultimately be reached. Similarly, if the price of the good is lowered from its equilibrium value, consumers will demand more of it, but firms will supply less. The increased demand will cause the price to rise, hopefully, again to an ultimate equilibrium. Let us concern ourselves with this equilibrium point first and examine uniqueness and stability considerations later.

Figure 4.2 shows a typical curve of supply versus price. From solving the problem of the firm, we have an equation relating the quantity produced (supplied) to its price. This is a curve such as that shown in Fig. 4.2. It might seem that the axes of the curve are rotated 90° from that which would represent common engineering practice. This is the convention in economics, however, and it is a useful convention in that it ultimately equates supply and demand.

Example 4.1: In Example 2.11, we found that the maximum profit to a firm in the long run resulted when the input factor was $\hat{\mathbf{x}} = -\mathbf{B}^{-1}(\mathbf{a} - \mathbf{w}/p)$ and that this resulted in a production level given by $\hat{q} = -0.5\mathbf{a}^{\mathrm{T}}\mathbf{B}^{-1}\mathbf{a} + (0.5\mathbf{w}^{\mathrm{T}}\mathbf{B}^{-1}\mathbf{w})/p^2$ (Eq. 2.106). We use the numerical values given by Equation

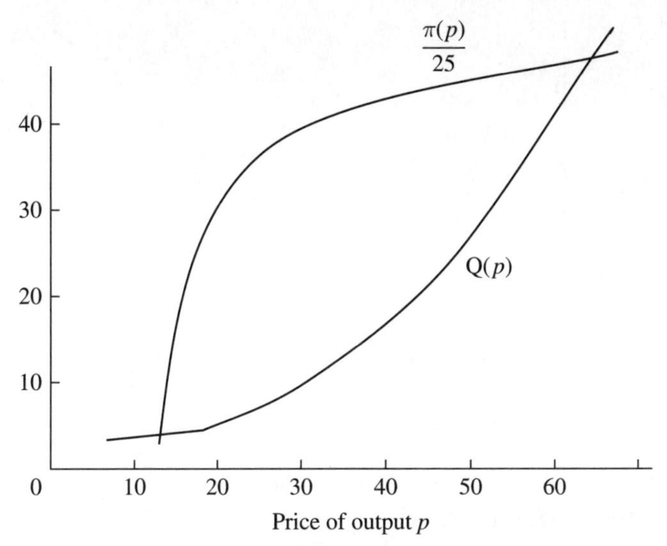

Figure 4.3. Commodities Produced and Profit as a Function of Price.

2.142 and obtain the following equation for the optimum production and profit (excluding fixed costs):

$$\hat{q} = 34.75 - \frac{2850}{p^2}$$

$$\hat{\Pi}(\hat{q}) = p\hat{q} - \mathbf{w}^T\hat{\mathbf{x}} = 34.75p + \frac{2850}{p} - 620$$

For $p < 9.38$, we purchase a negative amount of factor x_3. If we restrict ourselves to prices given by $p \geq 9.38$, then all input factors are positive, and the quantity produced and the profit are all positive. Figure 4.2 is actually a sketch of this production function, and Fig. 4.3 illustrates both the production function and the profit as functions of price. Although this is a specific example, the results are indeed generally typical of supply curve results. ■

Figure 4.4 shows a typical curve of demand versus price. As price decreases, demand for a commodity or good increases. The demand curve is obtained from consumer maximization of utility. This maximization is accomplished with the prices of all goods but one fixed, as well as with a fixed consumer's (budget) income for good purchases.

Example 4.2: We reconsider Example 3.7 for the case considered, except that we will regard p_1 as a free parameter. Equations 3.52–3.54 become, for the assumed $\mathbf{a}$, $\mathbf{B}$, and $\boldsymbol{I}$ parameters and $p_2 = 5$,

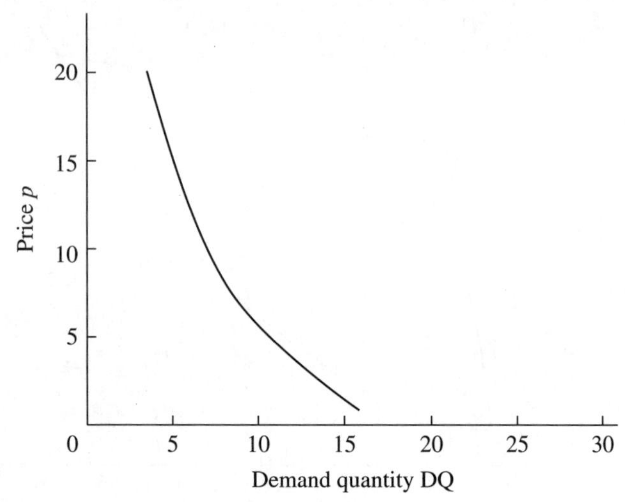

Figure 4.4. Typical Demand Curve.

$$\hat{x}_1 = \frac{75 + 5p_1}{5 + p_1 + 0.2p_1^2} \tag{4.3}$$

$$\hat{x}_2 = \frac{100 + 5p_1 + 3p_1^2}{5 + p_1 + 0.2p_1^2} \tag{4.4}$$

Our simple consumer demand equation is just the relation between p_1 and the quantity x_1 of commodity 1 purchased. The relation does concern itself directly with the fact that the consumer trades off purchase of commodity 1 for purchase of commodity 2 as the price of commodity 1 changes. The consumer does this, of course, to maintain maximum utility. We have for this utility:

$$U(p_1) = \frac{1062.5 + 387.5p_1 + 100p_1^2 + 11.5p_1^3 + 0.9p_1^4}{(5 + p_1 + 0.2p_1^2)^2} \tag{4.5}$$

The utility function does not go to zero as p_1 becomes infinite, since the consumer will spend the entire budget on commodity 2 if the cost of commodity 1 is infinite. The amount of commodity 1 purchased does not become infinite for zero price because the quadratic utility function used here actually shows consumer disutility for sufficiently large x_1. Figure 4.5 shows the demand quantity and utility for this example. This consumer demand curve is illustrated in Fig. 4.4, which shows the price on the vertical axis, as is a common economic practice. ■

Our examples up to this point have considered a single firm and a single consumer only. Our second assumption concerning a perfectly competitive economy

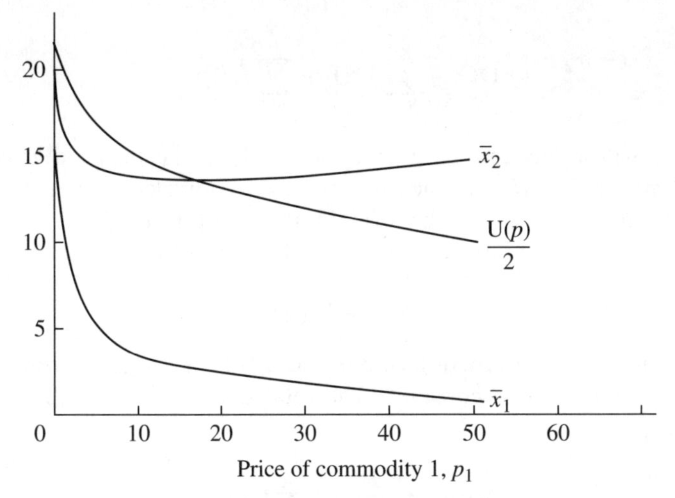

Figure 4.5. Commodities Purchased and Consumer Utility Versus Price p_1.

requires that there exist a number of firms and consumers.[2] Although individual firms and consumers have no direct influence on prices and wages, their numbers in the market at any given time certainly do determine supply–demand relationships and the prices of commodities on the marketplace.

We assume a number F of firms that, for the purposes of this section, will be assumed to be single-product firms. There will be H households. We assume a competitive economy and therefore that the firms and households each will accept wages $\mathbf{w}$ for factor inputs and prices $\mathbf{p}$ as given and do not influence them in a monopolistic, monopsonistic, oligopolistic, oligopsonistic, or some other nonperfectly competitive fashion. We consider only consumer demand for one good with price p_i. Each of the consumers maximizes their utility for all the commodities or goods available, subject to their individual budget constraints, which are assumed to be fixed here. Only the price of the ith good is assumed to be variable. Thus the demand of the jth consumer for the ith product, which is

$$\hat{x}_i^j = \mathrm{DQ}_i^j = f_d^j(p_i, \mathbf{p}_i^-, I^j)$$

can be written as

$$\mathrm{DQ}_i^j = f_d^j(p_i) \tag{4.6}$$

since the prices of all goods in the commodity bundle except for that of the ith good, $\mathbf{p}_i^-$, are assumed to be fixed and the income of the jth consumer, I^j, is also assumed to be fixed. The total demand for the ith good or service in the commodity bundle is just the sum of the individual demands; thus,

[2]If there were but a single firm, that firm would surely operate as a monopoly, since it could control the total number of products delivered to the community market.

$$\mathrm{DQ}_i = \sum_{j=1}^{H} \mathrm{DQ}_i^j = \sum_{j=1}^{H} f_{\mathrm{d}}{}^j(p_i) \tag{4.7}$$

The jth firm will supply a product or service level q that is the product or service level maximizing the profit of the jth firm (or the equivalent statement if we are considering the problem of the firm in the short run). The result of this optimization is that we have a supply quantity given by

$$\mathrm{SQ}_i^j = f_{\mathrm{s}}{}^j(p_i) \tag{4.8}$$

for the jth firm that is a function of the price p_i of the ith good produced since wages are assumed to be constant. The total quantity of the ith commodity supplied is just

$$\mathrm{SQ}_i = \sum_{j=1}^{F} \mathrm{SQ}_i^j = \sum_{j=1}^{F} f_{\mathrm{s}}{}^j(p_i) \tag{4.9}$$

Market equilibrium is established when supply equals demand. Thus we equate Equations 4.7 and 4.9 and solve for the resulting price of the ith commodity. In a given geographic region, which could perhaps be the whole world, we presumably obtain the consumer demand for the entire population in that region. Each consumer will purchase an amount of a given commodity, depending on the price. Firms making this commodity may make a profit if the price of the commodity is sufficiently high. This will encourage more firms to enter the market. Thus for a given price the supply quantity will necessarily increase. This will necessarily result in a lowering of the price for the quantity, as it is the intersection of the supply–demand curve that determines the market price for a commodity. This lowering of the price will result in a lower profit for the firms making the commodity. As long as the profit is positive, however, more firms would enter the market until the price has been reduced to a sufficiently low level such that the profit is zero. No additional firms will then subsequently enter the market, as the attractiveness for production of this commodity is insufficient owing to the zero profit. In a similar way, if the number of firms in the market with respect to production of a given commodity is so large that the price is driven sufficiently low such that profits are negative, then firms will be driven out of production and the price and profit will therefore rise. Thus our assumption of a perfectly competitive market necessarily results in a number of firms in the market for a given commodity such that the profit is zero. We can easily include a *normal* profit term with each profit function such that profit can be interpreted to mean excess profit. When the excess profit is zero, we will say that the firm is earning the minimum remuneration necessary for continued operation. This minimum remuneration will be a function of fixed costs, interest rates, payment for advertising costs and management salaries, risks, etc.

Example 4.3: Suppose that the jth individual consumer utility function is given by

$$U^j(\mathbf{x}) = \sum_{i=1}^{n} a_i^j \log x_i^j$$

and that the budget constraint is

$$\sum_{i=1}^{n} p_i^j x_i^j \le I^j$$

This is precisely Example 3.9, and so we have for the optimum commodity bundle for the jth consumer the following relationship given by Equation 3.73:

$$\hat{x}_i^j = \frac{I^j a_i^j}{\sum_{i=1}^{n} a_i^j} \frac{1}{p_i} = \frac{c_i^j}{p_i}, \qquad i = 1, 2, \ldots, n; \quad j = 1, 2, \ldots, H$$

where

$$c_i^j = \frac{I^j a_i^j}{\sum_{i=1}^{n} a_i^j}, \qquad i = 1, 2, \ldots, n; \quad j = 1, 2, \ldots, H$$

represents the total monetary expenditure of the jth consumer for the ith commodity. This c_i^j will generally vary from consumer to consumer. For example, if $a_i^j = 0$, then consumer j has zero utility for commodity i. Thus consumer j will not purchase any of commodity i, regardless of the price.

The total consumer demand for product or commodity i is given by

$$\mathrm{DQ}_i = \sum_{j=1}^{H} \hat{x}_i^j = \frac{1}{p_i} \sum_{j=1}^{H} c_i^j = \frac{C_i}{p_i}, \qquad C_i = \sum_{j=1}^{H} c_i^j$$

where C_i is the total price paid by all consumers for commodity i. As the number of consumers increases, C_i will increase, and the demand at a given price will increase. Alternately stated, the price for a given demand will increase as the demand increases. From the point of view of the firm, this will result in increased profits if there are a given fixed number of firms in the market. ■

Example 4.4: We presume a logarithmic production function for each firm as follows:

$$q^j = \sum_{i=1}^{n} a_i^j \log x_i^j, \quad j = 1, 2, \ldots, F \tag{4.10}$$

We assume that each firm desires to maximize profit:

$$\Pi^j(\mathbf{x}^j) = pq^j - \mathbf{w}^{\mathrm{T}}\mathbf{x}^j, \quad j = 1, 2, \ldots F \tag{4.11}$$

The optimum factor input is given for factor i by

$$\frac{\partial \Pi^j(x^j)}{\partial x_i^j} = 0 = \frac{a_i^j p}{x_i^j} - w_i \tag{4.12}$$

as

$$x_i^j = \frac{a_i^j}{w_i} p, \quad j = 1, 2, \ldots F; i = 1, 2, \ldots, n \tag{4.13}$$

The optimum production level for each firm is then of the form

$$\hat{q}^j = \sum_{i=1}^{n} a_i^j \log\left(\frac{a_i^j}{w_i} p\right) = \alpha^j + \beta^j \log p \tag{4.14}$$

where

$$\alpha^j = \sum_{i=1}^{n} a_i^j \log\left(\frac{a_i^j}{w_i}\right) \tag{4.15}$$

$$\beta^j = \sum_{i=1}^{n} a_i^j \tag{4.16}$$

The profit of the jth firm is given by

$$\Pi^j(\hat{\mathbf{x}}) = p(\alpha^j - \beta^j + \beta^j \log p) \tag{4.17}$$

In competitive equilibrium, this profit will be at some threshold value T^j. Above this value a firm will either enter or remain in the marketplace, and below it a firm will leave the marketplace or refuse to enter it. In equilibrium the profit will be T, and we will have

$$\Pi^j(\hat{\mathbf{x}}) = T^j = p(\alpha^j - \beta^j + \beta^j \log p)$$

Suppose that for a 20-input factor production process we have $w_i = a_i^j = 1$ and $T^j = 0$. We obtain $\beta^j = 20$ and $\alpha^j = 0$ from Equations 4.15 and 4.16. From Equation 4.17 we obtain the equilibrium price as $p = e = 2.718$. The quantity produced by each firm is given by Equation 4.14 as $\hat{q}^j = 20$.

Suppose further that the consumer demand curve is given by Example 4.3 for $C_i = 5436.56$. We then have the demand quantity

$$\mathrm{DQ} = \frac{5436.56}{p} \tag{4.18}$$

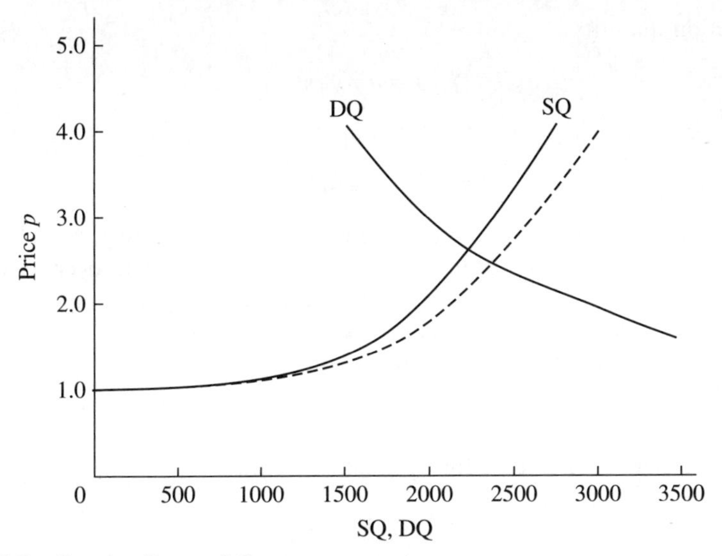

Figure 4.6. Supply–Demand Curves.

and the supply quantity of Equation 4.14 for 100 firms

$$SQ = 2000 \log p \tag{4.19}$$

We equate supply and demand and obtain $p = 2.718$ as the equilibrium price and a supply or demand quantity SQ = DQ = 2000. A total of 20 items, $q^j = 20$, are supplied by each of the 100 firms.

Figure 4.6 illustrates the supply–demand curve for this example. The equilibrium condition is SQ = DQ = 2000 and $p = 2.718$. The profit to each firm is zero at equilibrium. If any additional firms enter the market, the supply curve moves to the right, as shown in Fig. 4.6, since a greater quantity of the commodity will be supplied at any given price. This will result in a new equilibrium point at a lower price. The firms will now be losing money, or making a negative profit, even though each firm is operating optimally. Thus firms will leave the marketplace, and the supply curve will shift to the left such as to establish equilibrium at $p = 2.718$ and DQ = SQ = 2000. Of course, new technologies will be introduced in realistic situations. These can reduce costs, some of which can be shared with consumers, yielding benefits to both producers and consumers. ■

Our approach in the previous examples of this chapter, in which input factors to the production process were used to discern profit, is a bit cumbersome for the analysis and assessments generally desired here. A more convenient approach consists of relating profit to produced quantities and fortunately this is possible, as we have seen in the analysis leading to Equations 2.51 and 2.52. We take this approach here. We desire to maximize the output production quantity

$$J = q = f(\mathbf{x}) \tag{4.20}$$

subject to a constraint on the production costs

$$\mathrm{PC} = C(\mathbf{x}) = \mathbf{w}^{\mathrm{T}}\mathbf{x} = C_f \tag{4.21}$$

We assume that we operate within the relevant economic region such that at optimum production the output will necessarily decrease with decreased input factor expenditures. Thus we do not introduce an inequality constraint into Equation 4.21. We adjoin the equality constraint of Equation 4.21 to Equation 4.20, using a Lagrange multiplier, to obtain

$$J = f(\mathbf{x}) + \lambda(C_f - \mathbf{w}^{\mathrm{T}}\mathbf{x}) \tag{4.22}$$

The first-order requirements for optimality are that

$$\frac{\partial J}{\partial \mathbf{x}} = \frac{\partial f(\mathbf{x})}{\partial \mathbf{x}} - \lambda \mathbf{w} = 0 \tag{4.23}$$

$$C_f = \mathbf{w}^{\mathrm{T}}\mathbf{x} \tag{4.24}$$

Let us assume that it is possible to express the production function as

$$l(q) = (\mathbf{x}^{\mathrm{T}}\mathbf{A}\mathbf{x})^{1/2} = l[f(\mathbf{x})] \tag{4.25}$$

where $\mathbf{A}$ is a symmetric positive semidefinite matrix and $l(q)$ is some function of q. Then we have as the requirements for optimality:

$$\frac{\partial l(q)}{\partial \mathbf{x}} = \frac{\partial l(q)}{\partial q}\frac{\partial q}{\partial \mathbf{x}} = (\mathbf{x}^{\mathrm{T}}\mathbf{A}\mathbf{x})^{-1/2}\mathbf{A}\mathbf{x} \tag{4.26}$$

$$\frac{\partial q}{\partial \mathbf{x}} = \frac{\partial f(\mathbf{x})}{\partial \mathbf{x}} = \frac{(\mathbf{x}^{\mathrm{T}}\mathbf{A}\mathbf{x})^{-1/2}\mathbf{A}\mathbf{x}}{\partial l(q)/\partial q} \tag{4.27}$$

Equation 4.23 yields, with use of the foregoing equations,

$$\mathbf{x} = (\mathbf{x}^{\mathrm{T}}\mathbf{A}\mathbf{x})^{-1/2}\lambda\frac{\partial l(q)}{\partial q}\mathbf{A}^{-1}\mathbf{w} \tag{4.28}$$

By premultiplying this expression by $\mathbf{w}^{\mathrm{T}}$ and using Equation 4.24, we have

$$C_f = (\mathbf{x}^{\mathrm{T}}\mathbf{A}\mathbf{x})^{1/2}\lambda\frac{\partial l(q)}{\partial q}\mathbf{w}^{\mathrm{T}}\mathbf{A}^{-1}\mathbf{w} \tag{4.29}$$

Substitution of the value of λ from Equation 4.29 into Equation 4.28 leads to the optimum factor input as

$$\hat{\mathbf{x}} = \frac{\mathbf{A}^{-1}\mathbf{w}C_f}{\mathbf{w}^{\mathrm{T}}\mathbf{A}^{-1}\mathbf{w}} \tag{4.30}$$

Use of Equations 4.25 and 4.30 leads to the following desired expression between production level and input factor production costs to maximize output production:

$$\mathrm{PC} = C(q) = C_f = (\mathbf{w}^{\mathrm{T}}\mathbf{A}^{-1}\mathbf{w})^{1/2} l(q) \tag{4.31}$$

Thus we see that we can obtain an explicit solution for the optimum production costs in the form of Equation 4.31 if we can model the production function in the form of Equation 4.25. This is a fairly general model in that the functional form of l is unspecified.

The long-term optimization problem of the firm can now be stated, in terms of production level q, as that of maximizing the difference between revenue, or pq, and total costs. For the case where the total costs are the production costs only, we have

$$\Pi(q) = pq - C(q) \tag{4.32}$$

In supply–demand equilibrium, the profit of each firm is zero, and so we see that the *supply–demand equilibrium price is just the average cost of the output product*. This statement is obtained in mathematical form by setting Equation 4.32 equal to zero, to reflect zero profit:

$$p = \mathrm{AC} = \frac{C(q)}{q} = \frac{\mathrm{PC}}{q} \tag{4.33}$$

The firm will optimize its profits by selecting an optimum production level. From the *marginal profit equals zero* equation, which is given by

$$\frac{\partial \Pi(q)}{\partial q} = 0 = p - \frac{\partial C(q)}{\partial q}$$

we see that *the firm's equilibrium price is the marginal cost of the product of the firm*:

$$p = \mathrm{MC} = \frac{\partial C_p(q)}{\partial q} \tag{4.34}$$

The rate of change of profit with price may be obtained from Equation 4.32, which becomes

$$\frac{\partial \Pi(q)}{\partial p} = q + p\frac{\partial q}{\partial p} - \frac{\partial C_p(q)}{\partial q}\frac{\partial q}{\partial p} = q$$

Here we have used Equation 4.34 to simplify this relationship. Thus we see that *optimality ensures that price equals marginal cost* and *equilibrium is ensured by the relationship "price equals average cost."*

Using Equation 4.34, we can rewrite the profit relationship of Equation 4.32 as

$$\Pi(q) = q[\mathrm{MC} - \mathrm{AC}] \tag{4.35}$$

We see that the profit is positive whenever the firms' supplied quantity is such that the marginal cost is greater than the average cost, and is negative whenever the marginal cost is less than the average cost. A firm will generally not operate with negative profit, and so $\mathrm{MC} < \mathrm{AC}$ is an infeasible region. $\mathrm{MC} = \mathrm{AC}$ is the equilibrium condition as we have previously indicated.

We can easily extend these results to include other elements that influence production costs, such as rent, transportation, and taxes. As we have mentioned previously, elements of costs such as these are simply included in the profit equation prior to its optimization. Generally they do not modify the consumer demand curve. Since the price to the consumer for a given quantity of a commodity rises because of rent, transportation, and other costs and taxes to the firm, the consumer demand is reduced from what it would be if these elements acting to reduce profits did not exist. We should also mention the costs of capital. Those costs are generally independent of number of units sold. Also, we could consider differing costs among producers due to, for example, different manufacturing technologies.

Transportation costs, due to the physical separation of the firm from the marketplace and the need to transport products to market, are one element that reduces profits. Energy costs would appear in much the same way as transportation costs, as would land rental costs. These could be accounted for by considering them as input factors to the production process, elements of distribution, or external or nonproduction costs. For the jth firm, these might be of the form

$$C_e^j(q^j) = a^j + b^j q^j \tag{4.36}$$

Thus we may represent a constant or fixed cost plus a cost that varies linearly with output production level. If we wish to determine the effect of costs added after a basic production process has been established, then use of Equation 4.36 to reduce profit will often be a preferred approach to restructuring the production model.

Specific or unit sales taxes in which a firm must pay a tax equal to an amount of money per unit commodity sold take much the same form as transportation costs, except that the tax rate would be the same for all firms. The external costs due to a specific or unit sales tax will be of the form

$$C_e^j(q^j) = tq^j \tag{4.37}$$

Ad valorem sales taxes are the taxes a firm must pay; these are a percentage of the money received from the sales of a commodity. The *ad valorem* sales tax for the jth firm is given by

$$C_e^j(q^j) = vpq^j \tag{4.38}$$

where the tax is $100v\%$ of the firm's price p of a commodity. These taxes may easily be computed as *add-on* taxes to the consumer (see Problem 13). The two forms of taxation are essentially equivalent, as our examples will show. The procedure for determining economic equilibrium is again that of determining the most economic operation of the firm by setting price equal to marginal total costs and determining equilibrium by setting price equal to average total costs. *Total* is used here to emphasize that external factors to production, such as taxes, are included. In general, these added costs, such as costs due to transportation energy and taxes, will shift the supply curve upward, in that the total price a firm will charge for a given commodity quantity will increase. The new equilibrium point will reflect a higher price paid by consumers and a lower production quantity. The price received by the firm will be reduced because of the payment due to the government for the taxes, or the payment due to the transportation firm or energy firm for that factor. It is not possible to say whether it is the consumer or the firm that pays the majority of the increased product cost without examination of those specific supply–demand curves that are applicable to the commodity being considered. Elasticity concepts, such as those discussed in Section 3.5, are of value for this purpose.

Example 4.5: We consider an average cost of production equation given by

$$\mathrm{AC}_1(q) = a(q-q_0)^2 + \mathrm{AC}_0 \tag{4.39}$$

This relation indicates, for positive a, that production costs increase quadratically as the output differs from the value q_0 at which average production costs equal the minimum value AC_0. The production cost is then given by

$$\mathrm{PC} = C(q) = q\mathrm{AC}_1(q) = aq^3 - 2aq_0q^2 + (\mathrm{AC}_0 + aq_0^2)q \tag{4.40}$$

This appears to be a very simple production cost function, and it has neither zero nor infinite equilibrium production levels. The profit of the firm is given by

$$\Pi(q) = pq - C(q) \tag{4.41}$$

This expression is maximum whenever price equals the marginal production cost, or

$$\frac{\partial \Pi}{\partial q} = 0 = p - \frac{\partial C(q)}{\partial q} \tag{4.42}$$

or where, using Equation 4.40,

$$p = \mathrm{MC}(q) = 3aq^2 - 4aq_0q + \mathrm{AC}_0 + aq_0^2 \tag{4.43}$$

The supply curve for the firm is obtained by solving the foregoing equation for q to obtain

$$\mathrm{SQ} = q = 0, \qquad p < \mathrm{AC}_0$$

$$\mathrm{SQ} = q = \frac{2}{3}q_0 + \frac{1}{3}\left[q_0^2 + \frac{3(p-\mathrm{AC}_0)}{a}\right]^{1/2}, \quad p \geq \mathrm{AC}_0 \tag{4.44}$$

If the price equals the minimum average cost, then $q = q_0$. This serves as a check on this calculation. The price must be greater than or equal to the average cost m for the firm to operate, as we know, since $p = \text{MC} > \text{AC}$. Figure 4.7 illustrates the optimum profit, marginal costs, and average costs for the case where $\text{AC}_0 = 50$, $a = 0.10$, and $q_0 = 20$ (case 1). The profit relationship for maximum profit is obtained from Equations 4.41 and 4.43 as

$$\hat{\Pi}(q) = 2a(q^3 - q_0q^2) \tag{4.45}$$

The marginal costs and average costs may be obtained from Equations 4.39 and 4.43. Figure 4.8 illustrates the curves of these relations obtained for the case where $\text{AC}_0 = 100$, $a = 0.10$, and $q = 10$ (case 2). Let us now suppose that there are 100 case 1 firms in operation and no case 2 firms. The consumer demand curve is first presumed to be given by $\text{DQ}_1 = 3000 - 20p$. Figure 4.9 illustrates the supply–demand curves for this case and, together with Fig. 4.7, indicates that we have a zero profit equilibrium for the case 1 firms at $p = 50$, $\text{DQ} = \text{SQ} = 2000$.

Now suppose that there are 50 case 1 firms in operation and 50 case 2 firms, and that we have the same consumer demand relationship. The optimum equilibrium production conditions for the case 2 firms indicate that they should charge a price of \$100, and that each of the firms should deliver 10 units of the commodity to the marketplace for a total of 500 units. The consumer demand at a price $p = 100$ is 1000 units. This amount of production will only be supplied by case 2 firms at a price greater than $p = 100$. Supply–demand equilibrium is $q = 716$ and $p = 114.2$. But case 1 firms can surely supply the needed 500 or 716 units. Surely they would do this at the going market price of $p = 100$ or 114.2. In fact, at $p = 100$ the case 1 firms are quite willing to supply 28 units each for

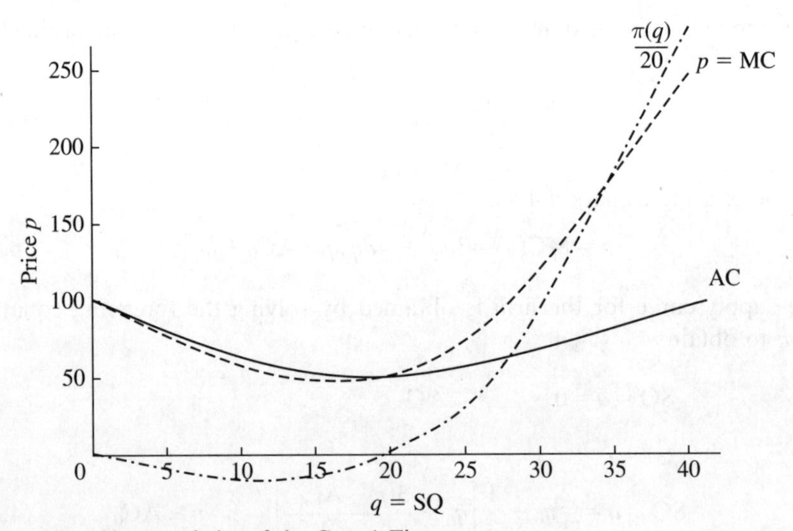

Figure 4.7. Characteristics of the Case 1 Firm.

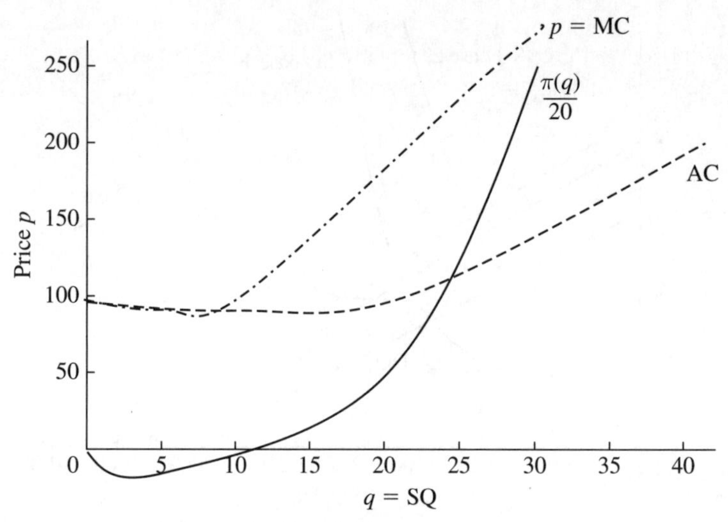

Figure 4.8. Characteristics of the Case 2 Firm.

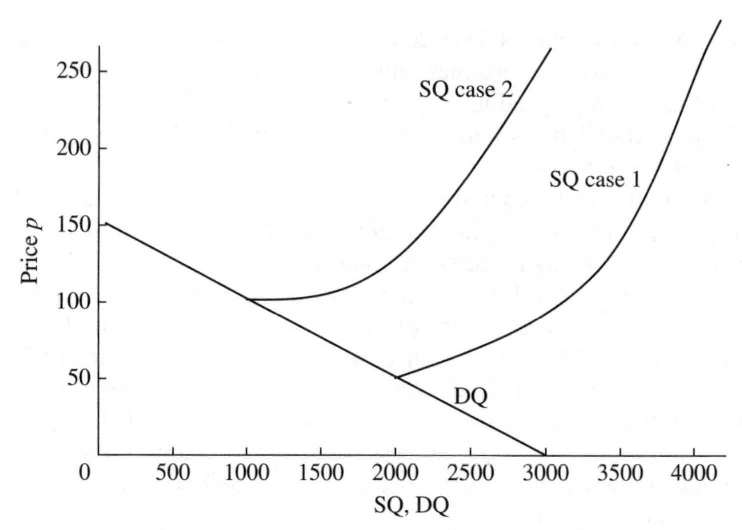

Figure 4.9. Supply–Demand Curve: 100 Case 1 Firms or 100 Case 2 Firms.

a total of 1400 commodities. At the price of 100, the consumers do not desire a total of 1900 commodities. Thus the market pressures will cause the case 1 firms to reduce prices, since they have a surplus commodity production at the price of \$100. Any reduction of prices will cause the high-production-cost firms of case 2 to leave the marketplace.

A new equilibrium will now be established. Figure 4.10 shows the supply–demand curves for the two firms and the consumers in this case where there are 50 case 1 firms and 50 case 2 firms. In this particular case, the market equilibrium price of $p = 84.8$ and the supply of 26 units per firm result in a

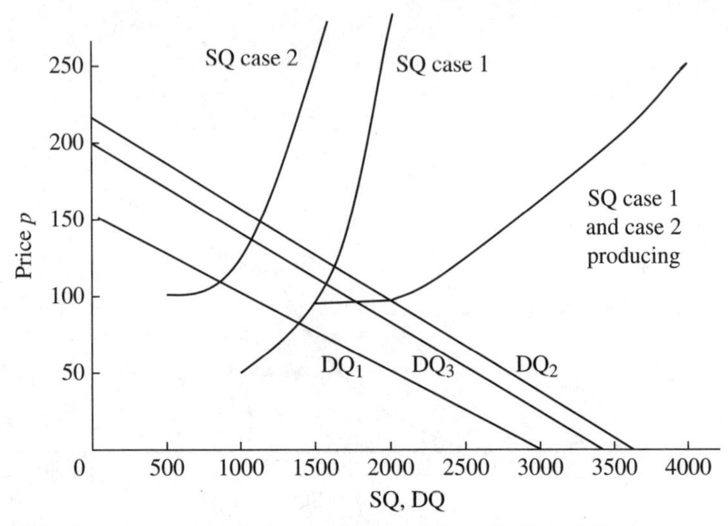

Figure 4.10. Supply–Demand Curves: 50 Case 1 and 50 Case 2 Firms.

profit for the case 1 firms of \$811.20 each, as we see from Fig. 4.7 or Equation 4.45. Case 2 firms will not produce under these conditions. Their selling price is less than their average product cost to produce.

Why do case 1 firms make a profit here? It must be because they have more of some resource, such as superior technology or more fertile land, than case 2 firms that attempt to make the same product. This fact should act as an incentive for others to adopt the superior technology or offer to purchase some of the fertile land so as to enter the marketplace with some of the superior characteristics of the case 1 firm. In the long run, these market pressures will act to bring about equilibrium at zero profit and to somehow capture the superior resource of the case 1 firms, which is worth \$811.20 in the present case. Of course, if each case 1 firm owned the scarce resource, then the profit is really just part of the operating costs for the case 1 firms. Each has a valuable resource and should not produce unless they recover an amount of profit equal to the value of the resource; otherwise, they would be better off renting out the valuable resource to others.

To illustrate this further, suppose that the consumer demand curve shifts to $DQ_2 = 3600 - 17p$. Now the demand curve intersects the supply curve at a point where both case 1 and case 2 firms will supply products. At $p = 100$, we have $DQ_2 = 1900$. From Fig. 4.10, we see that the case 1 firms will produce a total of 1400 units, or 28 units each, and the case 2 firms will produce 500 units, or 10 units each. The case 2 firms will earn no profit, whereas each case 1 firm will earn a profit, from Fig. 4.7 or Equation 4.45, of \$1254. This is interpreted as the payment to firms in case 1 for the rent of a valuable resource.

If the demand curve shifts to DQ_3, we may show, by considering a price reduced slightly below \$100, that some case 2 firms will leave the market and that each case 1 firm will produce 28 units. ■

Example 4.6: We will now illustrate the difference between the effects of a unit or specific sales tax and an *ad valorem* sales tax. We assume a consumer demand function of the form

$$\mathrm{DQ} = \mathrm{DQ}_0\left(l - \frac{p}{p_\mathrm{m}}\right) \tag{4.46}$$

Here DQ_0 is the demand at zero price and p_m is the maximum price at which the demand is reduced to zero. The cost of production for the firm is assumed to be given by

$$C_\mathrm{p}(q) = 0.5aq^2 \tag{4.47}$$

The unit tax is given by tq, and the total operating costs by

$$C(q) = 0.5aq^2 + tq + r \tag{4.48}$$

where r is the rent or royalty income due to the profit, if any, of the firm. The profit of the firm, which includes this rent, is given by

$$\Pi(q) = pq - 0.5aq^2 - tq - r \tag{4.49}$$

The firm's profit is maximized by

$$\frac{\partial \Pi(q)}{\partial q} = 0 = p - aq - t \tag{4.50}$$

which results, with a unit sales tax, in the price $p = \mathrm{MC} = aq + t$, or

$$\mathrm{SQ} = \frac{p - t}{a} \tag{4.51}$$

Equilibrium is determined when $\mathrm{DQ} = \mathrm{SQ}$ or at a price and demand given by

$$p = \frac{(\mathrm{DQ}_0 a + t)p_\mathrm{m}}{\mathrm{DQ}_0 a + p_\mathrm{m}} \tag{4.52}$$

$$\mathrm{DQ} = \mathrm{SQ} = \frac{\mathrm{DQ}_0(p_\mathrm{m} - t)}{\mathrm{DQ}_0 a + p_\mathrm{m}} \tag{4.53}$$

The quantity supplied is reduced and the price increased by imposition of a unit tax. The consumer expenditure with tax is given by

$$(p\mathrm{DQ})_\mathrm{t} = \frac{\mathrm{DQ}_0 p_\mathrm{m}(\mathrm{DQ}_0 a + t)(p_\mathrm{m} - t)}{(\mathrm{DQ}_0 a + p_\mathrm{m})^2} \tag{4.54}$$

and this may be more or less than the consumer expenditure with no tax, which is

$$(p\mathrm{DQ})_{\mathrm{nt}} = \frac{\mathrm{DQ}_0^2 p_{\mathrm{m}}^2 a}{(\mathrm{DQ}_0 a + p_{\mathrm{m}})^2} \tag{4.55}$$

The ratio of these expenditures is given by

$$\frac{(p\mathrm{DQ})_{\mathrm{t}}}{(p\mathrm{DQ})_{\mathrm{nt}}} = \left(1 + \frac{t}{\mathrm{DQ}_0 a}\right)\left(1 - \frac{t}{p_{\mathrm{m}}}\right) \tag{4.56}$$

and this shows us that

$$(p\mathrm{DQ})_{\mathrm{t}} > (p\mathrm{DQ})_{\mathrm{nt}}, \quad \text{if} p_{\mathrm{m}} - \mathrm{DQ}_0 a > t\left(\text{i.e.}, 1 - \frac{a\mathrm{DQ}_0}{p_{\mathrm{m}}} > \frac{t}{p_{\mathrm{m}}}\right)$$

$$(p\mathrm{DQ})_{\mathrm{t}} < (p\mathrm{DQ})_{\mathrm{nt}}, \quad \text{if} p_{\mathrm{m}} - \mathrm{DQ}_0 a < t\left(\text{i.e.}, 1 - \frac{a\mathrm{DQ}_0}{p_{\mathrm{m}}} < \frac{t}{p_{\mathrm{m}}}\right) \tag{4.57}$$

From Equation 3.138, we can compute the price demand elasticity as

$$\varepsilon_{\mathrm{d}}^{\mathrm{p}} = \frac{p}{\mathrm{DQ}} \frac{\partial \mathrm{DQ}}{\partial p} \tag{4.58}$$

Evaluating this at $t = 0$ by use of Equations 4.46, 4.52, and 4.53, we obtain

$$\varepsilon_{\mathrm{d}}^{\mathrm{p}} = -\frac{a\mathrm{DQ}_0}{p_{\mathrm{m}}} \tag{4.59}$$

We see that consumer expenses, due to the unit tax, increase whenever the demand price elasticity magnitude is less than $1 - t/p_{\mathrm{m}}$ and decrease whenever the demand price elasticity magnitude is greater than $1 - t/p_{\mathrm{m}}$. Thus if the demand is inelastic, such that $|\varepsilon_{\mathrm{d}}^{\mathrm{p}}|$ is small, then the consumer will expend more with imposition of a unit tax. Necessary commodities for human survival are inelastic, and taxing them in this fashion will result in greater consumer expenditure. If the demand is price elastic, such that $|\varepsilon|$ is large ($|\varepsilon| > 1$), then the consumer will actually reduce total expenditures due to imposition of a unit sales tax, as Equation 4.57 clearly shows.

The profit of the firm with the unit sales tax is, at supply–demand equilibrium, from Equations 4.49, 4.52, and 4.53, given by

$$\Pi_{\mathrm{t}}(q) = 0.5aq^2 - r_{\mathrm{t}} = \frac{\mathrm{DQ}_0^2(p_{\mathrm{m}} - t)^2 a}{2(\mathrm{DQ}_0 a + p_{\mathrm{m}})^2} - r_{\mathrm{t}} \tag{4.60}$$

To reduce this profit to zero, the rent or royalty is set at the value

$$r_{\mathrm{t}} = \frac{\mathrm{DQ}_0^2(p_{\mathrm{m}} - t)^2 a}{2(\mathrm{DQ}_0 a + p_{\mathrm{m}})^2} \tag{4.61}$$

With no unit sales tax, the firm's rent or royalty payment, which results in zero profit, is necessarily larger. It is given by

$$r_{nt} = \frac{DQ_0^2 p_m^2 a}{2(DQ_0 a + p_m)^2} \tag{4.62}$$

The ratio of the foregoing two quantities is

$$\frac{r_t}{r_{nt}} = \frac{(p_m - t)^2}{p_m^2} = \left(1 - \frac{t}{p_m}\right)^2 \tag{4.63}$$

With an *ad valorem* sales tax vpq, where the sales tax is $100v\%$, the profit of the firm is, if we assume the same production cost, given by

$$\Pi(q) = pq + 0.5aq^2 - vpq - r \tag{4.64}$$

The optimum production quantity is given from

$$\frac{\partial \Pi(q)}{\partial q} = 0 = p - aq - vp \tag{4.65}$$

as

$$SQ = q = \frac{p(1 - v)}{a} \tag{4.66}$$

Equilibrium is determined when DQ = SQ or when

$$p = \frac{DQ_0 a p_m}{DQ_0 a + p_m(1 - v)} \tag{4.67}$$

and

$$SQ = DQ = \frac{DQ_0 p_m (1 - v)}{DQ_0 a + p_m(1 - v)} \tag{4.68}$$

Again, we see that the price is increased and the quantity purchased reduced by imposition of a sales tax. The consumer expenditures with and without tax, respectively, are

$$(pDQ)_{vt} = \frac{DQ_0^2 a p_m^2 (1 - v)}{[DQ_0 a + p_m(1 - v)]^2} \tag{4.69}$$

$$(pDQ)_{nvt} = \frac{DQ_0^2 a p_m^2}{(DQ_0 a + p_m)^2} = (pDQ)_{nt} \tag{4.70}$$

and the ratio of these expenditures is given by

$$\frac{(p\text{DQ})_{\text{vt}}}{(p\text{DQ})_{\text{nvt}}} = \frac{(1-v)(\text{DQ}_0 a + p_{\text{m}})^2}{[\text{DQ}_0 a + p_{\text{m}}(1-v)]^2} = \frac{(1-v)(1-\varepsilon^{\text{P}})^2}{(1-\varepsilon_{\text{d}}^{\text{P}}-v)^2} \tag{4.71}$$

We see that

$$\begin{aligned} (p\text{DQ})_{\text{vt}} &> (p\text{DQ})_{\text{nvt}}, \quad \text{if} |\varepsilon_{\text{d}}^{\text{P}}| < (1-v)^{1/2} \\ (p\text{DQ})_{\text{vt}} &< (p\text{DQ})_{\text{nvt}}, \quad \text{if} |\varepsilon_{\text{d}}^{\text{P}}| > (1-v)^{1/2} \end{aligned} \tag{4.72}$$

Again, the total consumer expenditures will increase or decrease, depending on whether the price demand elasticity is large or small. The profit of the firm with the value-added tax is, from Equations 4.64 and 4.66, 4.67, and 4.68, given by

$$\Pi_{\text{vt}}(q) = \frac{p^2(1-v)^2}{2a} - r_{\text{vt}} = \frac{\text{DQ}_0^2 a(1-v)^2}{2(1-\varepsilon_{\text{d}}^{\text{P}}-v)^2} - r_{\text{vt}} \tag{4.73}$$

The rent or royalties to the firm for zero profit is

$$r_{\text{vt}} = \frac{\text{DQ}_0^2 a(1-v)^2}{2(1-\varepsilon_{\text{d}}^{\text{P}}-v)^2} \tag{4.74}$$

and the ratio of the royalties with taxes to those without taxes is

$$\frac{r_{\text{vt}}}{r_{\text{nvt}}} = \frac{(1-v)^2(1-\varepsilon_{\text{d}}^{\text{P}})^2}{(1-\varepsilon_{\text{d}}^{\text{P}}-v)^2} \tag{4.75}$$

We see that if $\varepsilon_{\text{d}}^{\text{P}} < 0$, the presence of the *ad valorem* sales tax reduces the excess profit to the firm. The particular form of our supply–demand function is such that $\varepsilon_{\text{d}}^{\text{P}}$ is necessarily negative, although this is not true in general.

A quantity of interest here is the price paid by the consumer for products as a function of the equilibrium price with no tax. This is given from Equations 4.59 and 4.68, with $v = 0$, as

$$p_{\text{nt}} = \frac{\text{DQ}_0 a p_{\text{m}}}{\text{DQ}_0 a + p_{\text{m}}} = \frac{-\varepsilon_{\text{d}}^{\text{P}}}{1-\varepsilon_{\text{d}}^{\text{P}}} p_{\text{m}} \tag{4.76}$$

From Equation 4.52, which gives the price with a unit sales tax, we have

$$p_{\text{t}} = \left(1 - \frac{t}{p_{\text{m}}\varepsilon_{\text{d}}^{\text{P}}}\right) p_{\text{nt}} \tag{4.77}$$

From Equation 4.67, which gives the price with an *ad valorem* tax, we have

$$p_{\text{vt}} = \frac{p_{\text{nt}}}{1 - [v/(1-\varepsilon_{\text{d}}^{\text{P}})]} \tag{4.78}$$

In each case we see that if the price demand elasticity is large, there will be little change in price to the consumer due to either of the sales taxes. To yield the same apparent equilibrium tax revenue for either tax, the government would set $tq = vp_{nt}q$. Combining this with Equation 4.76 results in

$$t = \frac{-\varepsilon_d^p p_m v}{1 - \varepsilon_d^p} \tag{4.79}$$

This is precisely the relationship we obtain from comparing Equation 4.77 with a Taylor's series expansion of Equation 4.78 truncated after first-order terms in v. The actual prices to the consumer are increased just a bit more using *ad valorem* taxes than using unit taxes. Both are taxes that extract more of a penalty from the consumer than the firm for inelastic demand commodities that include the necessities of life.

The actual government revenue for the unit sales tax is SQt, and for the *ad valorem* sales tax it is $vpSQ$. Thus we have for the unit sales tax

$$G_t = \frac{DQ_0(p_m - t)t}{DQ_0 a + p_m} = \frac{DQ_0(p_m - t)t}{(1 - \varepsilon_d^p)p_m} \tag{4.80}$$

and for the *ad valorem* sales tax

$$G_{vt} = \frac{DQ_0^2 a p_m^2 v(1 - v)}{[DQ_0 a + p_m(1-v)]^2} = \frac{DQ_0^2 a v(1 - v)}{(1-\varepsilon_d^p - v)^2} \tag{4.81}$$

where we have used actual prices and supply quantities for each tax.

To illustrate these results numerically, we consider that $a = 0.01$ and examine the following three cases:

1. Case 1: inelastic demand ($\varepsilon_d^p = -0.1$); $DQ_0 = 1100$; $p_m = 110$
2. Case 2: unitary demand ($\varepsilon_d^p = -1$); $DQ_0 = 2000$; $p_m = 20$
3. Case 3: elastic demand ($\varepsilon_d^p = -10$); $DQ_0 = 11{,}000$; $p_m = 11$

It is a simple matter to show that the no-tax supply–demand equilibrium is \$10, $SQ = DQ = 1000$ for all cases. Suppose that the government desires a total apparent return of \$1000 or 10% of the equilibrium revenue for commodities sold. Thus we may use $t = 1$ and $v = 0.1$ to yield this return.

With the tax imposed, the price the consumer must pay for commodities at the new equilibrium condition is the following:

1. Case 1: inelastic demand $p_t = 1.09p_{nt} = \$10.90$; $p_{vt} = 1.10p_{nt} = \$11.00$
2. Case 2: unitary demand $p_t = 1.05p_{nt} = \$10.590$; $p_{vt} = 1.053p_{nt} = \$10.53$
3. Case 3: elastic demand $p_t = 1.009p_{nt} = \$10.09$; $p_{vt} = 1.00p_{nt} = \$10.09$

We see, again, that the price rise to the consumer is nearly equal to the full tax for the inelastic demand case. It is only one-tenth of the tax for these commodities where the demand is elastic. The price to the firm is equal to

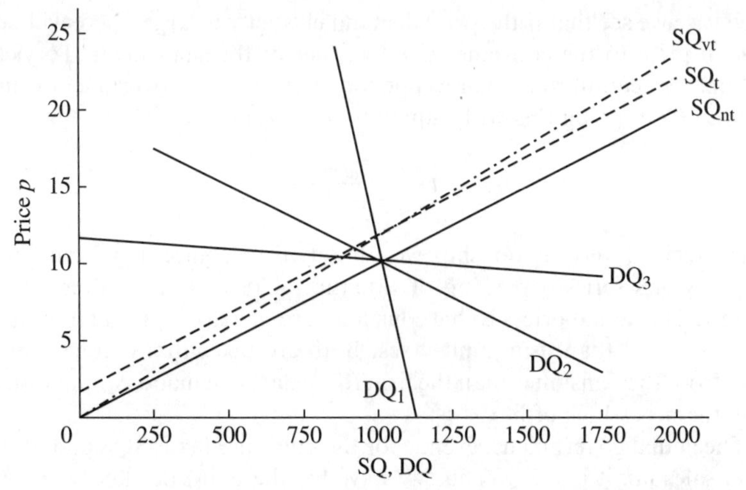

Figure 4.11. Supply–Demand Curves (DQ_1, Inelastic Demand; DQ_2, Unitary Demand; DQ_3, Elastic Demand; SQ_{nt}, No-Tax Supply; SQ_t, Unitary Tax Supply; SQ_{vt}, *Ad Valorem* Tax Supply).

the consumer's price less the tax per commodity. Thus the firms pick up almost all the tax increase for elastic demand goods. They are able to pass on the whole tax to the consumer for inelastic demand goods.

The supply–demand equilibrium shifts under the burden of the tax. Figure 4.11 shows these results for the no-tax case and for the unit and *ad valorem* sales taxes where supply curves are given by Equations 4.51 and 4.66. The ratio of revenues to the firm, that is to say the ratio of royalties for zero profit, with and without taxes is given, from Equation 4.63, as follows:

1. Case 1: inelastic demand ($r_t/r_{nt} = 0.98$)
2. Case 2: unitary demand ($r_t/r_{nt} = 0.90$)
3. Case 3: elastic demand ($r_t/r_{nt} = 0.83$)

The ratios for the *ad valorem* taxes are essentially the same as these and are not shown. The actual revenue to the government is given by Equation 4.80 as follows:

1. Case 1: inelastic demand $G_t = \$990.91$
2. Case 2: unitary demand $G_t = \$950.00$
3. Case 3: elastic demand $G_t = \$909.09$ ■

Market equilibrium, under perfect competition considerations, will exist if we have at least one nonnegative price at which supply and demand amounts are equal and also nonnegative. That is, the supply–demand curves must intersect at least once in the first quadrant (nonnegative quadrant) if there is to be a market equilibrium. We can easily construct supply–demand curves that do not intersect owing to the finite lengths of these curves. However, it is generally always possible to meaningfully extend the length of a finite supply–demand curve in a direction of either constant price or constant quantity such that equilibrium can be established. Figures 4.12 and 4.13 indicate two situations in which this can be accomplished. In Fig. 4.12, supply always exceeds demand and extension of the supply curve along the line $p = 0$ is meaningful, since a firm

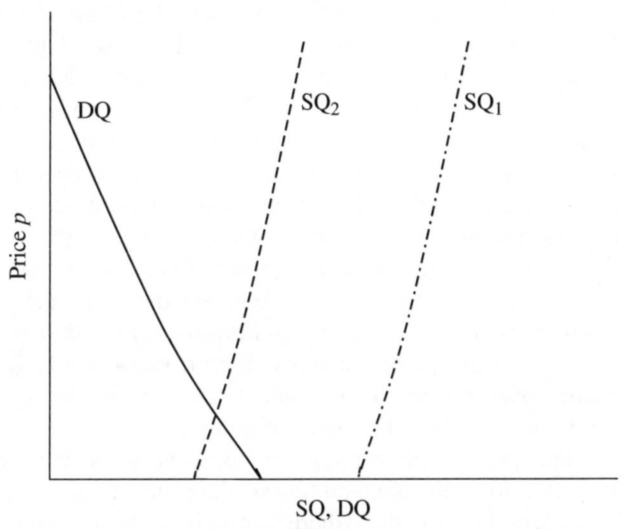

Figure 4.12. Supply–Demand Curve for a Normally Free Commodity (Case 1) That Becomes Nonfree Owing to a Negative Externality (Case 2).

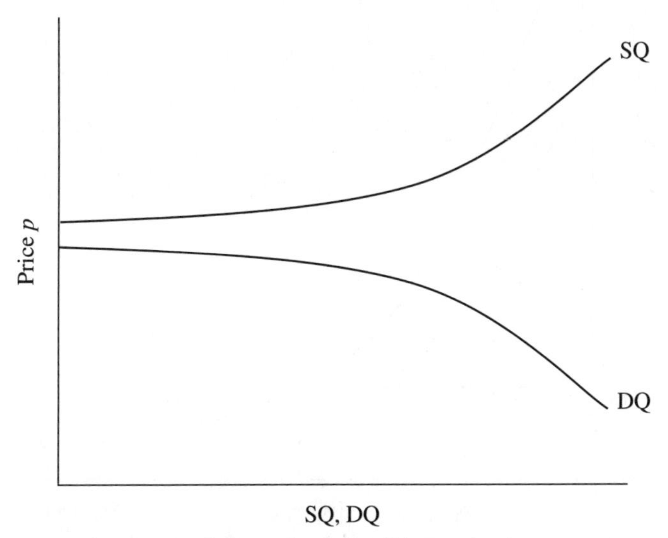

Figure 4.13. Supply–Demand Curve for Infeasible Production.

should always supply a quantity $q < Q_0$ at a price of zero if it is willing to supply Q_0 at a price of zero.

Figure 4.12 represents a classic case of a free commodity such as air. However, if something, such as pollution, reduces the quantity of *good* air available for free, the supply curve will shift to the left. The consumer now becomes willing to pay for pollution removal efforts. Figure 4.13 shows a case in which the supply price is always greater than the demand price. We can extend the demand curve upward along the $q = 0$ axis, since if consumers demand nothing at price $p_{\min}$, they will surely demand nothing at a price higher than $p_{\min}$. This situation might apply, for example, to solid gold taxi cabs. Production costs are far greater than what the consumer is willing to pay, and the supply is zero, as is the demand. Zero cost commodities and zero supply and demand commodities are meaningful, as is equilibrium under these conditions. Other cases such as those in which demand always exceeds supply for every price are not meaningful in a perfectly competitive economy. In a perfectly competitive economy, equilibrium will exist; if it does not, this is often due to poor modeling. In a nonperfect economy it is possible, for example, to have demand exceed supply at a fixed price controlled, say, by the government. Figure 4.14 shows such a case where an equilibrium point would exist in a free-market situation, but where price controls to limit prices at $p = p_{\max}$ result in an excess of demand over supply. This would lead to consumer shortages and possibly rationing to distribute the scarce commodity.

If we assume that supply curves have positive slope for all positive q (i.e., $\partial \text{SQ}/\partial q > 0$) and that demand curves are negatively sloped for all positive q (i.e., $\text{dDQ}/\text{d}q < 0$), then there can exist only a single equilibrium

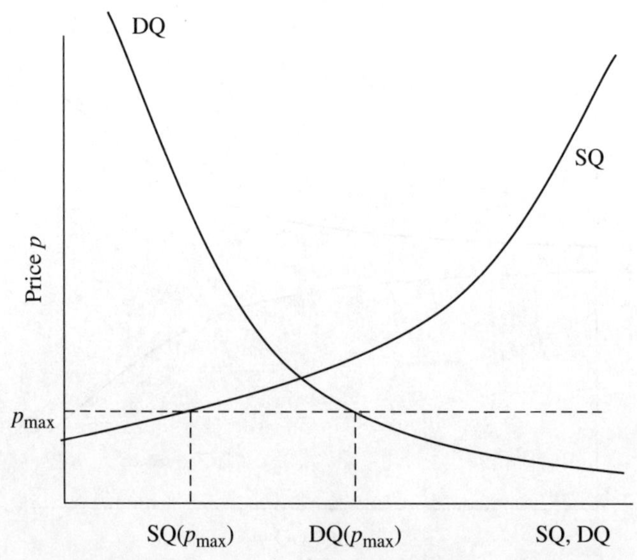

Figure 4.14. Case Where Demand Exceeds Supply Owing to Price Controls.

point. Usually supply and demand curves will have positive and negative slopes, respectively; there can be exceptions, however. These would appear to be more prevalent in factor markets, such as labor, than in product or commodity markets. It is conceivable, for example, for the labor supply to expand as wages increase up to a given maximum, but further increases in wages reduce the labor supply by making increased leisure possible and desirable. Although nonunique supply–demand equilibrium conditions are possible, they are not normally occurring phenomena.

In our work thus far in this chapter, we have determined static equilibrium points where supply equals demand. It is important to examine conditions under which this static equilibrium is stable in the sense that the market conditions will not cause a divergence from the equilibrium point(s). At price p, there will be a quantity $\mathrm{SQ}(p)$ produced or supplied and a quantity $\mathrm{DQ}(p)$ demanded. The *excess demand*[3] is defined as the difference between demand at price p and supply at price p:

$$\mathrm{ED}(p) = \mathrm{DQ}(p) - \mathrm{SQ}(p) \tag{4.82}$$

If the excess demand is positive, then consumers will tend to raise the price offered for commodities. Conversely, if the excess demand is negative, they will tend to lower prices. Thus a price increase should reduce the excess demand and a price reduction should increase the excess demand; hence,

$$\frac{\mathrm{dED}(p)}{\mathrm{d}p} = \frac{\mathrm{dDQ}(p)}{\mathrm{d}p} - \frac{\mathrm{dSQ}(p)}{\mathrm{d}p} < 0 \tag{4.83}$$

This is known as the *Walras stability condition.*

Alternately, the *excess price* for a commodity is defined as the difference between what consumers are willing to pay and what firms are willing to charge for a commodity at a given quantity produced or supplied. For the consumer price

$$p_c(q) = \mathrm{DQ}^{-1}(q) \tag{4.84}$$

and the firm's price

$$p_f(q) = \mathrm{SQ}^{-1}(q) \tag{4.85}$$

the excess price at demand q is

$$\mathrm{EP}(q) = p_c(q) - p_f(q) = \mathrm{DQ}^{-1}(q) - \mathrm{SQ}^{-1}(q) \tag{4.86}$$

When the excess price is positive and consumers are willing to pay more for a commodity, producers will generally increase the quantity supplied and raise prices. This will reduce the excess demand. However, if the excess price is

[3]See Problems 25 and 26 for a suggested example involving the excess demand approach to general economic equilibrium.

negative and consumers are tending to lower the prices paid, the firms will reduce the supply of goods. This will lower prices; hence

$$\frac{\mathrm{dEP}(q)}{\mathrm{d}q} = \frac{\mathrm{dDQ}^{-1}(q)}{\mathrm{d}q} - \frac{\mathrm{dSQ}^{-1}(q)}{\mathrm{d}q} < 0 \tag{4.87}$$

This is known as the *Marshall stability condition.*

These two stability conditions can be compared by recasting Equation 4.87, the Marshall stability condition, in terms of the more familiar supply–demand functions $q = \mathrm{DQ}(p)$ and $q = \mathrm{SQ}(p)$. By substituting

$$\frac{\mathrm{dDQ}^{-1}(q)}{\mathrm{d}q} = \frac{\mathrm{d}p}{\mathrm{dDQ}(p)} = \frac{1}{\mathrm{dDQ}(p)/\mathrm{d}p}$$

and

$$\frac{\mathrm{dSQ}^{-1}(q)}{\mathrm{d}q} = \frac{\mathrm{d}p}{\mathrm{dSQ}(p)} = \frac{1}{\mathrm{dSQ}(p)/\mathrm{d}p}$$

in Equation 4.87 we obtain

$$\frac{1}{\mathrm{dDQ}(p)/\mathrm{d}p} - \frac{1}{\mathrm{dSQ}(p)/\mathrm{d}p} < 0 \tag{4.88}$$

which is another statement of the *Marshall stability condition.* In the usual case where $\mathrm{dDQ}(p)/\mathrm{d}p < 0$ and $\mathrm{dSQ}(p)/\mathrm{d}p > 0$, we multiply both sides of the foregoing relation by the product of these two quantities and reverse the sign of the inequality, since we are multiplying by a negative quantity, to obtain

$$\frac{\mathrm{dSQ}(p)}{\mathrm{d}p} - \frac{\mathrm{dDQ}(p)}{\mathrm{d}p} > 0$$

which is equivalent to Equation 4.83. Thus under normal conditions the Walras and Marshall stability conditions are equivalent.[4] In those unusual cases in which these stability conditions yield different results, we cannot determine which result is correct on the basis of our effects thus far. We must examine additional information concerning market behavior that will allow us to associate dynamics with supply–demand behavior.

We conclude this section with a brief discussion of dynamic equilibrium. We again consider the very restrictive case of only a single good. We assume that firms bring a quantity q of goods to the market to sell them. They determine this quantity based on their anticipation of the price of the goods in the market and their supply curve. The anticipated price will be the price at which the goods were sold during the previous period. Thus we will say that

[4]This is not the case in general, particularly if $\mathrm{dDQ}(p)/\mathrm{d}p < 0$ and $\mathrm{dSQ}(p)/\mathrm{d}p < 0$ (see Problem 16).

at time t the firms bring to the marketplace a quantity of goods SQ(t) that is based on the price in the previous period:

$$\mathrm{SQ}(t) = f_s[p(t-1), t] = \mathrm{SQ}[p(t-1), t] \tag{4.89}$$

The firms anticipate selling these goods at time t at a price $p(t-1)$.

The consumers arrive at the marketplace at time t and demand a quantity of goods DQ(t). We assume that actual sales and purchases at time t are initially fluid but eventually take place at a specific price such that all goods are sold. This price will be the equilibrium price for time t such that the consumers purchase the entire quantity of goods brought to the marketplace.[5] Thus we have

$$\mathrm{DQ}(t) = f_d[p(t), t] = \mathrm{DQ}[p(t), t] \tag{4.90}$$

as the quantity demanded. Since we assume that the quantity is the supply quantity to bring about zero excess demand for every day, we have

$$\mathrm{DQ}(t) = \mathrm{SQ}(t) = f_s[p(t-1), t] = f_d[p(t), t] \tag{4.91}$$

We have just obtained a nonlinear difference equation in the marketplace price that we can, in theory, solve.[6]. Also, we could examine the stability conditions for this nonlinear differential equation, but it is not the primary objective of this chapter. From Equation 4.91 we have the difference equation

$$p(t) = f_d^{-1} f_s[p(t-1), t] = F[p(t-1), t] \tag{4.92}$$

If we are to reach a stable equilibrium point $\bar{p} = p(t) = p(t-1)$, we must have

$$p(t) = F[p(t), t] = F[p(t)] \tag{4.93}$$

and

$$\left| \frac{\partial F(\bar{p})}{\partial \bar{p}} \right| < 1 \tag{4.94}$$

This condition is necessary and sufficient for the establishment of equilibrium. It ensures that perturbations $\Delta p(t)$ given by

$$\Delta p(t) = \frac{\partial F[p(t-1)]}{\partial p(t-1)} \Delta p(t-1)$$

will be smaller than the cause of the perturbations, which is $\Delta p(t-1)$. If we use the definition of F, given by Equation 4.92, in Equation 4.94, we obtain

[5]We could consider cases of unfilled consumer demand or unsold goods, but we will not do this here.

[6]This is *but* one of the several models of time behavior of firms and consumers, and a very simple one.

$$\left| \frac{\partial f_{\mathrm{d}}^{-1}(\overline{q})}{\partial \overline{q}} \frac{\mathrm{d}f_{\mathrm{s}}(\overline{p})}{\mathrm{d}\overline{p}} \right| < 1 \tag{4.95}$$

where $\overline{q}$ is the equilibrium demand.

Since

$$\frac{f_{\mathrm{d}}^{-1}(q)}{q} = \left(\frac{\mathrm{DQ}(p)}{p}\right)^{-1}$$

we have as the requirement for stable equilibrium

$$\left| \frac{\mathrm{dSQ}(\overline{p})}{\mathrm{d}\overline{p}} \right| < \left| \frac{\mathrm{dDQ}(\overline{p})}{\mathrm{d}\overline{p}} \right| \tag{4.96}$$

Since $\mathrm{DQ}/p = \mathrm{SQ}/p$ at equilibrium, we can rewrite the foregoing relation as

$$\left| \frac{p}{\mathrm{SQ}} \frac{\mathrm{dSQ}}{\mathrm{d}p} \right| < \left| \frac{p}{\mathrm{DQ}} \frac{\mathrm{dDQ}}{\mathrm{d}p} \right| \tag{4.97}$$

Thus we see that equilibrium stability exists whenever the absolute value of the supply price elasticity is less than the absolute value of the demand price elasticity. When the inequality of Equation 4.96 holds over a range of supply–demand values, then any initial price starting within this range will result in convergence. However, when Equation 4.97 does not hold and the magnitude of the supply elasticity is greater than the demand elasticity, then instability results and there is no equilibrium price (and quantity). Figure 4.15 illustrates *cobweb* diagrams for stable and unstable supply–demand equilibria. Of course, many modifications to these curves are possible; for instance, we can have limit cycle behavior.

Example 4.7: We consider the linear supply–demand curves

$$\mathrm{SQ}(p) = q = ap - b \tag{4.98}$$

and

$$\mathrm{DQ}(p) = q = c - dp \tag{4.99}$$

For this example, the Walras stability condition of Equation 4.83 is given by

$$\frac{\mathrm{dED}(p)}{\mathrm{d}p} = \frac{\mathrm{dDQ}(p)}{\mathrm{d}p} - \frac{\mathrm{dSQ}(p)}{\mathrm{d}p} = -d - a < 0 \tag{4.100}$$

and this is satisfied for

$$a + d > 0 \tag{4.101}$$

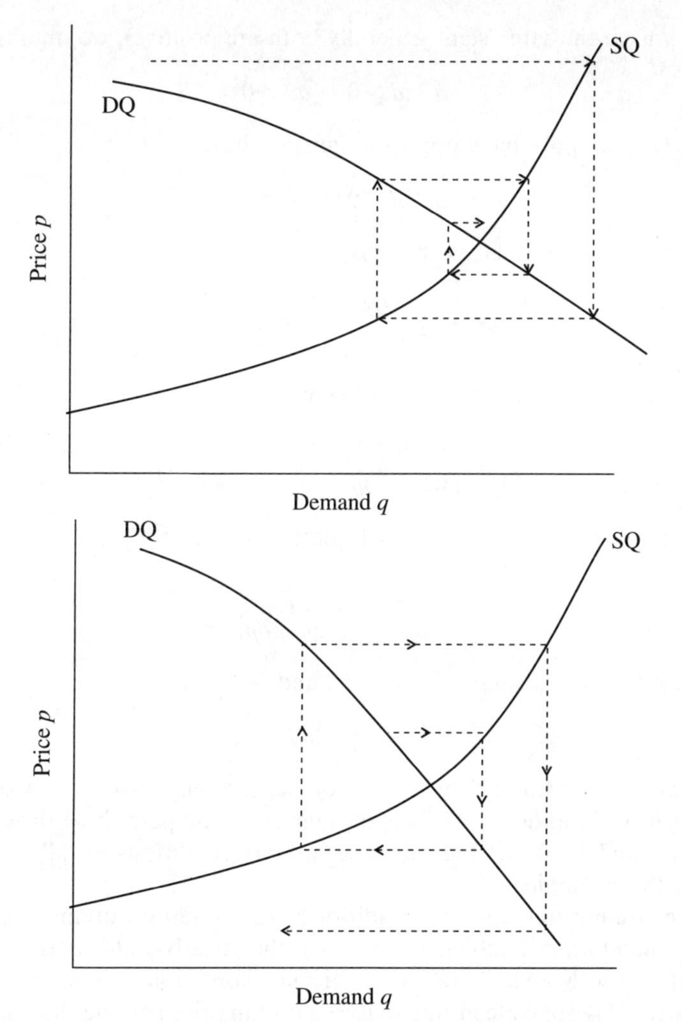

Figure 4.15. Typical Cobweb Diagrams: (a) Stable and (b) Unstable.

Since a and d are usually positive, we see that the supply–demand curves of Equations 4.98 and 4.99 are stable in the sense (of Walras) that excess demand, or $\mathrm{ED}(p) = \mathrm{DQ}(p) - \mathrm{SQ}(p)$, has a negative price derivative. A price increase will reduce the excess demand and a price reduction will increase the excess demand, as we see from

$$\Delta \mathrm{ED}(p) = \frac{\mathrm{dED}(p)}{\mathrm{d}p} \Delta p = -(d + a)\Delta p$$

The Marshall stability condition of Equation 4.88 becomes

$$\frac{1}{\mathrm{dDQ}(p)/\mathrm{d}p} - \frac{1}{\mathrm{dSQ}(p)/\mathrm{d}p} = -\frac{1}{d} - \frac{1}{a} < 0 \qquad (4.102)$$

If a and d have the same sign (generally both are positive), we obtain

$$a + d > 0 \quad (ad > 0) \tag{4.103}$$

Conversely, if a and d have opposite signs, we have

$$a + d < 0 \quad (ad < 0) \tag{4.104}$$

From supply price elasticity expression

$$\varepsilon_s^p = \frac{p}{SQ}\frac{\partial SQ(p)}{\partial p} = \frac{p}{ap - b}a$$

and the demand price elasticity expression

$$\varepsilon_s^p = \frac{p}{DQ}\frac{\partial DQ(p)}{\partial p} = \frac{p}{c - dp}(-d)$$

we obtain the stability condition of Equation 4.96 or 4.97 as

$$\left|\frac{pa}{ap - b}\right| < \left|\frac{pd}{c - dp}\right|$$

At equilibrium we have $ap - b = c - dp$, and so

$$|a| < |d| \tag{4.105}$$

is the stability condition. This is, of course, the relationship we would have obtained from Equation 4.96. It is of interest to compare these three stability conditions, and Table 4.1 indicates the stability conditions for all the possible cases for this example.

The equilibrium stability condition is the condition upon which supply equals demand after a sufficient time such that steady-state conditions result. The Walras and Marshall conditions are just conditions where the change in excess demand is opposite in sign to the change in price and the change in excess price is opposite in sign to the change in quantity purchased or supplied. These changes may well be so great that time instability may occur, even though the change in excess demand or excess price is in the *correct* direction.

For the case where the supply–demand curves are linear and of the form of Equations 4.98 and 4.99, the price equilibrium equation (Eq. 4.91 or 4.92) becomes

TABLE 4.1. Results of Different Stability Conditions

a	d	Walras condition	Marshall condition	Equilibrium stability condition
Positive	Positive	Always satisfied	Always satisfied	$a < d$
Positive	Negative	$a + d > 0$	$a + d < 0$	$a + d < 0$
Negative	Positive	$a + d > 0$	$a + d < 0$	$a + d > 0$
Negative	Negative	Never satisfied	Never satisfied	$a > d$

$$ap(t-1) - b = c - dp(t) \tag{4.106}$$

or

$$p(t) = \frac{-a}{d}p(t-1) + \frac{c+b}{d} \tag{4.107}$$

At time $t = 0$, we assume that market supply–demand is reached with price $p(0) = p_0$. At time t the price is, from the solution of Equation 4.107, given by

$$p(t) = \left(\frac{-a}{d}\right)^t p_0 + \frac{b+c}{d+a}\left[1 + \left(\frac{-a}{d}\right)^t\right]$$

If $|a/d| < 1$, the $(-a/d)^t$ term in the foregoing relation will decrease to zero as t increases, and we will approach the equilibrium solution, $(b+c)/(a+d)$. If $|a/d| > 1$, the magnitude of the $(-a/d)^t$ term becomes larger and larger as t increases and the solution becomes unstable. Converging or diverging oscillations in price will exist whenever a and d have the same sign. For cases where a and d have opposite signs, the price will either converge or diverge monotonically. Figure 4.16 illustrates the four possible time–response types. We could now illustrate the eight possible cobweb time paths and indicate whether stability is achieved for the three stability conditions mentioned here. We must not interpret these equilibrium stability conditions to be always true. They are equilibrium stability conditions only for the particular discrete time model of supply and demand that has been assumed here. ■

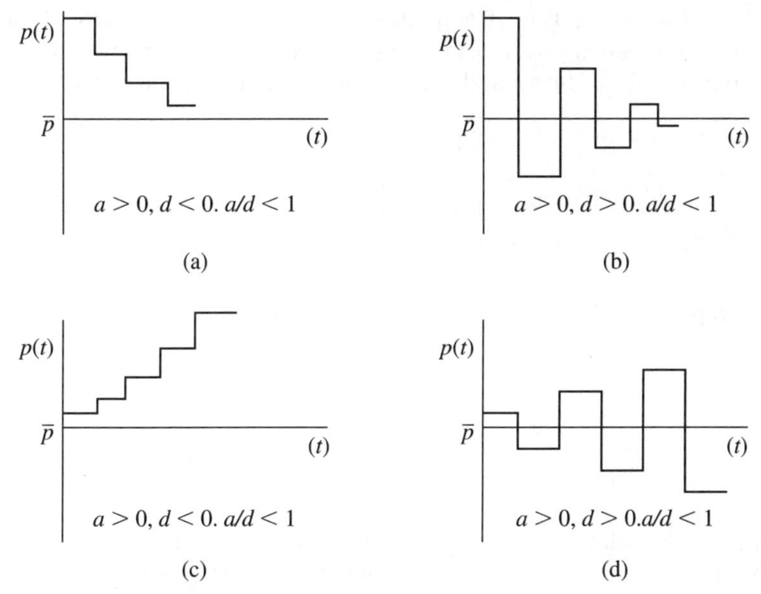

Figure 4.16. Four Price Response Behaviors: (a) Stable; (b) Stable and Oscillatory; (c) Unstable; and (d) Unstable and Oscillatory.

We could now proceed to examine other discrete or continuous time models for supply–demand equilibrium. The cobweb model assumed here forces zero excess demand (i.e., supply and demand are in equilibrium) at each and every time period. It would probably be more realistic to assume imperfect market conditions prior to equilibrium. Then each firm would bring a different quantity to the marketplace at different prices, depending on each firm's differing temporal assumption concerning future demand. We could model a situation such as this and determine equilibrium conditions. Since we are dealing with but a single good here, our interests would be better served by an examination of multiple-commodity markets. We will now turn to this much more general case. Unfortunately, in so doing we will lose much of the graphic insight possible in the single-good case.

4.3 GENERAL SUPPLY–DEMAND EQUILIBRIUM CONDITIONS

We will now extend the simple supply–demand equilibrium concepts of Section 4.2 to the case where there are F firms, H households or consumers, a total of n goods or commodities, and a total of m input factors to production. We assume the factor and commodity markets of Fig. 4.1. We will consider that a production equation of the form

$$\Phi^f(\mathbf{q}^f, \mathbf{x}^f), \quad f = 1, 2, \ldots, F \tag{4.108}$$

is associated with each firm. The quantity Φ^f represents the functional form of the production function for firm f and is a scalar. $\mathbf{q}^f$ represents the vector production output of firm f and consists of n components given by

$$\mathbf{q}^f = \begin{bmatrix} q_1^f \\ q_2^f \\ \vdots \\ q_n^f \end{bmatrix}, \quad f = 1, 2, \ldots, F \tag{4.109}$$

Here $\mathbf{x}^f$ represents the input factor m-vector to firm f:

$$\mathbf{x}^f = \begin{bmatrix} x_1^f \\ x_2^f \\ \vdots \\ x_m^f \end{bmatrix}, \quad f = 1, 2, \ldots, F \tag{4.110}$$

We will let the sales price vector for a unit commodity bundle—the price at which the firm sells or at which a household purchases one unit of each component in the commodity bundle—be expressed as

$$\mathbf{p} = \begin{bmatrix} p_1 \\ p_2 \\ \vdots \\ p_n \end{bmatrix} \tag{4.111}$$

It is assumed that prices will be given. The wages or prices paid for the input factor $\mathbf{x}$, that is, the wage or price paid for one unit of each component of the input factor vector, will be denoted by

$$\mathbf{w} = \begin{bmatrix} w_1 \\ w_2 \\ \vdots \\ w_m \end{bmatrix} \tag{4.112}$$

The profit of each firm will be the difference between product sales revenue and input factor cost. Thus we have

$$\Pi^f = \mathbf{p}^T \mathbf{q}^f - \mathbf{w}^T \mathbf{x}^f, \quad f = 1, 2, \ldots, F \tag{4.113}$$

We will assume a total of H households. Each household will purchase an n-vector commodity bundle $\mathbf{y}^h$ of the products of the firms, where

$$\mathbf{y}^h = \begin{bmatrix} y_1^h \\ y_2^h \\ \vdots \\ y_n^h \end{bmatrix}, \quad h = 1, 2, \ldots, H \tag{4.114}$$

Each household, through its labor or work, will supply an m-vector of factors h to the firms:

$$\mathbf{r}^h = \begin{bmatrix} r_1^h \\ r_2^h \\ \vdots \\ r_m^h \end{bmatrix}, \quad h = 1, 2, \ldots, H \tag{4.115}$$

The utility of each household is a function of the labor factor $\mathbf{r}^h$ that is sold on the factor markets at wages $\mathbf{w}$, and the number of products $\mathbf{y}^h$ each household purchases in a commodity market at prices $\mathbf{p}$. We will let the utility of each household be denoted by

$$U^h = U^h(\mathbf{y}^h, \mathbf{r}^h), \quad h = 1, 2, \ldots, H \tag{4.116}$$

Each household will determine the optimum factor vector $\mathbf{r}^h$ to sell in the factor market and the optimum commodity bundle $\mathbf{y}^h$ to purchase in the commodity market. The vectors $\mathbf{y}^h$ and $\mathbf{r}^h$ will maximize the consumer utility subject to a budget constraint for each household. This constraint will be given by

$$\mathbf{I}^h = \mathbf{p}^{\mathrm{T}}\mathbf{y}^h = \mathbf{w}^{\mathrm{T}}\mathbf{r}^h + \mathbf{\Pi}^{\mathrm{T}}\mathbf{s}^h, \quad h = 1, 2, \ldots, H \tag{4.117}$$

The $\mathbf{p}^{\mathrm{T}}\mathbf{y}^h$ expression represents the total expenditure on the commodity bundle $\mathbf{y}^h$ and is assumed to be equal to the total household income $\mathbf{I}^h$. The expression $\mathbf{w}^{\mathrm{T}}\mathbf{r}^h$ represents the income that the consumers derive from sales of their labor (factors) in the factor market. The quantity $\mathbf{\Pi}$ is the profit F-vector of the F firms and is given by

$$\mathbf{\Pi} = \begin{bmatrix} \Pi^1 \\ \Pi^2 \\ \vdots \\ \Pi^F \end{bmatrix} \tag{4.118}$$

The expression $\mathbf{s}^h$ represents the fractional share of ownership of the firms by consumer h:

$$\mathbf{s}^h = \begin{bmatrix} s_1^h \\ s_2^h \\ \vdots \\ s_F^h \end{bmatrix} \tag{4.119}$$

Since firms purchase factors $\mathbf{x}^f$ from households and households supply factors $\mathbf{r}^h$ to the firms, there must be a *factor market clearing equation* that will indicate that the sum of all factors received by all firms, $\mathbf{x}$, is the sum of all factors delivered by all households, $\mathbf{R}$. Thus we have

$$\mathbf{x} = \sum_{f=1}^{F} \mathbf{x}^f = \sum_{h=1}^{H} \mathbf{r}^h = \mathbf{R} \tag{4.120}$$

Also, since the products delivered to the market by the firms $\mathbf{q}^f$ are the commodities purchased in the market by the households $\mathbf{y}^h$, we must have a *commodity market clearing equation* to indicate this. Here we have

$$\mathbf{Q} = \sum_{f=1}^{F} \mathbf{q}^f = \sum_{h=1}^{H} \mathbf{y}^h = \mathbf{y} \tag{4.121}$$

as the market clearing equation. The general equilibrium problem for the firm and the household consists of the following three activities:

1. Maximization of the profit equation (4.113), subject to the constraint equation (4.108) of the production in terms of goods and factor inputs
2. Maximization of the household utility equation (4.116), subject to the budget income constraint equation (4.117)
3. Simultaneous solution of the relations obtained from the above two operations, subject to the factor clearing equation (4.120) and the market clearing equation (4.121)

TABLE 4.2. Defining Relations and Problems of General Economic Equilibrium

Element	Symbol or equation
Product of the *f*th firm	$\mathbf{q}^f$ (*n*-vector)
Factor input of the *f*th firm	$\mathbf{x}^f$ (*m*-vector)
Production of the *f*th firm	$\Phi^f(\mathbf{q}^f, \mathbf{x}^f) = 0$
Profit of the *f*th firm	$\Pi^f = \mathbf{p}^{\mathrm{T}}\mathbf{q}^f - \mathbf{w}^{\mathrm{T}}\mathbf{x}^f$
Sales price of products of the firm	$\mathbf{p}$ (*n*-vector)
Wages for factor inputs from consumers	$\mathbf{w}$ (*m*-vector)
Commodity bundle from *h*th household	$\mathbf{y}^h$ (*n*-vector)
Factor input from *h*th household	$\mathbf{r}^h$ (*m*-vector)
Economic utility of *h*th household	$U^h = U^h(\mathbf{y}^h, \mathbf{r}^h)$
Firm fractional ownership by *h*th household	$\mathbf{S}^h$ (*F*-vector), $\sum_{h=1}^{H} \mathbf{s}^h = \mathbf{1}$
Budget of *h*th household	$\mathbf{p}^{\mathrm{T}}\mathbf{y}^h = \mathbf{w}^{\mathrm{T}}\mathbf{r}^h + \mathbf{\Pi}^{\mathrm{T}}\mathbf{s}^h = I^h$
Factor market clearing equation	$\mathbf{x} = \sum_{f=1}^{F} \mathbf{x}^f = \sum_{h=1}^{H} \mathbf{r}^h = \mathbf{R}$
Commodity market clearing equation	$\mathbf{Q} = \sum_{f=1}^{F} \mathbf{q}^f = \sum_{h=1}^{H} \mathbf{y}^h = \mathbf{y}$
Problem of the *f*th firm	Maximize Π^f with constraint on Φ^f
Problem of *h*th household	Maximize U^h with constraint on I^h

Table 4.2 summarizes the problem of the general economic equilibrium of supply and demand.

One of the basic relations in general economic equilibrium theory, known to economists as *Walras' law*, states that for any given vector of prices and wages the total economic value of the supply must be equal to the total economic value of demand. This can be proved as follows: We can sum the household budget income equation (4.117) over all households to yield

$$I = \sum_{h=1}^{H} I^h = \sum_{h=1}^{H} \mathbf{p}^{\mathrm{T}}\mathbf{y}^h = \sum_{h=1}^{H} \mathbf{w}^{\mathrm{T}}\mathbf{r}^h + \sum_{h=1}^{H} \mathbf{\Pi}^{\mathrm{T}}\mathbf{s}^h \tag{4.122}$$

Now all the profit of the firms must be distributed over the households. The profit can go nowhere else except the people sector, since we have not assumed that the government will impose any taxes. Thus we must have

$$\sum_{h=1}^{H} \mathbf{s}^h = \mathbf{1} \tag{4.123}$$

Here $\mathbf{1}$ is an *F*-vector of all 1's. We can therefore rewrite the last term in Equation 4.122 as

$$\sum_{h=1}^{H} \mathbf{\Pi}^{\mathrm{T}}\mathbf{s}^h = \sum_{f=1}^{F} \Pi^f \tag{4.124}$$

The total income equation (Eq. 4.122), using the foregoing relation and Equation 4.113, becomes

$$I = \sum_{h=1}^{H} \mathbf{p}^{\mathrm{T}}\mathbf{y}^h = \sum_{h=1}^{H} \mathbf{w}^{\mathrm{T}}\mathbf{r}^h + \sum_{f=1}^{F} (\mathbf{p}^{\mathrm{T}}\mathbf{q}^f - \mathbf{w}^{\mathrm{T}}\mathbf{x}^f)$$

We group together the terms involving **p** and **w** to obtain

$$\mathbf{p}^{\mathrm{T}}\left(\sum_{h=1}^{H} \mathbf{y}^h - \sum_{f=1}^{F} \mathbf{q}^f\right) = \mathbf{w}^{\mathrm{T}}\left(\sum_{h=1}^{H} \mathbf{r}^h - \sum_{f=1}^{F} \mathbf{x}^f\right) \tag{4.125}$$

This expression is known as *Walras' law*. We recognize that Equations 4.120 and 4.121 are embedded in this relation, which can be written in several different ways. Also, that there is at least one redundant equation in Equation 4.125 and this allows establishment of a numeraire.

Use of basic optimization theory for the general problem of the firm, profit maximization with the constraint that the production equation be satisfied, results in the following Lagrangian, from the results summarized in Table 4.2:

$$L^f = \Pi^f + \lambda^f \Phi^f = \mathbf{p}^{\mathrm{T}}\mathbf{q}^f - \mathbf{w}^{\mathrm{T}}\mathbf{x}^f + \lambda^f \Phi^f(\mathbf{q}^f, \mathbf{x}^f) \tag{4.126}$$

Here the necessary conditions for optimality are, from Table 2.2, given by

$$\frac{\partial L^f}{\partial \mathbf{q}^f} = \mathbf{0} = \mathbf{p} + \lambda^f \frac{\partial \Phi^f(\mathbf{q}^f, \mathbf{x}^f)}{\partial \mathbf{q}^f} \tag{4.127}$$

$$\frac{\partial L^f}{\partial \mathbf{x}^f} = \mathbf{0} = -\mathbf{w} + \lambda^f \frac{\partial \Phi^f(\mathbf{q}^f, \mathbf{x}^f)}{\partial \mathbf{x}^f} \tag{4.128}$$

$$\frac{\partial L^f}{\partial \lambda^f} = \mathbf{0} = \Phi^f(\mathbf{q}^f, \mathbf{x}^f) \tag{4.129}$$

These relations are solved for each firm, $f = 1, 2, \ldots, F$. This results in the need to solve $n + m + 1$ scalar equations for each firm, for a total of $(n + m + l)F$ scalar equations.

The Lagrangian for the general problem of the consumer, to maximize utility subject to a budget constraint, is, from Table 4.2, given by

$$L^h = U^h + \gamma^h(I^h - \mathbf{p}^{\mathrm{T}}\mathbf{y}^h) = U^h(\mathbf{y}^h, \mathbf{r}^h) + \gamma^h(\mathbf{w}^{\mathrm{T}}\mathbf{r}^h + \mathbf{\Pi}^{\mathrm{T}}\mathbf{s}^h - \mathbf{p}^{\mathrm{T}}\mathbf{y}^h) \tag{4.130}$$

The necessary conditions for optimality are, from Table 2.2 where we assume that the entire consumer budget is used, given by

$$\frac{\partial L^h}{\partial \mathbf{y}^h} = \mathbf{0} = \frac{\partial U^h(\mathbf{y}^h, \mathbf{r}^h)}{\partial \mathbf{y}^h} + \gamma^h \mathbf{p} \tag{4.131}$$

$$\frac{\partial L^h}{\partial \mathbf{r}^h} = \mathbf{0} = \frac{\partial U^h(\mathbf{y}^h, \mathbf{r}^h)}{\partial \mathbf{r}^h} + \gamma^h \mathbf{w} \tag{4.132}$$

$$\frac{\partial L^h}{\partial \gamma^h} = 0 = \mathbf{w}^{\mathrm{T}}\mathbf{r}^h + \mathbf{\Pi}^{\mathrm{T}}\mathbf{s}^h - \mathbf{p}^{\mathrm{T}}\mathbf{y}^h \tag{4.133}$$

This results in $n + m + 1$ scalar equations for each of the H firms for a total of $(n + m + l)H$ scalar equations.

The factor market clearing equation (Eq. 4.120) results in m scalar equations, and the commodity market clearing equation (Eq. 4.121) results in n scalar equations. Table 4.3 shows the necessary conditions for optimality of all firms and all consumers that we have just derived. There are a total of $(n + m + 1)(F + H) + n + m$ equations that need to be solved here. We obtain the n-vector of products delivered ($\mathbf{q}^f$) for each of the F firms, m-vector of factors purchased ($\mathbf{x}^f$) for each of the F firms, the n-vector commodities or products purchased ($\mathbf{y}^h$) for each of the H households, the m-vector ($\mathbf{r}^h$) of factors delivered by each of the H households, the Lagrange multipliers (λ^f), the H Lagrange multipliers (γ^h), and the $m + n$ market clearing conditions. Because of Walras' law (Eq. 4.124), we see that one of the equations in the set of equations in Table 4.3 is not necessary in that there are only $(n + m + 1)$ $(F + H) + n + m - 1$ independent equations in this table. This is actually quite

TABLE 4.3. Necessary Conditions for General Economic Market Equilibrium

Relation	Equation
Optimization of the fth firm	
Optimum *production supply* quantity	$\lambda^f \frac{\partial \varphi^f(\mathbf{q}^f, \mathbf{x}^f)}{\partial \mathbf{q}^f} = -\mathbf{p}$
Optimum input *factor demand* quantity	$\lambda^f \frac{\partial \varphi^f(\mathbf{q}^f, \mathbf{x}^f)}{\partial \mathbf{x}^f} = \mathbf{w}$
Production constraint	$\varphi^f(\mathbf{q}^f, \mathbf{x}^f) = 0$
Optimization of the hth household	
Optimum *commodity* market *demand*	$\frac{\partial U^h(\mathbf{y}^h, \mathbf{r}^h)}{\partial \mathbf{y}^h} = \gamma^h \mathbf{p}$
Optimum amount of *factor supply*	$\frac{\partial U^h(\mathbf{y}^h, \mathbf{r}^h)}{\partial \mathbf{r}^h} = -\gamma^h \mathbf{w}$
Budget constraint	$\mathbf{w}^{\mathrm{T}}\mathbf{r}^h + \mathbf{\Pi}^{\mathrm{T}}\mathbf{s}^h = \mathbf{p}^{\mathrm{T}}\mathbf{y}^h$
Market clearing equations	
Factor market clearing	$\sum_{f=1}^{F} \mathbf{x}^f = \sum_{h=1}^{H} \mathbf{r}^h = \mathbf{R}$
Commodity market clearing	$\sum_{f=1}^{F} \mathbf{q}^f = \sum_{h=1}^{H} \mathbf{y}^h = \mathbf{Q}$

obvious because maximizing a multiple of Π^f is the same as maximizing Π^f in terms of the optimum $\mathbf{q}^f$, $\mathbf{x}^f$, and λ^f obtained. If we multiply all prices $\mathbf{p}$ and wages $\mathbf{w}$ by the same amount a, we also multiply all profits and the consumer budget by a. The utility function is not a function of a, and so we can use a new Lagrange multiplier $\gamma^h = \gamma^h/a$. Thus we see that the optimum equilibrium point is unaffected by increasing all prices and wages by the same multiplicative factor a. Thus we conclude that all optimum demand and supply functions are homogeneous of degree zero in all prices. The price and wage vectors can be normalized by selecting one price or wage component as a numeraire or anchor and then determining all prices and wages relative to it. That is, the vectors

$$\begin{bmatrix} \mathbf{p} \\ \mathbf{w} \end{bmatrix}, \quad \begin{bmatrix} \dfrac{\mathbf{p}}{p_i} \\ \dfrac{\mathbf{w}}{p_i} \end{bmatrix}, \quad \begin{bmatrix} \dfrac{\mathbf{p}}{w_j} \\ \dfrac{\mathbf{w}}{w_j} \end{bmatrix}, \quad i = 1, 2, \ldots, n; \quad j = 1, 2, \ldots, m$$

are all equivalent in the sense that they all result in the same supply–demand equilibrium. In Section 4.2, we have indicated that zero profit for each firm will be a consequence of long-run equilibrium. This will occur here as well if we allow the number of firms F producing a product to be a variable.

Example 4.8: We consider a relatively comprehensive example to illustrate general market equilibrium. The production function of firm f is presumed to be the very general quadratic

$$[(\mathbf{b}^f)^{\mathrm{T}}\mathbf{q}^f + 0.5(\mathbf{q}^f)^{\mathrm{T}}\mathbf{C}^f\mathbf{q}^f]^2 = 0.5(\mathbf{x}^f)^{\mathrm{T}}\mathbf{A}^f\mathbf{x}^f = \Gamma^2(\mathbf{q}^f) \tag{4.134}$$

where $\mathbf{C}^f$ and $\mathbf{A}^f$ are positive definite matrices. Each of the F firms is presumed to produce according to the structural model represented by this equation. The profit of firm f is given by

$$\Pi^f = \mathbf{p}^{\mathrm{T}}\mathbf{q}^f - \mathbf{w}^{\mathrm{T}}\mathbf{x}^f \tag{4.135}$$

Table 4.3 provides the following necessary conditions for optimality of firm f:

$$\lambda^f \frac{\partial \varphi^f(\mathbf{q}^f, \mathbf{x}^f)}{\partial \mathbf{q}^f} = 2\lambda^f \Gamma(\mathbf{q}^f) \frac{\partial \Gamma(\mathbf{q}^f)}{\partial \mathbf{q}^f} = -\mathbf{p} \tag{4.136}$$

$$\lambda^f \frac{\partial \varphi^f(\mathbf{q}^f, \mathbf{x}^f)}{\partial \mathbf{x}^f} = -\mathbf{A}^f\mathbf{x}^f\lambda^f = \mathbf{w} \tag{4.137}$$

$$\varphi^f(\mathbf{q}^f, \mathbf{x}^f) = \Gamma^2(\mathbf{q}^f) - 0.5(\mathbf{x}^f)^{\mathrm{T}}\mathbf{A}^f\mathbf{x}^f = 0 \tag{4.138}$$

From Equation 4.137, we obtain

$$\mathbf{x}^f = \frac{-(\mathbf{A}^f)^{-1}\mathbf{w}}{\lambda^f}$$

We insert this relation in Equation 4.138 and obtain

$$(\lambda^f)^2 = -\frac{0.5\mathbf{w}^{\mathrm{T}}(\mathbf{A}^f)^{-1}\mathbf{w}}{\Gamma^2(\mathbf{q}^f)} \tag{4.139}$$

$$\mathbf{x}^f = \frac{(\mathbf{A}^f)^{-1}\mathbf{w}\Gamma(\mathbf{q}^f)}{[0.5\mathbf{w}^{\mathrm{T}}(\mathbf{A}^f)^{-1}\mathbf{w}]^{1/2}} \tag{4.140}$$

Here we assume that λ^f is negative as required by the sufficiency condition that the matrix of second partials be negative definite. From Equations 4.136 and 4.139 we obtain

$$[2\mathbf{w}^{\mathrm{T}}(\mathbf{A}^f)^{-1}\mathbf{w}]^{1/2}\frac{\partial\Gamma(\mathbf{q}^f)}{\partial(\mathbf{q}^f)} = \mathbf{p} \tag{4.141}$$

Equations 4.140 and 4.141 represent the solution to the problem of the firm for any given value of $\Gamma(\mathbf{q}^f)$. For the specific $\Gamma(\mathbf{q}^f)$ assumed in Equation 4.134, we obtain

$$\mathbf{q}^f = (\mathbf{C}^f)^{-1}\left(\frac{\mathbf{p}}{[2\mathbf{w}^{\mathrm{T}}(\mathbf{A}^f)^{-1}\mathbf{w}]^{1/2}} - \mathbf{b}^f\right) \tag{4.142}$$

$$\mathbf{x}^f = \frac{(\mathbf{A}^f)^{-1}\mathbf{w}}{[2\mathbf{w}^{\mathrm{T}}(\mathbf{A}^f)^{-1}\mathbf{w}]^{1/2}}\left(\frac{\mathbf{p}^{\mathrm{T}}(\mathbf{C}^f)^{-1}\mathbf{p}}{2\mathbf{w}^{\mathrm{T}}(\mathbf{A}^f)^{-1}\mathbf{w}} - (\mathbf{b}^f)^{\mathrm{T}}(\mathbf{C}^f)^{-1}\mathbf{b}^f\right) \tag{4.143}$$

as the optimum production and input factor levels for firm f. These levels depend only on the firm's production characteristics—$\mathbf{A}^f$, $\mathbf{b}^f$, and $\mathbf{C}^f$—and the prices and wages in the commodity and factor markets $\mathbf{p}$ and $\mathbf{w}$. We note that unless the inequality

$$\mathbf{p} > \mathbf{b}^f[2\mathbf{w}^{\mathrm{T}}(\mathbf{A}^f)^{-1}\mathbf{w}]^{1/2} \tag{4.144}$$

holds, the production of the firm $\mathbf{q}^f$ and the input factor $\mathbf{x}^f$ are each zero. Unless the inequality of Equation 4.144 is satisfied, the firm will also not make a profit and will not remain in business. The profit for firm f is given by Equations 4.135, 4.142, and 4.143 as

$$\Pi^f = \frac{0.5\{\mathbf{p}^{\mathrm{T}} - [2\mathbf{w}^{\mathrm{T}}(\mathbf{A}^f)^{-1}\mathbf{w}]^{1/2}(\mathbf{b}^f)^{\mathrm{T}}\}(\mathbf{C}^f)^{-1}\{\mathbf{p} - [2\mathbf{w}^{\mathrm{T}}(\mathbf{A}^f)\mathbf{w}]^{1/2}(\mathbf{b}^f)\}}{[2\mathbf{w}^{\mathrm{T}}(\mathbf{A}^f)^{-1}\mathbf{w}]^{1/2}} \tag{4.145}$$

We assume that each household has a utility function in the form of a general quadratic

$$U^h(\mathbf{y}^h, \mathbf{r}^h) = (\mathbf{g}^h)^{\mathrm{T}}\mathbf{y}^h - 0.5(\mathbf{y}^h)^{\mathrm{T}}\mathbf{D}^h\mathbf{y}^h - 0.5(\mathbf{r}^h)^{\mathrm{T}}\mathbf{E}^h\mathbf{r}^h \tag{4.146}$$

where $\mathbf{g}^h > \mathbf{0}$, and $\mathbf{D}^h$ and $\mathbf{E}^h$ are each symmetric positive definite matrices. The consumer likes commodities and dislikes labor. In the absence of any constraints, the consumer would choose $\mathbf{y}^h = (\mathbf{D}^h)^{-1}\mathbf{g}^h$ and $\mathbf{r}^h = \mathbf{0}$. Table 4.3 leads us to the following necessary conditions for consumer optimality with an income constraint:

$$\frac{\partial U(\mathbf{y}^h, \mathbf{r}^h)}{\partial \mathbf{y}^h} = \mathbf{g}^h - \mathbf{D}^h \mathbf{y}^h = \gamma^h \mathbf{p} \tag{4.147}$$

$$\frac{\partial U^h(\mathbf{y}^h, \mathbf{r}^h)}{\partial \mathbf{r}^h} = -\mathbf{E}^h \mathbf{y}^h = -\gamma^h \mathbf{w} \tag{4.148}$$

$$\mathbf{w}^{\mathrm{T}} \mathbf{r}^h + \Pi^{\mathrm{T}} \mathbf{s}^h = \mathbf{p}^{\mathrm{T}} \mathbf{y}^h \tag{4.149}$$

From Equations 4.147 and 4.148, on solving for $\mathbf{y}$ and $\mathbf{r}$, we obtain

$$\mathbf{y}^h = (\mathbf{D}^h)^{-1}(\mathbf{g}^h - \gamma^h \mathbf{p}) \tag{4.150}$$

$$\mathbf{r}^h = (\mathbf{E}^h)^{-1} \mathbf{w} \gamma^h \tag{4.151}$$

Then, from Equation 4.149 and the foregoing equations, we obtain

$$[\mathbf{p}^{\mathrm{T}}(\mathbf{D}^h)^{-1}\mathbf{p} + \mathbf{w}^{\mathrm{T}}(\mathbf{E}^h)^{-1}\mathbf{w}]\gamma^h = -\mathbf{\Pi}^{\mathrm{T}}\mathbf{s}^h + \mathbf{p}^{\mathrm{T}}(\mathbf{D}^h)^{-1}\mathbf{g}^h$$

Thus we have the optimum consumer behavior and the optimum commodity demand bundle and factor supply of Equations 4.150 and 4.151, where the Lagrange multiplier of the hth household is

$$\gamma^h = \frac{\mathbf{p}^{\mathrm{T}}(\mathbf{D}^h)^{-1}\mathbf{g}^h - \mathbf{\Pi}^{\mathrm{T}}\mathbf{s}^h}{\mathbf{p}^{\mathrm{T}}(\mathbf{D}^h)^{-1}\mathbf{p} + \mathbf{w}^{\mathrm{T}}(\mathbf{E}^h)^{-1}\mathbf{w}} \tag{4.152}$$

The expression γ^h can be determined if we know the profit vector for all firms $\mathbf{\Pi}$ and the percentage ownership of household h in all firms. The commodity demand $\mathbf{y}^h$ and factor supply $\mathbf{r}^h$ must be positive for our solution to be meaningful. For $\mathbf{r}^h$ to be positive, we must also have γ^h positive. Thus we require

$$\mathbf{p}^{\mathrm{T}}(\mathbf{D}^h)^{-1}\mathbf{g}^h > \mathbf{\Pi}^{\mathrm{T}}\mathbf{s}^h$$

$$\mathbf{g}^h > \frac{\mathbf{p}^{\mathrm{T}}(\mathbf{D}^h)^{-1}\mathbf{g}^h - \mathbf{\Pi}^{\mathrm{T}}\mathbf{s}^h}{\mathbf{p}^{\mathrm{T}}(\mathbf{D}^h)^{-1}\mathbf{p} + \mathbf{w}^{\mathrm{T}}(\mathbf{E}^h)^{-1}\mathbf{w}}\mathbf{p}$$

If these requirements are not satisfied, consumer h will purchase zero commodities $\mathbf{y}$ and supply zero factor inputs $\mathbf{r}^h$. We recall our discussion in Chapter 3 concerning the restrictions we need to ensure meaningful solutions. Here these restrictions require $\mathbf{y} \geq 0$, $\mathbf{r}^h \geq 0$, and $(\mathbf{y})^{\mathrm{T}}\mathbf{p} \leq I^h = \mathbf{w}^{\mathrm{T}}\mathbf{r}^h + \mathbf{\Pi}^{\mathrm{T}}\mathbf{s}^h$. It is interesting to note that the commodity bundle purchased by the hth consumer, given by Equation 4.150, is necessarily smaller than that which would be purchased if there were no budget restrictions.

Each firm will supply products $\mathbf{q}^f$, $f = 1, 2, \ldots, F$, and demand factor inputs $\mathbf{x}^f$, $f = 1, 2, \ldots, F$, according to the relations in Equations 4.142 and 4.143. Households demand commodities $\mathbf{y}^h$ and supply factors $\mathbf{r}^h$ according to the relations in Equations 4.150–4.152. If the profit of all firms $\mathbf{\Pi}$ has been given and the percentage ownership vector $\mathbf{s}^h$ is known, γ^h of Equation 4.152 and the households or consumers commodity demand and factor supply can be computed. The share distribution $\mathbf{s}^h$ is presumably known and the profit vector $\mathbf{\Pi}$ can be determined from the solution of Equation 4.147 for $\mathbf{\Pi}^f$ for $f = 1, 2, \ldots, F$. For the consumer and the producers to come together in the factor and commodity markets, the market clearing equations of Table 4.3 must hold. Table 4.4 summarizes the result of this example. Even though we have obtained explicit closed-form expressions for the supply and demand functions, we cannot simply equate them to obtain explicit analytic equilibrium wages, prices, and input factors from each household and to each firm, and the

TABLE 4.4. General Equilibrium Equations for Example 4.8

Quantity	Expression
Production function of firm f	$[(\mathbf{b}^f)^T\mathbf{q}^f + 0.5(\mathbf{q}^f)^T\mathbf{C}^f\mathbf{q}^f]^2 = 0.5(\mathbf{x}^f)^T\mathbf{A}^f\mathbf{x}^f$
Profit of firm f	$\Pi^f = \mathbf{p}^T\mathbf{q}^f - \mathbf{w}^T\mathbf{x}^f$
Utility of consumer h	$U^h(\mathbf{y}^h,\mathbf{r}^h) = (\mathbf{g}^h)^T\mathbf{y}^h - 0.5(\mathbf{y}^h)^T\mathbf{D}^h\mathbf{y}^h - 0.5(\mathbf{r}^h)^T\mathbf{E}^h\mathbf{r}^h$
Budget of consumer h	$\mathbf{w}^T\mathbf{r}^h + \mathbf{\Pi}^T\mathbf{s}^h = \mathbf{p}^T\mathbf{y}^h$
Definitions	$\mathbf{q}^f$: production of firm f $\mathbf{x}^f$: input factor demand by firm f $\mathbf{y}^h$: commodity demand by consumer h $\mathbf{r}^h$: factor supply by consumer h $\mathbf{\Pi}$: firm profit vector = $(\Pi^1, \Pi^2, \ldots, \Pi^F)^T$ $\mathbf{s}^h$: firm fractional ownership by household h
Optimum production or commodity supply of firm f	$\mathbf{q}^f = (\mathbf{C}^f)^{-1}\left(\dfrac{\mathbf{p}}{[2\mathbf{w}^T(\mathbf{A}^f)^{-1}\mathbf{w}]^{1/2}} - \mathbf{b}^f\right)$
Optimum input factor demand by firm f	$\mathbf{x}^f = \dfrac{(\mathbf{A}^f)^{-1}\mathbf{w}}{[2\mathbf{w}^T(\mathbf{A}^f)^{-1}\mathbf{w}]^{1/2}}\left[\dfrac{\mathbf{p}^T(\mathbf{C}^f)^{-1}\mathbf{p}}{2\mathbf{w}^T(\mathbf{A}^f)^{-1}\mathbf{w}} - (\mathbf{b}^f)^T(\mathbf{C}^f)^{-1}\mathbf{b}^f\right]$
Maximum profit of firm f	$\Pi^f = 0.5\dfrac{(\mathbf{C}^f)^{-1}\{\mathbf{p} - [2\mathbf{w}^T(\mathbf{A}^f)^{-1}\mathbf{w}]^{1/2}\mathbf{b}^f\}}{[2\mathbf{w}^T(\mathbf{A}^f)^{-1}\mathbf{w}]^{1/2}}\{\mathbf{p}^T - [2\mathbf{w}^T(\mathbf{A}^f)^{-1}\mathbf{w}]^{1/2}(\mathbf{b}^f)^T\}$
Optimum demand of consumer h	$\mathbf{y}^h = (\mathbf{D}^h)^{-1}\left\{\mathbf{g}^h - \dfrac{[\mathbf{p}^T(\mathbf{D}^h)^{-1}\mathbf{g}^h - \mathbf{\Pi}^T\mathbf{s}^h]\mathbf{p}}{\mathbf{p}^T(\mathbf{D}^h)^{-1}\mathbf{p} + \mathbf{w}^T(\mathbf{E}^h)^{-1}\mathbf{w}}\right\}$
Optimum factor supply of consumer h	$\mathbf{r}^h = \dfrac{(\mathbf{E}^h)^{-1}\mathbf{w}[\mathbf{p}^T(\mathbf{D}^h)^{-1}\mathbf{g}^h - \mathbf{\Pi}^T\mathbf{s}^h]}{\mathbf{p}^T(\mathbf{D}^h)^{-1}\mathbf{p} + \mathbf{w}^T(\mathbf{E}^h)^{-1}\mathbf{w}}$
Factor market clearing	$\sum_{f=1}^{F}\mathbf{x}^f = \sum_{h=1}^{H}\mathbf{r}^h = \mathbf{R}$
Commodity market clearing	$\sum_{h=1}^{H}\mathbf{q}^f = \sum_{h=1}^{H}\mathbf{y}^h = \mathbf{Q}$

production of each firm and the commodity bundle of each consumer. We will now proceed to examine several cases that will add further insight into the solution obtained here. ■

Example 4.9: The simplest possible specific case in applying the results of Example 4.8 occurs when there is a single firm supplying a single product and a single consumer supplying a single factor. Alternately, we could consider F identical firms producing the same one product from a single factor input, and H identical consumers each demanding a single product, the same product, and each supplying the same factor input. We assume further that $b = 0$, such that the equilibrium requirements of Table 4.4 become

$$q = \left(\frac{a^{1/2}}{2^{1/2}c}\right)\frac{p}{w} \tag{4.153}$$

$$x = \left(\frac{a^{1/2}}{2^{3/2}c}\right)\frac{p^2}{w^2} \tag{4.154}$$

$$\Pi = \left(\frac{a^{1/2}}{2^{3/2}c}\right)\frac{p^2}{w} \tag{4.155}$$

$$y = \frac{gw^2 + \Pi pe}{p^2e + w^2d} \tag{4.156}$$

$$r = \frac{(pg - \Pi d)w}{p^2e + w^2d} \tag{4.157}$$

$$x = r \tag{4.158}$$

$$q = y \tag{4.159}$$

We combine Equations 4.153, 4.155, and 4.156 to obtain

$$\frac{a^{1/2}e}{2^{3/2}c}\frac{p^3}{w} + \frac{a^{1/2}d}{2^{1/2}c}pw - gw^2 = 0$$

Combination of Equations 4.154, 4.155, and 4.157 results in

$$\left(\frac{a^{1/2}e}{2^{3/2}c}\right)\frac{p^4}{w^2} + \left(\frac{a^{1/2}d}{2^{1/2}c}\right)p^2 - gpw = 0$$

and we see that these two relations are not independent. We recall our previous discussions concerning this issue and realize that we are completely free to pick any single price or wage quantity as the numeraire. If we let $w = 1$, we then have from the foregoing equation:

$$\left(\frac{a^{1/2}e}{2^{3/2}c}\right)p^3 + \left(\frac{a^{1/2}d}{2^{1/2}c}\right)p - g = 0 \tag{4.160}$$

This is a relation whose solution yields the equilibrium price p. Unfortunately, the general solution to this equation is quite cumbersome. Thus we consider the specific case where $a = c = d = g = 1$ such that Equation 4.160 becomes

$$ep^3 + 2p - 2^{3/2} = 0 \tag{4.161}$$

The specific problem being considered here reduces to the long-term problem of the firm,

$$q = 2^{1/4} \quad x^{1/2} = 0$$

$$\Pi = pq - wx$$

and the consumer problem,

$$U(y, r) = y - 0.5y^2 - 0.5r^2e$$

$$wr + \Pi = py$$

Now, we have as the optimum equilibrium solution, in terms of the unspecified parameter e, with the numeraire $w = 1$,

$$y = q = 2^{-1/2}p$$

$$r = x = 2^{-3/2}p^2$$

$$\Pi = 2^{3/2}p^2$$

$$U(y, r) = 2^{-1/2}p - 2^{-2}p^2 - 2^{-4}p^4e$$

where p is obtained from Equation 4.161 for specific values of e.

Table 4.5 presents relevant results for this example. As e approaches zero, the consumer has no aversion to labor and will supply whatever labor results in maximum consumer utility. For $e = 0$, we obtain $p = 2^{1/2}$ from Equation 4.161 and also $y = q = 1.0$, $\Pi = r = x = 2^{-1/2}$, and $U = 2^{-1}$. The consumer utility is just that which we would obtain for no budget constraint at all, since the consumer has no aversion to labor. The firm will purchase its best factor input to maximize profit. As e becomes infinite and the consumer demonstrates great aversion to labor, the price of goods drops to zero, but so do the demand and profit and we obtain what most would regard as a very backward economy. ■

TABLE 4.5. Price Versus Consumer Aversion to Labor

E	p
0	1.414
0.2	1.229
1	0.966
2	0.834
5	0.669
10	0.556
50	0.349
100	0.283
∞	0

Example 4.10: The previous example is somewhat unrealistic in that not all consumers will own the same percentage of shares in a firm and not all will have the same inclination toward labor. We again assume that there is a single firm producing a single product from a single factor input, but assume that there are four households to supply this labor and purchase the product of the firm. We assume $b = 0$, $a = c = d = g = 1$, and that household 1 owns 50% of the firm and has infinite aversion to labor ($e^1 = \infty$), household 2 owns 50% of the firm and has zero aversion to labor, household 3 owns none of the firm and has infinite aversion to labor, and household 4 owns none of the firm and has an aversion to labor denoted by e.

For this example, the equilibrium requirements of Table 4.4 become

$$q = \frac{p}{2^{1/2}w}$$

$$x = \frac{p^2}{2^{3/2}w^2}$$

$$\Pi = \frac{p^2}{2^{3/2}w}$$

$$y^1 = \frac{0.5\Pi}{p}, \qquad r^1 = 0$$

$$y^2 = 1, \qquad r^2 = \frac{p - 0.5\Pi}{w}$$

$$y^3 = 0, \qquad r^3 = 0$$

$$y^4 = \frac{w^2}{p^2e + w^2}, \qquad r^4 = \frac{pw}{p^2e + w^2}$$

$$x = \sum_{h=1}^{4} r^h = r^1 + r^2 + r^3 + r^4 = R$$

$$Q = \sum_{h=1}^{4} y^h = y^1 + y^2 + y^3 + y^4$$

Each of the four households or consumers functions just as they should according to their individual utility functions and ownership participation in the profit of the firm. Consumer 1 has infinite aversion to labor and naturally will not supply any factor inputs. This consumer will purchase as many products as 50% ownership of the profits of the firm will purchase. Household 2 has no aversion to labor and so will work as much as necessary to purchase the optimum commodity bundle ($y^2 = 1$) with no budget constraints. Part of this purchase price (0.5Π) will come from ownership of the firm. The rest of the purchase price for one unit commodity ($p - 0.5\Pi$) will come from input factor supply at wage w. Household 3 owns no part of the firm and has infinite aversion to labor. Thus the income to this household is zero and the household is effectively inert. Household 4 owns none of the firm and has an aversion coefficient e.

We are free to pick one numeraire and we pick unit wages $w = 1$. The commodity market equilibrium equation is given by

$$0.265ep^3 - 0.5ep^2 + 0.265p - 1 = 0$$

As a check on the possibility of errors, we can obtain the factor market clearing equation, although, because of the dependence of one of the equilibrium equations, this will result in no new information. Solution of the foregoing equation for specific values of e determines the price p of goods in the commodity market. Table 4.6 illustrates the behavior of the firm and the households as a function of the labor aversion factor of the fourth household. From the foregoing equation, we see that we have $p = 3.77$ for $e = 0$. This leads to a producer supply of $q = 2.67$ products made from $x = 5.03$ input factors to the firm. The firm earns a profit of $\Pi = 5.03$ units. The product of the firm is distributed to the households as $y^1 = 0.67$ units, $y^2 = 1$, $y^3 = 0$, and $y^4 = 1$.

TABLE 4.6. Price as a Function of Aversion to Labor of Household 4

e	p
0	3.774
0.2	2.667
1	2.209
2	2.083
5	1.980
10	1.936
50	1.897
100	1.893

The input factors supplied by the consumers are $r^1 = 0$, $r^2 = 1.26$, $r^3 = 0$, and $r^4 = 3.77$. The utilities of the consumers are $U^1 = 0.45$, $U^2 = 0.50$, $U^3 = 0$, and $U^4 = 0.50$.

When household 4 has infinite aversion to labor, $e = \infty$, the equilibrium price becomes, from the foregoing equation, $p = 1.89$. This lower price results in a lowered production of $q = 1.33$ units from $x = 1.26$ input factors, and a profit of 1.26 monetary units. The 1.33 commodities are subject to consumer demands of $y^1 = 0.33$, $y^2 = 1$, $y^3 = 0$, and $y^4 = 0$. All consumers supply zero input factors, except for consumer 2, who supplies the full amount of 1.26 factors. The utilities of the consumers are $U^1 = 0.11$, $U^2 = 0.5$, $U^3 = 0$, and $U^4 = 0$. As consumer 4 increases its labor aversion factor, not only does the utility of this consumer decrease, but so does the utility of consumer 1, who depends only on the profits of the firm for utility satisfaction. Consumer 2 is willing to work whatever amount is needed to obtain utility satisfaction. However, it turns out that consumer 2 must supply precisely the same input factor as when consumer 4 had zero aversion to labor. Although prices have decreased, the decreased profits of firms have also decreased the shareholder earnings of household 2. ■

4.4 EXTENSIONS TO GENERAL EQUILIBRIUM TO INCLUDE MARKET INTERDEPENDENCIES

A sometimes unrealistic feature of our efforts in Section 4.3 is that we assumed that firms use only factor inputs from consumers and that consumers use only the products of the firms. More often than not, a firm will use inputs not only from the factor market but also from the consumer market. For example, corn production depends not only on labor but also on corn seed and fertilizer. Each of these is sold on a commodity market. In fact, corn seed will be sold by the firms that produce corn. Similarly, consumers desire not only the goods or products sold on the commodity market but also the services sold on the factor market, and there exist items such as electricity that cannot be classified entirely as a commodity (good) or factor (service).

Our efforts in Section 4.3 can easily be extended to this more general case. The problem summarized in Table 4.2 is applicable, with just a few modifications, to the case where firms and consumers purchase both factors and commodities. The product output of firm f is still an n-vector $\mathbf{q}^f$. The factor output of consumer h is still an m-vector denoted by $\mathbf{r}^h$, but the factor input to the production of firm f, $\mathbf{x}^f$, should now be an $(n+m)$-vector, since it contains both the m-vector input factor $\mathbf{R}^f = \mathbf{C}^f\mathbf{R}$ and the n-vector commodity input $\mathbf{Q}^f$, where $\mathbf{Q}^f$ is a fraction $\mathbf{A}^f\mathbf{Q}$ of the total commodity production of all firms. Similarly, the items demanded by consumers $\mathbf{y}^h$ should now be an $(n+m)$-vector, since these items contain commodities $\mathbf{Q}^h = \mathbf{D}^h\mathbf{Q}$ as well as services or factors $\mathbf{R}^h$, where $\mathbf{R}^h$ is a fraction $\mathbf{B}^h\mathbf{R}$ of the total factor supply of all consumers. Here the matrices $\mathbf{A}^f$, $\mathbf{B}^h$, $\mathbf{C}^f$, and $\mathbf{D}^h$ may be functions of supply and demand. We will later consider the case where these are constant matrices.

The market clearing equations must now be modified to include the fact that products and services flow to firms and consumers. Thus we replace the market clearing equations of Section 4.3 with a commodity market clearing equation that indicates that the sum of all commodities produced by all firms $\mathbf{Q}$ must be equal to the commodities delivered to consumers, $\sum_{h=1}^{H} \mathbf{Q}^h$, plus the quantities consumed by the firms, $\sum_{f=1}^{F} \mathbf{Q}^f$. We also use a factor market clearing equation that indicates that the sum of all factors supplied by consumers, $\mathbf{R}$, must be equal to all factors purchased by consumers, $\sum_{h=1}^{H} \mathbf{R}^h$, plus the sum of all the factors purchased by the firms, $\sum_{f=1}^{F} \mathbf{R}^f$.

As in Section 4.3, we obtain the equilibrium or optimality conditions for each of the firms and the equilibrium conditions for each of the consumers. Then we combine these by use of the market clearing equations to obtain general supply–demand market equilibrium conditions. Tables 4.7–4.9 present the necessary equilibrium conditions for producer optimality, consumer optimality, and general market equilibrium. As in Section 4.3, the firm may make a profit in the short run.

As we should expect, in the long run new firms will enter the market and ultimately the profit will be reduced to zero.

TABLE 4.7. Necessary Conditions for Optimality of Firm *f*

Relation	Equation
Optimum *production*: *supply* quantity of firm f	$\lambda^f \dfrac{\partial \varphi^f(\mathbf{q}^f, \mathbf{R}^f, \mathbf{Q}^f)}{\partial \mathbf{q}^f} = -\mathbf{p}$
Optimum input *factor*: *demand* quantity by firm f	$\lambda^f \dfrac{\partial \varphi^f(\mathbf{q}^f, \mathbf{R}^f, \mathbf{Q}^f)}{\partial \mathbf{R}^f} = \mathbf{w}$
Optimum *input*: *commodity* demand by firm f	$\lambda^f \dfrac{\partial \varphi^f(\mathbf{q}^f, \mathbf{R}^f, \mathbf{Q}^f)}{\partial \mathbf{Q}^f} = \mathbf{p}$
Production constraint	$\varphi^f(\mathbf{q}^f, \mathbf{R}^f, \mathbf{Q}^f) = 0$

TABLE 4.8. Necessary Conditions for Optimality of Consumer or Household *h*

Relation	Equation
Optimum *factor demand* by consumer h	$\dfrac{\partial U^h(\mathbf{R}^h, \mathbf{Q}^h, \mathbf{r}^h)}{\partial \mathbf{R}^h} = \gamma^h \mathbf{w}$
Optimum *commodity demand* by consumer h	$\dfrac{\partial U^h(\mathbf{P}^h, \mathbf{Q}^h, \mathbf{r}^h)}{\partial \mathbf{Q}^h} = \gamma^h \mathbf{p}$
Optimum *factor supply* by consumer h	$\dfrac{\partial U^h(\mathbf{R}^h, \mathbf{Q}^h, \mathbf{r}^h)}{\partial \mathbf{r}^h} = \gamma^h \mathbf{w}$
Budget constraint of consumer h	$\mathbf{p}^T \mathbf{Q}^h + \mathbf{w}^T \mathbf{R}^h = \mathbf{w}^T \mathbf{r}^h + \mathbf{\Pi}^T \mathbf{s}^h$

TABLE 4.9. Market Clearing Equations for General Equilibrium with Market Interdependencies

Relation	Equation
Factor market clearing	$\mathbf{R} = \sum_{h=1}^{H} \mathbf{r}^h = \sum_{h=1}^{H} \mathbf{R}^h + \sum_{f=1}^{F} \mathbf{R}^f$
Commodity market clearing	$\mathbf{Q} = \sum_{f=1}^{F} \mathbf{q}^f = \sum_{f=1}^{F} \mathbf{Q}^f + \sum_{h=1}^{H} \mathbf{Q}^h$

Example 4.11: We consider the following production function:

$$0.25(\mathbf{q}^{\mathrm{T}}\mathbf{A}\mathbf{q})^2 = 0.5\mathbf{R}^{\mathrm{T}}\mathbf{B}\mathbf{R} + 0.5\mathbf{Q}^{\mathrm{T}}\mathbf{C}\mathbf{Q} \tag{4.162}$$

where **A**, **B**, and **C** are symmetric positive definite. We will momentarily drop the f superscript. Table 4.7 leads us to the following necessary conditions for optimality of the firm:

$$\lambda(\mathbf{q}^{\mathrm{T}}\mathbf{A}\mathbf{q})\mathbf{A}\mathbf{q} = -\mathbf{p} \tag{4.163}$$

$$\lambda\mathbf{B}\mathbf{R} = -\mathbf{w} \tag{4.164}$$

$$\lambda\mathbf{C}\mathbf{Q} = -\mathbf{p} \tag{4.165}$$

Solution of the last two equations for **R** and **Q** and substitution of the results into Equation 4.162 yields the relation

$$0.25(\mathbf{q}^{\mathrm{T}}\mathbf{A}\mathbf{q})^2 = 0.5(\mathbf{w}^{\mathrm{T}}\mathbf{B}^{-1}\mathbf{w} + \mathbf{p}^{\mathrm{T}}\mathbf{C}^{-1}\mathbf{p})\lambda^{-2}$$

So, we have as the Lagrange multiplier

$$\lambda = \frac{2^{1/2}(\mathbf{w}^{\mathrm{T}}\mathbf{B}^{-1}\mathbf{w} + \mathbf{p}^{\mathrm{T}}\mathbf{C}^{-1}\mathbf{p})^{1/2}}{\mathbf{q}^{\mathrm{T}}\mathbf{A}\mathbf{q}}$$

We have, as the optimum production equilibrium quantities for the firm, the total production

$$\mathbf{q}^f = \frac{\mathbf{A}^{-1}\mathbf{p}}{2^{1/2}(\mathbf{w}^{\mathrm{T}}\mathbf{B}^{-1}\mathbf{w} + \mathbf{p}^{\mathrm{T}}\mathbf{C}^{-1}\mathbf{p})^{1/2}}$$

the factor input to production

$$\mathbf{R}^f = \frac{(\mathbf{p}^{\mathrm{T}}\mathbf{A}^{-1}\mathbf{p})\mathbf{B}^{-1}\mathbf{w}}{2^{3/2}(\mathbf{w}^{\mathrm{T}}\mathbf{B}^{-1}\mathbf{w} + \mathbf{p}^{\mathrm{T}}\mathbf{C}^{-1}\mathbf{p})^{3/2}}$$

and the production input to production

$$\mathbf{Q}^f = \frac{(\mathbf{p}^{\mathrm{T}}\mathbf{A}^{-1}\mathbf{p})\mathbf{C}^{-1}\mathbf{p}}{2^{3/2}(\mathbf{w}^{\mathrm{T}}\mathbf{B}^{-1}\mathbf{w} + \mathbf{p}^{\mathrm{T}}\mathbf{C}^{-1}\mathbf{p})^{3/2}}$$

Only a portion of the output of the firm, the amount $\mathbf{q}^f - \mathbf{Q}^f$, is delivered to the consumer. The amount is used to supply the firm itself. The profit of the firm is

$$\Pi^f = \frac{\mathbf{p}^{\mathrm{T}}\mathbf{A}^{-1}\mathbf{p}}{2^{3/2}(\mathbf{w}^{\mathrm{T}}\mathbf{B}^{-1}\mathbf{w} + \mathbf{p}^{\mathrm{T}}\mathbf{C}^{-1}\mathbf{p})^{1/2}}$$

■

Example 4.12: We now consider a consumer utility maximization example in which the consumer utility is

$$U(\mathbf{R}^h, \mathbf{Q}^h, \mathbf{r}^h) = \mathbf{d}^{\mathrm{T}}\mathbf{R}^h - 0.5(\mathbf{R}^h)^{\mathrm{T}}\mathbf{E}\mathbf{R}^h + \mathbf{g}^{\mathrm{T}}\mathbf{Q}^h - 0.5(\mathbf{Q}^h)^{\mathrm{T}}\mathbf{J}\mathbf{Q}^h - 0.5(\mathbf{r}^h)^{\mathrm{T}}\mathbf{K}\mathbf{r}^h \tag{4.166}$$

where the consumer budget is given by

$$\mathbf{p}^{\mathrm{T}}\mathbf{Q}^h + \mathbf{w}^{\mathrm{T}}\mathbf{R}^h = \mathbf{w}^{\mathrm{T}}\mathbf{r}^h + \mathbf{\Pi}^{\mathrm{T}}\mathbf{s}^h \tag{4.167}$$

We assume $\mathbf{d} > \mathbf{0}$, $\mathbf{g} > \mathbf{0}$, and that $\mathbf{E}$, $\mathbf{J}$, and $\mathbf{K}$ are symmetric positive definite matrices.

It is straightforward to show that the optimum values for the commodity demand bundle $\mathbf{Q}^h$, the factor demand bundle $\mathbf{R}^h$, and the factor supply $\mathbf{r}^h$ are

$$\mathbf{R}^h = \mathbf{E}^{-1}(\mathbf{d} - \mathbf{w}\gamma^h) \tag{4.168}$$

$$\mathbf{Q}^h = \mathbf{J}^{-1}(\mathbf{g} - \mathbf{p}\gamma^h) \tag{4.169}$$

$$\mathbf{r}^h = \mathbf{K}^{-1}\mathbf{w}\gamma^h \tag{4.170}$$

where

$$\gamma^h = \frac{\mathbf{P}^{\mathrm{T}}\mathbf{J}^{-1}\mathbf{g} + \mathbf{w}^{\mathrm{T}}\mathbf{E}^{-1}\mathbf{d} - \mathbf{\Pi}^{\mathrm{T}}\mathbf{s}^h}{\mathbf{P}^{\mathrm{T}}\mathbf{J}^{-1}\mathbf{p} + \mathbf{w}^{\mathrm{T}}(\mathbf{E}^{-1} + \mathbf{K}^{-1})\mathbf{w}} \tag{4.171}$$

If there were no budget constraint, the consumer would demand $\mathbf{R}^h$, $\mathbf{Q}^h$, and $\mathbf{r}^h$ as given by Equations 4.168–4.170 with $\gamma^h = 0$. We require $\gamma^h \geq 0$ in order that $r^h \geq 0$. Table 4.8 displays these necessary conditions for consumer equilibrium. ■

Example 4.13: To illustrate market equilibrium conditions, we assume that there is a single homogeneous consumer group and a single homogeneous firm that produces all products. The firm and the consumer will be assumed to behave as in Examples 4.11 and 4.12. The market clearing equations of Table 4.9 becomes

$$\mathbf{R} = \mathbf{r}^h = \mathbf{R}^h + \mathbf{R}^f$$

$$\mathbf{Q} = \mathbf{q}^f = \mathbf{Q}^f + \mathbf{Q}^h$$

These equations and those of the previous two examples require solution to determine market equilibrium. Unfortunately, the resulting equations are very nonlinear in the price and wage vectors, which, in turn, determine the demand and *supply quantities for both* the factor and the commodity market. These equations have no general solution. For specific values of the various parameter vectors and matrices, we may determine numerical solutions to our problem. Insight into the solution can be obtained by determining the sensitivity of price, wage, and quantities in the commodity and factor markets to changes in these parameters. ■

A particular case in which the general equilibrium equations are quite tractable occurs when we have an economy with *linear input–output technology* such that all production functions are linear in input factors and input commodities to production. We have previously considered a simple linear input–output technology production function in Examples 2.1 and 2.2. The general case will have n commodity inputs to production and m factor inputs. We assume that the production of commodity i by firm f is given by

$$q_i^f = \min\left(\frac{Q_{1i}^f}{a_{1i}^f}, \frac{Q_{2i}^f}{a_{2i}^f}, \ldots, \frac{Q_{ni}^f}{a_{ni}^f}, \frac{R_{1i}^f}{b_{1i}^f}, \frac{R_{2i}^f}{b_{2i}^f}, \ldots, \frac{R_{mi}^f}{b_{mi}^f}\right) \tag{4.172}$$

Here q_i^f represents the production of the ith commodity by firm f in terms of the several inputs to production. Parameters a_{ji}^f and b_{ki}^f are the *production coefficients* that represent the (nonnegative) amounts of the jth commodity and the kth factor input needed to produce one unit amount of the ith commodity. The terms Q_{ji}^f and R_{ki}^f are the jth commodity and kth factor input to production of the ith commodity by firm f.

As we have indicated in Example 2.1, economic operation of the firm results in use of commodity and factor inputs such that all terms of the form $a_{ij}^{-1} Q_i^f$ and $b_{kj}^{-1} R_k^f$ in Equation 4.172 are equal. Thus we have the following for productive firm f, for $i, j = 1, 2, \ldots, n$ and $k = 1, 2, \ldots, m$:

$$Q_{ji}^f = a_{ji}^f q_i^f \tag{4.173}$$

$$R_{ki}^f = b_{ki}^f q_i^f \tag{4.174}$$

Thus we see that, as in Examples 2.1 and 2.2, the input commodities and factors needed to produce commodity q_i^f are linearly proportional to the output quantity produced.

The market clearing equations of Table 4.9 furnish the required conditions for economic equilibrium. For the lth produced commodity, the commodity market clearing equation of Table 4.9 is

$$Q_l = \sum_{f=1}^{F} q_l^f = \sum_{f=1}^{F} \sum_{j=1}^{n} Q_{lj}^f + \sum_{h=1}^{H} Q_l^h \tag{4.175}$$

It is convenient to define the *consumer aggregate commodity demand* as

$$\mathbf{C} = \sum_{h=1}^{H} \mathbf{Q}^h \tag{4.176}$$

such that, using Equations 4.173 and 4.176, Equation 4.175 becomes

$$Q_l = \sum_{f=1}^{F} q_l^f = \sum_{f=1}^{F} \sum_{j=1}^{n} a_{lj}^f q_j^f + C_l \tag{4.177}$$

The first term on the right-hand side of this equation can be written as

$$\sum_{f=1}^{F} \sum_{j=1}^{n} a_{lj}^f q_j^f = \sum_{j=1}^{n} A_{lj} Q_j \tag{4.178}$$

where

$$A_{lj} = \frac{\sum_{f=1}^{F} a_{lj}^f q_j^f}{\sum_{f=1}^{F} q_j^f} \tag{4.179}$$

Generally A_{lj} will be a function of the production levels of each firm. However, when the firms producing commodity l are homogeneous in the sense that a_{lj}^f, the production coefficient representing the amount of the lth commodity needed to produce one unit amount of commodity j, is the same for all firms, A_{lj} will not be a function of the production level. In most realistic cases, this will turn out to be a very useful as well as valid assumption. If a commodity can be made by two production processes with vastly different inputs to production, such a commodity could and should be considered as two commodities for the purpose of linear input–output production analysis. Butter and margarine represent examples of this, in that they could be considered as one commodity to serve a single end, but they are best considered as two commodities, owing to the very different inputs to production of these two commodities.

We may rewrite the linear input production equation in terms of the commodity market equilibrium equation (4.177) by using Equations 4.178 and 4.179:

$$Q_l = \sum_{j=1}^{n} A_{lj} Q_j + C_l \tag{4.180}$$

This equation states that the total amount of commodity l produced by all firms, Q_l, equals the total consumer demand for the lth commodity, C_i, plus the amount of commodity l needed as input to all production processes. We have assumed a special form of technology, such that the amount of commodity l needed for production process j is a linear function, $A_{lj}Q_j$, of the quantity of the jth produced commodity.

In vector form this equation becomes

$$\mathbf{Q} = \mathbf{AQ} + \mathbf{C} \tag{4.181}$$

and is known as the *Leontief input–output equation.* Here $\mathbf{Q}$ is the vector of total production of all firms in the economy, $\mathbf{A}$ is the *input–output commodity production coefficient matrix*, and $\mathbf{C}$ is the vector of consumer or *final demand.*

The total factor input of factor l to production can be obtained by summation of Equation 4.174 over all firms and output commodities:

$$\delta_l = \sum_{f=1}^{F} R_l^f = \sum_{f=1}^{F}\sum_{i=1}^{n} R_{li}^f = \sum_{f=1}^{F}\sum_{i=1}^{n} b_{li}^f q_i^f \tag{4.182}$$

as in Equation 4.178. Also, we let

$$\sum_{f=1}^{F}\sum_{i=1}^{n} b_{li}^f q_i^f = \sum_{i=1}^{n} B_{li} Q_i \tag{4.183}$$

where

$$B_{li} = \frac{\sum_{f=1}^{F} b_{li}^f q_i^f}{\sum_{f=1}^{F} q_i^f} \tag{4.184}$$

Generally B_{li} will be a function of the production levels of each firm. For nearly homogeneous firms we see that, as in the case with the commodity production coefficient A_{li}, B_{li} will not be a function of production level.

The factor market clearing equation of Table 4.9 becomes, for factor l,

$$R_l = \sum_{h=1}^{H} r_l^h = \sum_{f=1}^{F} R_l^f + \sum_{h=1}^{H} R_l^h \tag{4.185}$$

We denote the aggregate consumer demand for factor input l as

$$\Re_l = \sum_{h=1}^{H} R_l^h \tag{4.186}$$

and substitute this relation and Equations 4.182 and 4.183 in Equation 4.185 to obtain

$$R_l = \delta_l + \Re_l = \sum_{i=1}^{n} B_{li} Q_i + \Re_l \tag{4.187}$$

This equation states that the total supplied amount of factor l, R_l, is equal to the aggregate consumer demand for factor input l, $\Re_l$, plus the total demand of all firms for factor input l, δ_l. For the assumed linear input–output production technology, this is a linear function, $\sum_{i=1}^{n} B_{li}Q_i$, of the produced commodities.

In the vector case, Equation 4.187 is written as

$$\mathbf{R} = \boldsymbol{\delta} + \boldsymbol{\Re} = \mathbf{BQ} + \boldsymbol{\Re} \tag{4.188}$$

where $\mathbf{R}$ represents the total vector amount of factor demand, $\mathbf{B}$ is the factor production coefficient matrix, $\boldsymbol{\Re}$ is the total factor demand by consumers, and $\boldsymbol{\delta} = \mathbf{BQ}$ is the factor input to production supplied by consumers.

As a summary of this effort, we can state that for a linear input–output technology the aggregate production of commodities and factors can be described by the Leontief equations:

$$\mathbf{Q} = \mathbf{AQ} + \mathbf{C} \tag{4.181}$$

$$\mathbf{R} = \mathbf{BQ} + \boldsymbol{\Re} \tag{4.188}$$

where $\mathbf{A}$ and $\mathbf{B}$ are production coefficient matrices for commodities and factors, $\mathbf{C}$ is the consumer commodity demand vector, $\boldsymbol{\Re}$ is the consumer factor demand vector, $\mathbf{Q}$ is the commodity production vector or total demand vector, and $\mathbf{R}$ is the total factor supply vector.

It is important to note that the two foregoing equations simply describe the productive firm with linear input–output production technology. They do not represent optimum behavior of either the firm in maximizing profits or the consumer in maximizing utility. We again assume a price vector $\mathbf{p}$ and wage vector $\mathbf{w}$. The revenue to the firms, that is, the total revenue to all firms, is

$$\tau = \mathbf{p}^{\mathrm{T}}\mathbf{Q} \tag{4.189}$$

The cost to the firms of the input factors to production, $\mathbf{BQ}$, is $\mathbf{w}^{\mathrm{T}}\mathbf{BQ}$. The cost to the firm of input commodities to production is $\mathbf{p}^{\mathrm{T}}\mathbf{AQ}$. The total production cost is the sum of these two costs, and the profit of the firm is the difference between revenue and production costs. Thus we have for the profit of all firms

$$\Pi = \mathbf{p}^{\mathrm{T}}\mathbf{Q} - \mathbf{p}^{\mathrm{T}}\mathbf{AQ} - \mathbf{w}^{\mathrm{T}}\mathbf{BQ} = [\mathbf{p}^{\mathrm{T}}(\mathbf{I} - \mathbf{A}) - \mathbf{w}^{\mathrm{T}}\mathbf{B}]\mathbf{Q} = \mathbf{p}^{\mathrm{T}}\mathbf{C} - \mathbf{w}^{\mathrm{T}}\mathbf{BQ} \tag{4.190}$$

A useful interpretation can be given to the vector, where $\mathbf{I}$ is the identity matrix of dimension $n \times n$,

$$\mathbf{M\Pi}(\mathbf{Q}) = (\mathbf{I} - \mathbf{A}^{\mathrm{T}})\mathbf{p} - \mathbf{B}^{\mathrm{T}}\mathbf{w} = \frac{\partial \Pi}{\partial \mathbf{Q}} \tag{4.191}$$

in that it is the marginal profit vector for the economy or the partial derivative of the profit with respect to production. This quantity represents the vector profit for a unit vector production, $\mathbf{Q} = \mathbf{1}$. The marginal and average production cost vector is just $\mathbf{A}^{\mathrm{T}}\mathbf{p} + \mathbf{B}^{\mathrm{T}}\mathbf{w}$. The marginal and average profits $\mathbf{A\Pi}$ $(\mathbf{Q})$ are equal, where we define $\mathbf{A\Pi}_i(\mathbf{Q}) = Q_i^{-1}\Pi(Q_i, \mathbf{Q}\bar{i} = \mathbf{0})$.

It is not meaningful to speak of unconstrained maximization of the (national) product Π given by Equation 4.190. If the marginal profit of Equation 4.191 is positive, profits will increase linearly with increasing $\mathbf{Q}$, and we should, for maximum profit, produce as large an output as possible. Physically there must be constraints on maximum production $\mathbf{Q}_{\max} = \mathbf{c}$ and maximum factor input to productivity $(\mathbf{BQ})_{\max} = d$. So the problem of maximization of the profit of the firm is that of maximizing, by choice of a nonnegative $\mathbf{Q}$, the expression

$$\Pi = [\mathbf{p}^{\mathrm{T}}(\mathbf{I} - \mathbf{A}) - \mathbf{w}^{\mathrm{T}}\mathbf{B}]\mathbf{Q} \tag{4.192}$$

subject to the constraints

$$\mathbf{0} \leq \mathbf{Q} \leq \mathbf{Q}_{\max} = \mathbf{c} \tag{4.193}$$

$$\mathbf{BQ} \leq (\mathbf{BQ})_{\max} = \mathbf{d} \tag{4.194}$$

This is a problem in linear programming, and a variety of techniques are available for its solution. Often it turns out that the optimal solution that is obtained without using Equation 4.193 will satisfy $\mathbf{Q} < \mathbf{c}$. Alternately, we can incorporate Equation 4.193 into Equation 4.194. We will not consider Equation 4.193 further in our development and simply assume that it is satisfied.

Two other problem formulations are equivalent to this firm profit maximization formulation. The value of the consumer final demand, or gross national product (GNP) Y, is the product of the commodity price vector and the number of commodities sold:

$$Y = \mathbf{p}^{\mathrm{T}}\mathbf{C} \tag{4.195}$$

We may maximize Y, the value of the final demand or GNP, by choice of $\mathbf{C}$ subject to the equality constraint of the Leontief equation and the inequality constraints on commodities and factors:

$$\mathbf{Q} = \mathbf{AQ} + \mathbf{C} \tag{4.181}$$

$$\mathbf{BQ} \leq \mathbf{d} \tag{4.194}$$

Alternately, we can state this problem, by eliminating the consumer demand vector $\mathbf{C}$, as that of maximizing the value of consumer final demand or GNP by choice of $\mathbf{Q}$. Thus we maximize

$$Y = \mathbf{p}^{\mathrm{T}}(\mathbf{I} - \mathbf{A})\mathbf{Q} \tag{4.196}$$

subject to the inequality constraints

$$\mathbf{BQ} \leq \mathbf{d} \tag{4.194}$$

The marginal GNP is an important term. It is known as the *value added* and is given by

$$\mathbf{v} = \mathbf{MY} = \frac{\partial Y}{\partial \mathbf{Q}} = (\mathbf{I} - \mathbf{A}^{\mathrm{T}})\mathbf{p} \tag{4.197}$$

Again, either of these GNP formulations results in a linear programming problem that is solvable by any of the several approaches.

The mathematical dual of this final demand value maximization problem is of interest. In this dual problem, we minimize, by proper choice of the (national) wage vector $\mathbf{w}$, the input factor cost or national income given by

$$\tau = \mathbf{d}^{\mathrm{T}}\mathbf{w} \tag{4.198}$$

subject to the inequality constraints

$$(\mathbf{I} - \mathbf{A}^{\mathrm{T}})\mathbf{p} - \mathbf{B}^{\mathrm{T}}\mathbf{w} \leq \mathbf{0} \tag{4.199}$$

$$\mathbf{w} \geq \mathbf{0} \tag{4.200}$$

This is another linear programming problem.

That these three problems are equivalent can easily be seen by writing the Lagrangian for our problem of profit maximization. We adjoin, using a Lagrange multiplier, the constraint of Equation 4.194 to the cost function of Equation 4.192 and obtain the Lagrangian

$$L = \mathbf{p}^{\mathrm{T}}(\mathbf{I} - \mathbf{A})\mathbf{Q} - \mathbf{w}^{\mathrm{T}}\mathbf{BQ} + \boldsymbol{\lambda}^{\mathrm{T}}(\mathbf{d} - \mathbf{BQ}) \tag{4.201}$$

To obtain the GNP maximization formulation, we add the quantity $\mathbf{w}^{\mathrm{T}}\mathbf{d}$ to the Lagrangian and obtain

$$L' = L + \mathbf{w}^{\mathrm{T}}\mathbf{d} = \mathbf{p}^{\mathrm{T}}(\mathbf{I} - \mathbf{A})\mathbf{Q} + (\boldsymbol{\lambda} + \mathbf{w})^{\mathrm{T}}(\mathbf{d} - \mathbf{BQ}) \tag{4.202}$$

The value of $\mathbf{Q}$ that maximizes Equation 4.202 is surely that value which maximizes Equation 4.201. That is, the value of $\mathbf{Q}$ to maximize Π is also the value that maximizes $\Pi + \mathbf{w}^{\mathrm{T}}\mathbf{d}$. It is convenient to define a new Lagrange multiplier $\boldsymbol{\gamma} = \boldsymbol{\lambda} + \mathbf{w}$ and we have, from the foregoing equation,

$$L' = \mathbf{p}^{\mathrm{T}}(\mathbf{I} - \mathbf{A})\mathbf{Q} + \boldsymbol{\gamma}^{\mathrm{T}}(\mathbf{d} - \mathbf{BQ}) \tag{4.203}$$

The primal problem to which this Lagrangian L' corresponds is just that of maximization of GNP, given by Equation 4.196, subject to the inequality constraint on factor resource availability given by Equation 4.194.

To obtain the dual problem, which is applicable to linear cost and constraint equations only, corresponding to the Lagrangian of Equation 4.203, we interchange the role of the Lagrange multiplier and the dual variable to be optimized. Thus we let $\boldsymbol{\gamma} = \mathbf{w}$ and $\mathbf{Q} = \boldsymbol{\eta}$, where the wage vector $\mathbf{w}$ is the dual variable and the new Lagrange multiplier is $\boldsymbol{\eta}$. We thus have for the dual Lagrangian

$$L'' = \boldsymbol{\eta}^{\mathrm{T}}[(\mathbf{I}-\mathbf{A})^{\mathrm{T}}\mathbf{p} - \mathbf{B}^{\mathrm{T}}\mathbf{w}] + \mathbf{d}^{\mathrm{T}}\mathbf{w} \tag{4.204}$$

Thus we see that the dual problem is indeed that of Equations 4.198 through 4.200.

An interesting interpretation to Equation 4.199 can be given. We know that linear programming solutions generally operate on the boundary of the inequality constraint equation if this equation is of the minimum dimensionality necessary for a meaningful problem. We see that the expression in Equation 4.199 is just the marginal profit, and the average profit given by Equation 4.191 as well. Furthermore, we see from Equation 4.190 that the maximum profit $\Pi_{\max}$ is zero. Thus we see that at optimum, the maximized value of GNP and the minimized value of national income τ are the same, in that

$$Y = \mathbf{p}^{\mathrm{T}}\mathbf{C} = \tau = \mathbf{w}^{\mathrm{T}}\mathbf{d} \tag{4.205}$$

We have previously determined that the profit of firms in competitive equilibrium is zero, and here we come to this important conclusion again. Although we have demonstrated this result for the case where the inequality of Equation 4.193 is routinely satisfied by the optimum solution, it turns out that it is valid in conditions where the inequality commodity resource condition must be invoked to find a solution to the problem.

Example 4.14: We consider a simple two-commodity system with two factor inputs in which we assume

$$\mathbf{p} = \begin{bmatrix} 1.25 \\ 1.25 \end{bmatrix}, \quad \mathbf{A} = \begin{bmatrix} 0.1 & 0.2 \\ 0.1 & 0 \end{bmatrix}, \quad \mathbf{B} = \begin{bmatrix} 4 & 2 \\ 4 & 8 \end{bmatrix}, \quad \mathbf{d} = \begin{bmatrix} 20 \\ 40 \end{bmatrix}$$

To solve the GNP maximization problem of Equation 4.196 subject to the constraint of Equation 4.194, we maximize the expression

$$\gamma = \mathbf{p}^{\mathrm{T}}(\mathbf{I} - \mathbf{A})\mathbf{Q} = Q_1 + Q_2$$

subject to $Q_1 > 0$, $Q_2 > 0$, and $\mathbf{BQ} \leq \mathbf{d}$. This last vector inequality is equivalent to the two scalar inequalities

$$2Q_1 + Q_2 \leq 10$$
$$Q_1 + 2Q_2 \leq 10$$

Figure 4.17 illustrates the geometry for this single linear programming problem and shows that the optimum solution to maximize $\gamma = Q_1 + Q_2$ is $Q_1 = Q_2 = 3.33$. This leads to a GNP of $\gamma = 6.67$ and also to use of the maximum amount of factor resources available for production, $\boldsymbol{\delta} = \mathbf{d} = \mathbf{BQ} = [20 \;\; 40]^{\mathrm{T}}$. This is also the solution to the problem of profit

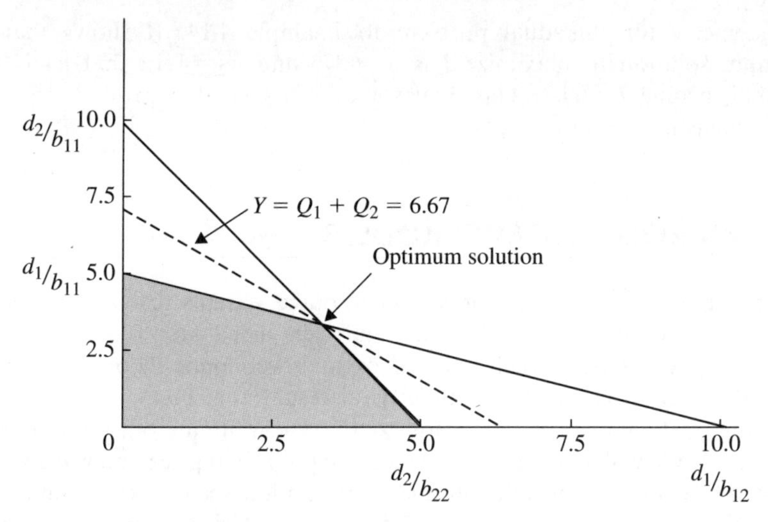

Figure 4.17. Geometry of the Primal Problem.

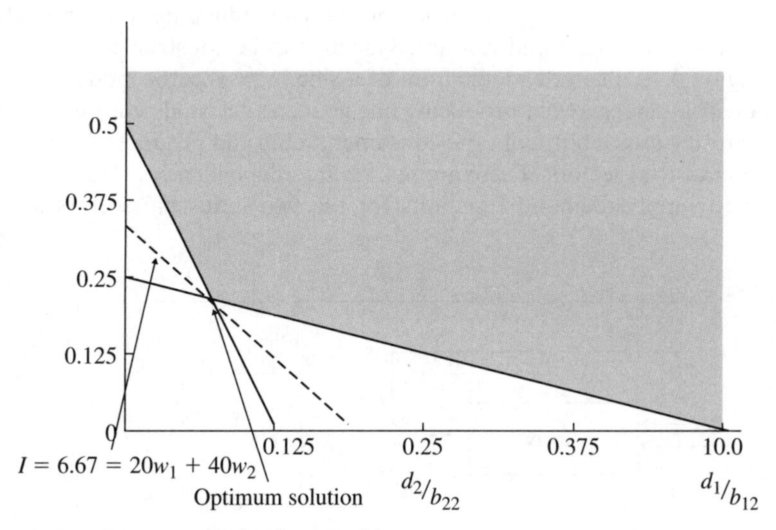

Figure 4.18. Geometry of the Dual Problem.

maximization, represented by Equations 4.192 and 4.194, for any positive wage vector $\mathbf{w}$ subject to the restriction that $(\boldsymbol{I} - \mathbf{A}^{\mathrm{T}})\mathbf{p} - \mathbf{B}^{\mathrm{T}}\mathbf{w} \geq \mathbf{0}$. ■

Example 4.15: The dual problem to Example 4.14 is given by Equations 4.198–4.200. We minimize national income given by

$$I = \mathbf{d}^{\mathrm{T}}\mathbf{w} = 20w_1 + 40w_2$$

subject to $w_1 \geq 0$, $w_2 \geq 0$, and $(\boldsymbol{I} - \mathbf{A}^{\mathrm{T}})\mathbf{p} - \mathbf{B}^{\mathrm{T}}\mathbf{w} \leq \mathbf{0}$. This last inequality becomes $1 - 4w_1 - 4w_2 \leq 0$ and $1 - 2w_1 - 8w_2 \leq 0$. Figure 4.18 illustrates the geometry for this dual problem to Example 4.14. It shows that the optimum solution to maximize I is $w_1 = 1/6$ and $w_2 = 1/12$ that leads to a national income $I = 6.67$. This dual solution is such that profit is zero and GNP equals national income. ■

4.5 MICROECONOMIC MODELS

A number of models of economic and resource systems result from the use of microeconomic theory. We have just discussed the fundamentals of input–output analysis, which is based on microeconomic theory. Our analysis in this section is based on our general results for economic equilibrium, including that analysis which leads to the linear input–output models of Section 4.4. We will assume the existence of equilibrium price and wage vectors **p** and **w**, equilibrium production and consumption vectors for commodities, and equilibrium supply and demand for factors. Using a simple two-sector model of economic equilibrium based on Cobb–Douglas or translog production and utility functions, we will conceptually illustrate how models of complex large economic and resource systems can be constructed.

Figure 4.19 illustrates a diagram of a simple two-sector model involving two industrial sectors, each producing one good, and a single consumer sector. Each industry uses labor from the consumer sector and products from each of the two industrial sectors. There are market interdependencies. Tables 4.7–4.9 lead us to supply–demand functions for the two industrial sectors and the

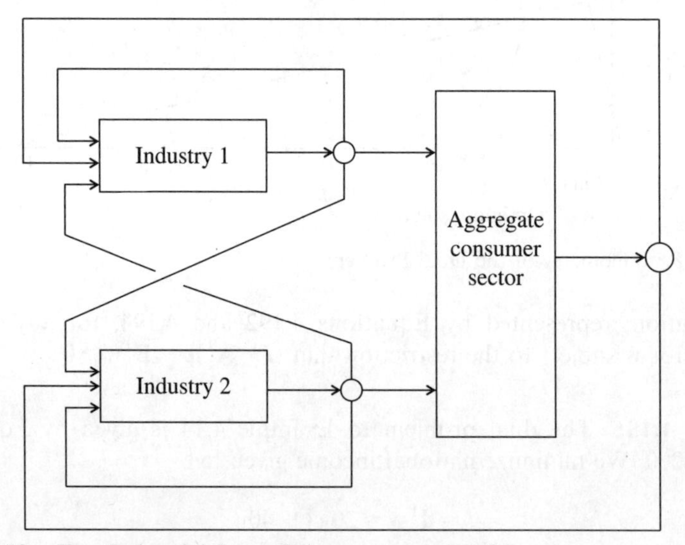

Figure 4.19. Two-Industrial-Sector Microeconomic Model.

single consumer sector of this simple economy. From these, we obtain six sets of important relations, where the notation is self-explanatory.

Two production supply equations:

$$q_1^1 = \mathrm{SQ}_1(p_1, p_2, w) \quad (4.206)$$

$$q_2^2 = \mathrm{SQ}_2(p_1, p_2, w) \quad (4.207)$$

Two factor demand equations:

$$R_1 = \mathrm{FD}_1(p_1, p_2, w) \quad (4.208)$$

$$R_2 = \mathrm{FD}_2(p_1, p_2, w) \quad (4.209)$$

Four production input to production demand equations:

$$Q_1^1 = \mathrm{PD}_1^1(p_1, p_2, w) \quad (4.210)$$

$$Q_2^1 = \mathrm{PD}_2^1(p_1, p_2, w) \quad (4.211)$$

$$Q_1^2 = \mathrm{PD}_1^2(p_1, p_2, w) \quad (4.212)$$

$$Q_2^2 = \mathrm{PD}_2^2(p_1, p_2, w) \quad (4.213)$$

Two commodity demand equations:

$$C_1 = Q_1^3 = \mathrm{DQ}_1^3(p_1, p_2, w) \quad (4.214)$$

$$C_2 = Q_2^3 = \mathrm{DQ}_2^3(p_1, p_2, w) \quad (4.215)$$

One factor (labor) supply equation:

$$r = \mathrm{FS}(p_1, p_2, w) \quad (4.216)$$

Three market clearing equations (at the summing junctions in Fig. 4.19):

$$q_1^1 = Q_1^1 + Q_1^2 + C_1 \quad (4.217)$$

$$q_2^2 = Q_2^1 + Q_2^2 + C_2 \quad (4.218)$$

$$r = R_1 + R_2 \quad (4.219)$$

We have 14 equations in 14 unknowns and should be able to solve these (setting one wage or price component as the numeraire) for the 14 unknowns. As our efforts thus far in this chapter have indicated, these equations are generally coupled. This sort of work is mainstream microeconomic analysis.

In this section, we have introduced the subject of microeconomic modeling. A generalized input–output modeling framework may be developed for specific situations and the input–output matrix coefficients calculated for a given production function. These nonlinear supply–demand relations could be used to implement a microeconomic model.

4.6 SUMMARY

In this chapter, we examined the supply–demand equilibrium conditions of microeconomic theory. We examined the simple case of a single commodity and considered problems of existence, uniqueness, and stability of economic equilibrium under conditions of perfect competition. We extended this analysis to the case of multiproduct firms that utilize both factor inputs and commodities in the production process. This chapter concluded with a discussion of linear input–output microeconomics. Several concepts from macroeconomic theory were useful in this general analysis. A number of advanced concepts were not presented. For example, we did not discuss economic equilibrium under conditions of imperfect competition because this would have necessitated a discussion of team theory in which various economic agents obtain information, such as that concerning consumer behavior, and transmit this information, according to various flow patterns, back to the firms. Also, we did not consider problems of existence and stability for general economic equilibrium, or problems of equity in which one consumer would not be allowed to increase utility if this increase would come at the expense of utility reduction for other consumers. We will turn our attention to considerations of equity or welfare economics in Chapter 5.

PROBLEMS

1. There are 500 firms in a perfectly competitive production firm, each of which has the production cost function given by $c_i = 0.1q_i^2$. The market demand function is $q = 10{,}000 - 500p$. What is the supply curve for an individual firm and for the entire set of firms? What is the equilibrium market price, quantity produced, and profit for each firm?
2. Suppose that there are opportunities for new firms to enter the market of Problem 1 and for old firms to leave. Production characteristics are the same as in Problem 1. What will be the new equilibrium market price, quantity produced, and profit?
3. The production function for a single firm is given by $q = 6K^{0.5} + 2L^{0.5} + 4M^{0.5}$, where the wages for the inputs are $w_K = 4$ (capital), $w_L = 1$ (labor), and $w_M = 5$ (natural resources). Find the supply function for the firm. Suppose that there are 250 firms. What is the market supply function?
4. The utility of a consumer for commodities is $U(\mathbf{x}) = 0.5\mathbf{x}^T\mathbf{A}\mathbf{x}$, where $\mathbf{A}$ is symmetric positive definite. Suppose that there are 10 consumers each with the same utility function with different budgets that is given by $(\mathbf{x}^j)^T\mathbf{p} \leq I^j$, $j = 1, 2, \ldots, 10$. Determine the demand function for the 10 consumers. What is the demand for x in terms of p_1? What do these results become if $\mathbf{A} = I$, $I^j = 10j$?
5. Repeat Problem 4 for the utility function $U(\mathbf{x}) = (\mathbf{x}^T\mathbf{A}\mathbf{x})^{1/2}$.
6. What is the supply function for a firm with an optimum production cost function given by $c(q) = e^q - 1$?
7. If a supply function is constrained to always having a positive slope, $\mathrm{d}SQ(p)/\mathrm{d}p > 0$, what must be the signs of the coefficients in the general production cost equation $C_p(q) = \sum_{j=1}^{n} a_j q^j$ for $n = 2,3,4$?

8. If supply and demand functions are $DQ = DQ_0(1-p/p_{max})$ and $SQ = SQ_0 + ap$, what are the market equilibria under
   a. normal conditions where $DQ_0 = 2000$, $p_{max} = 20$, $SQ_0 = 0$, $a = 100$?
   b. unusual supply conditions where $DQ_0 = 200$, $p_{max} = 20$, $SQ_0 = 1500$, and $a = -500$?
9. What is the effect of a \$1 unit sales tax on equilibrium for Problem 8?
10. What is the economic equilibrium and rent or royalty income to a firm with production cost function $C_p(q) = aq^2 + bq$ if the consumer demand function is $DQ = DQ_0(1 - p/p_{max})$?
11. What is the effect of a unit sales tax on equilibrium in Problem 8?
12. What is the effect on economic equilibrium in Example 4.5 of consumer demand curves where (a) $DQ_4 = 3000(1 - p/160)$ and (b) $DQ_5 = 3000(1 - p/3.57)$ considering (i) 100 case 1 firms and no case 2 firms, (ii) no case 1 firms and 100 case 2 firms, and (iii) 50 case 1 firms and 50 case 2 firms?
13. Develop the requirements for economic equilibrium for the case where the consumer pays the sales taxes rather than the firm. Show that these two cases yield essentially identical results for small tax rates.
14. What are the demand price elasticities for the equilibrium conditions of Problem 8? Plot a curve of demand price elasticity as a function of price and one as a function of the demand function of Problem 8.
15. Determine the Walras and Marshall equilibrium conditions for the supply–demand curves of Problem 8.
16. It has been speculated that the demand for labor in some countries is of the form $DQ = DQ_0(I - w/w_0)$, where $DQ_0 > 0$, $w_0 > 0$, and that the supply curve is of the form $SQ(w) = SQ_m - a(w - w_{max})^2$, where $SQ_m > 0$, $w_{max} > 0$, and $a > 0$. The term w represents wages, and it is speculated that the decrease in labor supply with increasing wage for large wages is due to an enhanced desire for leisure. Investigate static stability of the equilibrium points using the Walras and Marshall conditions. Which of these results appears most plausible?
17. Investigate static stability for supply $SQ = ap$ where the commodity is Giffen.
18. Investigate the effect of specific and *ad valorem* taxes added to the price of goods after they reach the market, that is, consider that the consumer budget is $I \geq \mathbf{x}^T\mathbf{p} + \text{tax}$. Show that the demand curve is now shifted, rather than the supply curve, but that the effects of this form of taxation are the same as when the tax is levied on the producer.
19. Subsidies appear to be an economic fact of life, and many instances of subsidies to producers and consumers may be cited. Government subsidy to low-income families for rental housing is one of many such examples. Consider a simple two-dimensional supply–demand relationship as shown in Fig. 4.20. We denote equilibrium with no subsidy as q_{NS} and p_{NS}. A subsidy will cause a shift in the demand curve, as shown in Fig. 4.20, but not in the supply curve. This occurs because the consumer will pay less for a given commodity and will demand more at a set price or pay less at a set demand. The new demand curve shows that at the new equilibrium point q_s, the total price subsidy (PS) by the government is

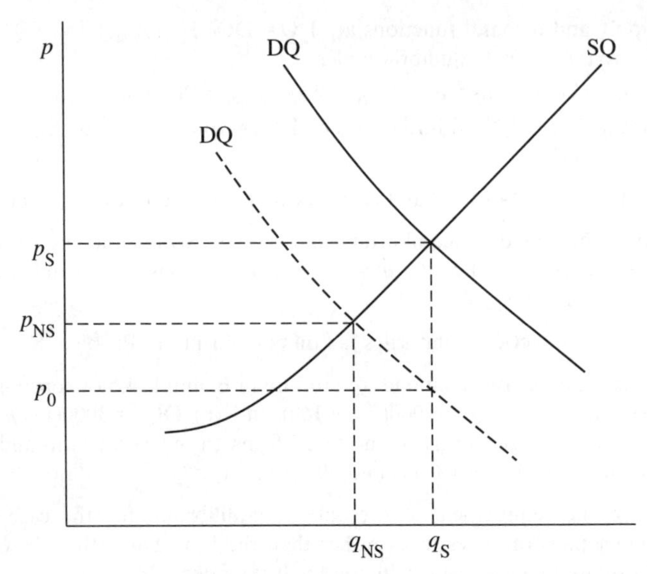

Figure 4.20. Supply and Demand Relations with Price Subsidy PS $= P_s - P_o$.

$P_s - P_o$, where P_s is the price paid to producers for one unit of the commodity and P_o is the price paid by consumers for one unit of the commodity at demand level q_s. Of this price subsidy PS $= p_s - p_0$, an amount PSP $= p_s - p_{NS}$ is the price subsidy to the producer or firm and PSC $= p_{NS} - p_o$ is the price subsidy to the consumer.

We can formulate three equations in the three unknowns p_s, q_s, and p_{NS}. We have the supply equation $q = \text{SQ}(p)$ and the consumer demand equation $q = \text{DQ}(p)$ that can be solved for the equilibrium values q_{NS} and p_{NS}. We introduce a subsidy PS and have $q_s = \text{SQ}(p_s)$ for the supply equation and $q_s = \text{DQ}(p_0)$ for the demand equation. The relation between prices is $P_S = P_s - p_0$ and this allows us, for fixed subsidy, to solve for the three unknowns.

a. Determine the relative economic distribution of the subsidy PS between consumers and firms as a function of the demand and supply elasticity.

b. Show that it does not matter, economically, whether consumers or producers receive the direct subsidy, but that the actual determinant of the benefits is specified entirely by the elasticities of demand and supply.

c. Illustrate your results in (a) and (b) for the specific case where $\text{SQ}(p) = p$, $\text{DQ}(p) = 10(1 + a) - ap$. Investigate this for a 10% subsidy where $P_S = 1$ and a 50% subsidy where $P_S = 5$.

20. Suppose that pollution control standards are enforced on a normally polluting firm. The effect of these standards is a reduction in pollution per unit output of the firm and an increase in costs such that the supply curve of the producers is altered. Determine the economic effect of pollution control on consumers and firms. How does the elasticity of supply and demand influence the fractional payment from pollution control paid by consumers and the fraction paid by producers?

21. Each of the three different firms produces under perfectly competitive conditions. The production functions for the firms are $q_1 = 10L_1 - L_1^2 + 3K$,

$q_2 = 2L_2 - 5L_2^2 + 4K$, and $q_3 = 40L_3 - 4L_3^2 + K$. The prices for the products are $p_1 = p_2 = p_3 = 1$. The firms compete in purchasing labor, and the supply curve that gives the total supply curve for labor is given by $L = L_1 + L_2 + L_3 = 2w_L$. The wages for capital are $w_K = 1$.

a. If there is a perfectly competitive labor market, what are the various equilibrium conditions?

b. If the firms form a monopsony with respect to hiring labor, what will be various equilibrium conditions?

c. What do the results for parts (a) and (b) become if new firms may enter the market?

22. This example concerns a simple economy with two consumers, two firms producing two different products, and a natural resource. Consumer 1 owns firm 1, and consumer 2 owns firm 2. Initially, each owns 20 units of the natural resource. The relevant equations are $q_1 = 10M_1$, $q_2 = 8M_2$, $u_1(q_1^1, q_2^1) = q_1^1 q_2^1 M^1$, $u_2(q_1^2, q_2^2) = 0.2\ \ln q_1^2 + 0.5\ \ln q_2^2 + 0.3\ \ln M^2$, and $M^1 + M^2 + M_1 + M_2 = 40 = M$. Find the economic equilibrium conditions for this simple economy. Please note that there are no exponents in this problem.

23. The concept of consumer and producer surplus can be used to advantage in determining the effect of government intervention in commodity and factor markets. This intervention may take various forms such as subsidies, price and wage controls, and rationing. The supply–demand curve of Fig. 4.21 illustrates *consumer surplus* and *producer surplus*. A number of consumers must be willing to pay more than the equilibrium price p_E to obtain market items that they desire, but market conditions allow them to pay a lower price. The aggregate savings to consumers is denoted the consumer surplus and is given by the indicated area in Fig. 4.21. Producer surplus occurs because some producers are willing to produce at prices lower than equilibrium p_E but existing market conditions allow them to

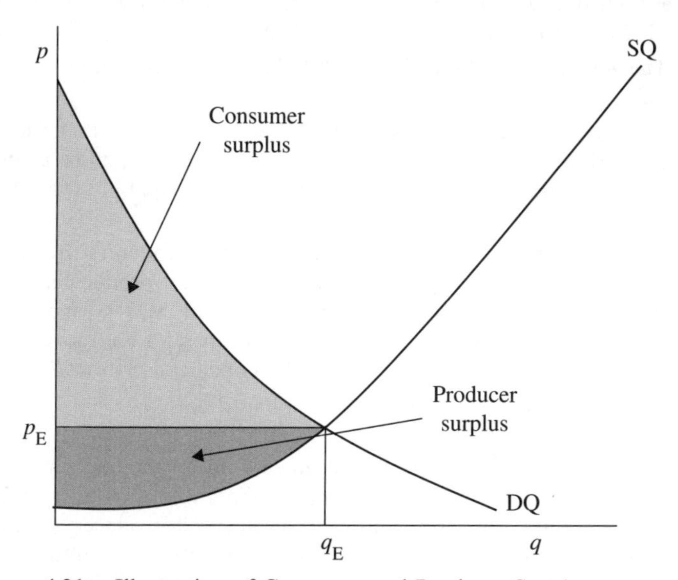

Figure 4.21. Illustration of Consumer and Producer Surplus.

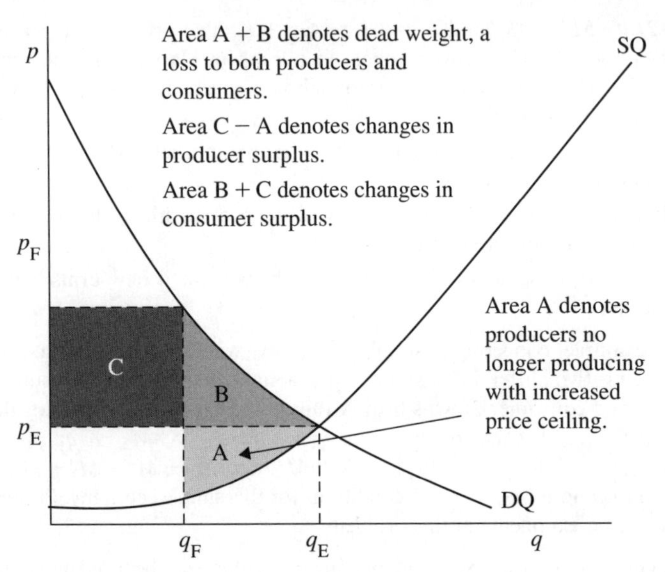

Figure 4.22. Illustration of Changes in Consumer and Producer Surplus due to Price Ceiling.

sell at the higher equilibrium price. The area indicated in Fig. 4.21 represents the aggregate amount of the producer surplus. We can determine changes in consumer and producer surplus due to policies, such as government fixed prices, as indicated in Fig. 4.22.

Suppose that we have an industry such as agriculture in which the supply curve is perfectly elastic such that $q = \mathrm{SQ}(p) = 10$. Suppose that we have the linear demand curve $q = \mathrm{DQ}(p) = 20 - 2p$. The agricultural producers believe the price of the farm product to be too low. Evaluate the following possible policies:

a. The government subsidizes agriculturalists to the amount of $\mathrm{PS} = 2$ for each unit of agricultural product sold.

b. The government imposes a fixed price support system that fixes the price charged to the consumer at $p_F = 7$ and agrees to purchase any surplus product at this price. It will sell the surplus purchased in export world pools at a price $p_{WP} = 4$.

c. The government subsidizes the agricultural producers by a soil bank arrangement in which payments are made for not producing. Suppose that the government decides to allow production $\mathrm{SQ} = 6$ and compensates producers such that their total revenue from sales and government is 70, the same as if they produced and sold 10 units at a price of 7.

d. The government does nothing.

Evaluate the costs and benefits of these four plans. Be sure to consider the following for each plan: equilibrium prices and quantities, products the government must purchase, consumer surplus and producer surplus, cost to the government and tax payers for subsidies, benefits to agriculturalists, and the ratios cost/(consumer surplus) and cost/(producer surplus) that would represent cost–benefit ratios for the public and the agriculturalists. Write a detailed case study report of your findings in this example.

24. Draw supply–demand curves to illustrate the effect on consumer and producer surplus of (a) a government-imposed fixed price p_F above the normal free-market equilibrium p_E and (b) a government-imposed fixed price ceiling p_c below the normal free-market equilibrium p_E. Illustrate your results for all four combinations of elastic and inelastic supply and demand curves. Indicate the relative costs and benefits to consumers and producers, and dead weight. Under which circumstances do consumers or producers benefit from price-fixing regulations? Figure 4.22 represents a typical curve for one of these cases. What is your interpretation of dead weight?

25. The concept of excess demand is sometimes introduced into problems of general economic equilibrium. In this formulation a vector of prices $\mathbf{p}$ is assumed. The price vector represents the price for both commodities and factors. An aggregate demand vector function $\mathbf{x}^{\mathrm{D}} = \mathbf{x}^{\mathrm{D}}(\mathbf{p})$ and an aggregate supply function $\mathbf{x}^{\mathrm{S}}(\mathbf{p})$ obtained by combining the individual demand and supply functions of all consumers and producers is assumed. The *excess demand* vector is defined as the difference between demand and supply, or $\mathbf{ED}(\mathbf{p}) = \mathbf{x}^{\mathrm{D}}(\mathbf{p}) - \mathbf{x}^{\mathrm{S}}(\mathbf{p})$. Economic equilibrium is defined as a nonnegative price vector $\mathbf{p}$ for which all components of the excess demand vector are nonpositive and for which the price of a commodity is zero if the excess demand for that commodity is negative. Under general economic equilibrium conditions, the aggregate market supply value will equal aggregate market demand value, in that $\mathbf{p}^{\mathrm{T}}\mathbf{ED}(\mathbf{p}) = 0$.

a. Reformulate the general economic equilibrium problem in terms of the excess demand approach. Determine the necessary conditions for equilibrium.

b. Under what conditions will the excess demand be homogeneous of degree zero such that for $a > 0$ we have $\mathbf{ED}(a\mathbf{p}) = \mathbf{ED}(\mathbf{p})$? What does this result indicate in terms of a numeraire quantity?

26. Suppose that in the particular case of a three-commodity system, we have the following. What are economic equilibrium prices?

$$\mathrm{ED}_1(\mathbf{p}) = -2p_1 + p_2 + p_3$$

$$\mathrm{ED}_2(\mathbf{p}) = p_1 - p_2 + p_3$$

$$\mathrm{ED}_3(\mathbf{p}) = 5p_1 - 2p_2 - 4p_3$$

27. In our input–output discussions we assumed that the required inputs to production are determined for a specified output level q_i as $x_{ji} = a_{ji}q_i$. Here x_{ji} is the jth input to the ith production process. The maximum output that could be obtained from a given input x_i is given by $q_i = \min[(x_{1i}/a_{1i}), (x_{2i}/a_{2i}), \ldots, (x_{ni}/a_{ni})] = \min(x_{ji}/a_{ji})$. Often it will occur that there will be a number m of different production activities that could be used to produce a given output. Let the total scalar output be the sum of the individual outputs, or $q = \sum_{i=1}^{m} q_i$, and the total input requirement for the jth input component to production is $x_j = \sum_{i=1}^{m} x_{ji} = \sum_{i=1}^{m} a_{ji}q_i$. For a unit output, $q = 1$, we must have from the foregoing equations, $I = \sum_{i=1}^{m} \gamma_i$ and $\beta_j = \sum_{i=1}^{m} a_{ji}\gamma_i$, where $\gamma_i = q_i/q$ and β_j is the input required for $q = 1$. The problem for the producer is to choose

$\gamma_i = 1, 2, \ldots, m$, to maximize the output that is given by $q = \min(x_j/\beta_j)$ by choice of γ_i that determine β_j. Suppose that the output can be produced using any combination of three processes, each of which uses two inputs. Suppose that $\mathbf{A} = \begin{bmatrix} 1 & 2 & 3 \\ 6 & 5 & 4 \end{bmatrix}$. Plot the isoquants of constant production for this case.

28. Generally linear input–output production activities may result in an n-vector output $\mathbf{q}$. Suppose that z_i denotes the level of the ith production activity and assume m activities. Assume that a_{ji} is the quantity of the jth output commodity produced by one unit of the ith activity such that $q_j = \sum_{i=1}^{m} a_{ji} z_i, j = 1, 2, \ldots, n$ or $\mathbf{q} = \mathbf{Az}$. Assume that b_{ji} is the quantity of the jth input required for a unit of the ith production activity such that $x_j = \sum_{i=1}^{m} b_{ji} z_i$, $j = 1, 2, \ldots, l$, or $x = \beta z$ represents the input to each activity at level z; x is an l-vector. We can define average or composite activities to produce an output as in Problem 22. We use $q_i = \min(x_{ji}/\beta_j)$, where $l = \sum_{i=1}^{l} \gamma_i$, and $\beta_j = \sum_{i=1}^{n} a_{ji} \gamma_i$. Suppose that each input quantity x_1 is fixed. Commodity market prices p_i are given.

a. What is the revenue to the firm?

b. What is the optimization problem that maximizes revenue to the firm?

c. What is the dual of (b)?

29. Two producers using Leontief technologies each make the same product. The first producer uses two units of labor, one unit of capital, and one unit of resources to make a single product. The second producer uses one unit of labor, three units of capital, and two units of resources to make a single product. Demand for the product is given by $p = 200 - q$; $w_L = 5$, $w_K = 1$, $w_M = 2$.

a. What are the production cost equations for the producers?

b. What are the economic equilibrium conditions?

c. Suppose that an unlimited number of firms with either production process can enter the market. What will be the new equilibrium conditions?

BIBLIOGRAPHY AND REFERENCES

The texts referenced in Chapters 2 and 3 also contain discussions concerning supply–demand equilibrium. Also, a number of advanced works that present existence and stability proofs for general economic equilibrium have not been discussed here. Included among these are the following:

Debreu G. Theory of value. New York: Wiley; 1959.

Leontief WW. Input–output economics. London: Oxford University Press; 1966.

Nikaido M. Convex structures and economic theory. New York: Academic; 1968.

Pindyck RS, Rubinfeld DL. Microeconomics. 6th ed. Englewood Cliffs, NJ: Prentice Hall; 2004.

Quirk J, Saposnik R. Introduction to general equilibrium theory and welfare economics. New York: McGraw-Hill; 1968.

Varian HR. Intermediate microeconomics: a modern approach. 7th ed. New York: W.W. Norton; 2005.

Von Neumann J. A model of general equilibrium. Rev Econ Stud 1945;13:1–9.

Wald A. On some systems of equations of mathematical economics. Econometrica 1957;19:368–403.

CHAPTER 5

NORMATIVE OR WELFARE ECONOMICS, DECISIONS AND GAMES, AND BEHAVIORAL ECONOMICS

5.1 INTRODUCTION

Our efforts thus far have been primarily devoted to perfect competition. When perfect competition conditions exist, it turns out that no individual consumer or firm can increase their objective function or payoff without decreasing the payoff for other firms and/or consumers. When imperfect competition (externalities, public goods, taxes and subsidies to selected firms or consumers, and other conditions) exist, the rules of perfect competition do not generally apply. Improvements in the accuracy of the results obtained by assuming perfect competition are then often possible, as well as improvements (or at least changes) in the results that accrue to firms and consumers. In this chapter we will examine concepts of optimality and economic efficiency of production and distribution under such conditions. We will focus on the allocation of land, capital, and resources in ways that generate the maximum benefit to society (or the community or nation). There are two fundamental issues underlying this: the allocation of the factors of production and the distribution of these products among different individuals. We will examine efficient, effective, and equitable ways in which we can accomplish this allocation. Then we will discuss the concept of social welfare and the conditions that must exist to determine optimum behavior under these social welfare conditions. This will lead us in the concluding part of this chapter to a discussion of behavioral economics. It is potentially important to be able to apply this thinking to markets where providers are not all identical and, less constraining, consumers are not all identical. Differences in production functions, perhaps in terms of differing fixed and variable cost structures, can be central to gaining competitive advantage. We will examine some of these issues in Chapters 6, 8, 9, and 10.

Economic Systems Analysis and Assessment,
by Andrew P. Sage and William B. Rouse

5.2 PARETO OPTIMALITY UNDER PERFECT COMPETITION CONDITIONS

Concepts of Pareto optimality serve as the basis for much of modern welfare economics. An allocation of resources is said to be *Pareto optimal* if there is no reallocation of production or distribution of the products of production that will increase the economic utility of one or more households without decreasing the economic utility of other households. We may also define a non-Paretian allocation of resources as one in which it is possible to increase the utility of one household without reducing the utility of any other. In most cases Pareto optimality cannot be achieved. But it is an obviously desirable criterion, in the sense that no one is harmed by a move in the direction of Pareto optimality. Often improvement in the utility of one household can occur only through reduction in the utility of others. Decisions that have to be made under such circumstances are difficult indeed. Thus when the conditions for Pareto optimality exist, we are fortunate; when they do not, issues such as income distribution and redistribution must be resolved.

We will now obtain the necessary conditions for the existence of Pareto optimal economic solutions. First we will consider Pareto optimal production and then Pareto optimal consumption. Finally we will combine production and consumption, and discuss Pareto optimal equilibrium.

We will first consider the case of two firms using two factor inputs to produce two commodities, and then generalize our arguments to higher order cases.

We assume the production functions

$$q_1 = f_1(x_1^1, x_2^1) \tag{5.1}$$

$$q_2 = f_2(x_1^2, x_2^2) \tag{5.2}$$

where the total factor input quantities are constrained, perhaps by availability concerns, such that

$$x_1^1 + x_1^2 = X_1 \tag{5.3}$$

$$x_2^1 + x_2^2 = X_2 \tag{5.4}$$

A problem of the firm in the short run is to maximize output. But *Pareto optimal production efficiency* requires a firm to maximize its output subject to the constraint that there is no reduction in the production output of any other firm. Thus the problem of firm 1 is to maximize its output such that the output of firm 2 remains at some fixed level Q_2. The Lagrangian for the problem of maximizing production output for firm 1 subject to the constraints of Equations 5.3 and 5.4 and the constraint that $q_2 = Q_2$ is given by

$$L = f_1(x_1^1, x_2^1) + \lambda[Q_2 - f_2(x_1^2, x_2^2)] + \gamma(X_1 - x_1^1 - x_1^2) + v(X_2 - x_2^1 - x_2^2) \tag{5.5}$$

where λ, γ, and v are Lagrange multipliers. The necessary conditions for optimality are given by

$$\frac{\partial L}{\partial x_1^1} = \frac{\partial f_1}{\partial x_1^1} - \gamma = 0 \tag{5.6}$$

$$\frac{\partial L}{\partial x_2^1} = \frac{\partial f_1}{\partial x_2^1} - \nu = 0 \tag{5.7}$$

$$\frac{\partial L}{\partial x_1^2} = -\lambda \frac{\partial f_2}{\partial x_1^2} - \gamma = 0 \tag{5.8}$$

$$\frac{\partial L}{\partial x_2^2} = -\lambda \frac{\partial f_2}{\partial x_2^2} - \nu = 0 \tag{5.9}$$

$$\frac{\partial L}{\partial \lambda} = Q_2 - f_2(x_1^2, x_2^2) = 0 \tag{5.10}$$

$$\frac{\partial L}{\partial \gamma} = X_1 - x_1^1 - x_1^2 = 0 \tag{5.11}$$

$$\frac{\partial L}{\partial \nu} = X_2 - x_2^1 - x_2^2 = 0 \tag{5.12}$$

We now combine Equations 5.6 and 5.8 and obtain the expression

$$\frac{\partial f_1}{\partial x_1^1} = -\lambda \frac{\partial f_2}{\partial x_1^2} = \mathrm{MP}_1^1 = -\lambda \mathrm{MP}_1^2 \tag{5.13}$$

We combine Equations 5.7 and 5.9 and obtain the expression

$$\frac{\partial f_1}{\partial x_2^1} = -\lambda \frac{\partial f_2}{\partial x_2^2} = \mathrm{MP}_2^1 = -\lambda \mathrm{MP}_2^2 \tag{5.14}$$

Here, MP_j^i is the marginal productivity of firm i with respect to the jth factor. Dividing these two relations results in the interesting and useful fact that the ratio of the marginal productivities of any one firm for two inputs x^i and x^j is the same for all firms. In doing this, we have the relation

$$\frac{\partial f_1/\partial x_1^1}{\partial f_1/\partial x_2^1} = \frac{\mathrm{MP}_1^1}{\mathrm{MP}_2^1} = \frac{\partial f_2/\partial x_1^2}{\partial f_2/\partial x_2^2} = \frac{\mathrm{MP}_1^2}{\mathrm{MP}_2^2} \tag{5.15}$$

This relation is important for *efficient production*. The ratio of marginal productivities is generally called the marginal rate of technical substitution or substitution, as we know from Equation 2.17. We may write Equation 5.15 as

$$\mathrm{MRS}_{12}^1 = \mathrm{MRS}_{12}^2 \tag{5.16}$$

Therefore, the marginal rate of technical substitution for the two firms must be equal if Pareto optimality exists. If this is not the case, then the produced output of one commodity can be increased without reducing the produced output of the other commodity. We have shown in Chapter 2 that the ratio of marginal productivities is the same as the ratio of wages for the factor involved if profit is to be maximized. For Pareto optimality, we require in addition that Equation 5.16 holds true. We can show that Pareto optimal production efficiency requires that the marginal rates of technical substitution for two input factors i and j be the same for all productive units that use these two factor inputs to production. Precisely these same results are obtained by maximizing the output of firm 2 in a Paretian fashion. Thus we see that

$$\frac{\mathrm{MP}_i^k}{\mathrm{MP}_j^k} = \frac{\mathrm{MP}_i^l}{\mathrm{MP}_j^l} = \frac{w_i}{w_j} \tag{5.17}$$

since the wages are the same for all firms for a given factor. These are very classic and very restrictive results.

Example 5.1: Suppose that the production equations for two firms are

$$q_1 = f_1(x_1^1, x_2^1) = (x_1^1)^{0.5}(x_2^1)^{0.5} \tag{5.18}$$

$$q_2 = f_2(x_1^2, x_2^2) = (x_1^2)(x_2^2) \tag{5.19}$$

The marginal rates of substitution are

$$\mathrm{MRS}_{12}^1 = \frac{x_2^1}{x_1^1} = -\left.\frac{\mathrm{d}x_2^1}{\mathrm{d}x_1^1}\right|_{\text{isoquant}} \tag{5.20}$$

$$\mathrm{MRS}_{12}^2 = \frac{x_2^2}{x_1^2} = -\left.\frac{\mathrm{d}x_2^2}{\mathrm{d}x_1^2}\right|_{\text{isoquant}} \tag{5.21}$$

Figure 5.1 illustrates the isoquants for the two firms. ■

Much of the preceding discussions concerning Pareto optimality of the firm apply directly to Pareto optimality of the household. Before considering a more general case, let us consider the case of two consumers who must share, in a Paretian fashion, two commodities that are in restricted supply. The consumers have the utility function $U_1(q_1^1, q_2^1)$ and $U_2(q_1^2, q_2^2)$. The total supply of commodity 1 is Q_1, so that

$$q_1^1 + q_1^2 = Q_1$$

The total supply of commodity 2 is Q_2, so that

$$q_2^1 + q_2^2 = Q_2$$

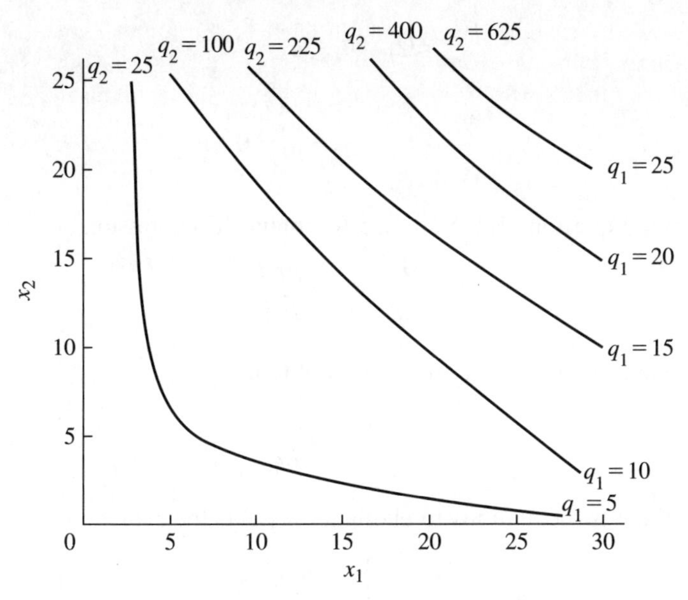

Figure 5.1. Isoquants for the two production functions of Example 5.1.

To maximize the utility of consumer or household 1, with the constraints that the utility of consumer 2 is fixed at U_2 and the commodity supply is restricted, we define the Lagrangian as

$$L = U_1(q_1^1, q_2^1) + \lambda[U_2 - U_2(q_1^2, q_2^2)] + \gamma(Q_1 - q_1^1 - q_1^2) + v(Q_2 - q_2^1 - q_2^2) \quad (5.22)$$

where λ, γ, and v are Lagrange multipliers. The necessary conditions for optimality are given by

$$\frac{\partial L}{\partial q_1^1} = \frac{\partial U_1}{\partial q_1^1} - \gamma = 0 \quad (5.23)$$

$$\frac{\partial L}{\partial q_2^1} = \frac{\partial U_1}{\partial q_2^1} - v = 0 \quad (5.24)$$

$$\frac{\partial L}{\partial q_1^2} = -\lambda\frac{\partial U_2}{\partial q_1^2} - \gamma = 0 \quad (5.25)$$

$$\frac{\partial L}{\partial q_2^2} = -\lambda\frac{\partial U_2}{\partial q_2^2} - v = 0 \quad (5.26)$$

$$\frac{\partial L}{\partial \lambda} = U_2 - U_2(q_1^2, q_2^2) = 0 \quad (5.27)$$

$$\frac{\partial L}{\partial \gamma} = Q_1 - q_1^1 - q_1^2 = 0 \tag{5.28}$$

$$\frac{\partial L}{\partial v} = Q_2 - q_2^1 - q_2^2 = 0 \tag{5.29}$$

We combine Equations 5.23 and 5.25 to obtain the expression

$$\frac{\partial U_1}{\partial q_1^1} = -\lambda \frac{\partial U_2}{\partial q_1^2} \tag{5.30}$$

We combine Equations 5.24 and 5.26 to obtain

$$\frac{\partial U_1}{\partial q_2^1} = -\lambda \frac{\partial U_2}{\partial q_2^2} \tag{5.31}$$

Dividing these two equations to eliminate λ yields the expressions

$$\frac{\partial U_1/\partial q_1^1}{\partial U_1/\partial q_2^1} = \frac{\partial U_2/\partial q_1^2}{\partial U_2/\partial q_2^2} \tag{5.32}$$

$$\frac{\mathrm{MU}_1^1}{\mathrm{MU}_2^1} = \frac{\mathrm{MU}_1^2}{\mathrm{MU}_2^2} \tag{5.33}$$

$$\mathrm{MRCS}_{12}^1 = \mathrm{MRCS}_{12}^1 \tag{5.34}$$

These three relations are important. Each is a statement of the requirement for *efficient consumption* under these fairly restrictive classic assumptions.

We have just derived the important result that under Paretian optimality of the household, the marginal rate of commodity substitution for the two households is the same. The marginal rate of commodity substitution is, as we have defined in Equation 3.7, the ratio of the marginal utilities for the commodities in question. If the relations of Equation 5.33 or 5.34 are not satisfied, it means we have not satisfied conditions for Pareto optimality. We recall from Chapter 3 that under conditions of optimality of the individual households, the ratio of marginal utilities is just the ratio of the prices for the commodities involved. Also, we can show that the results of Equation 5.33 are valid for any two consumers and any two commodities in the utility function of the two consumers. Thus we have the expression

$$\frac{\mathrm{MU}_i^k}{\mathrm{MU}_j^k} = \frac{\mathrm{MU}_i^l}{\mathrm{MU}_j^l} = \frac{p_i}{p_j} \tag{5.35}$$

since the price of a commodity is the same for one consumer as for any other consumer under the conditions assumed here.

Example 5.2: Suppose that the utility functions of the consumers are

$$U_1(q_1^1, q_2^1) = 20q_1^1 + q_2^1 \tag{5.36}$$

$$U_2(q_1^2, q_2^2) = q_1^2 q_2^2 \tag{5.37}$$

The marginal rates of commodity substitution for the two consumers are given by

$$\text{MRCS}_{12}^1 = 20 = -\left.\frac{dq_2^1}{dq_1^1}\right|_{\text{isoquant}} \tag{5.38}$$

$$\text{MRCS}_{12}^2 = \frac{q_2^2}{q_1^2} = -\left.\frac{dq_2^2}{dq_1^2}\right|_{\text{isoquant}} \tag{5.39}$$

Figures 5.2 and 5.3 illustrate the isoquants of constant utility for the two consumers.

From the equality of the MRCS under Pareto optimality, we see from Equations 5.38 and (5.39) that

$$q_2^2 = 20\, q_1^2 \tag{5.40}$$

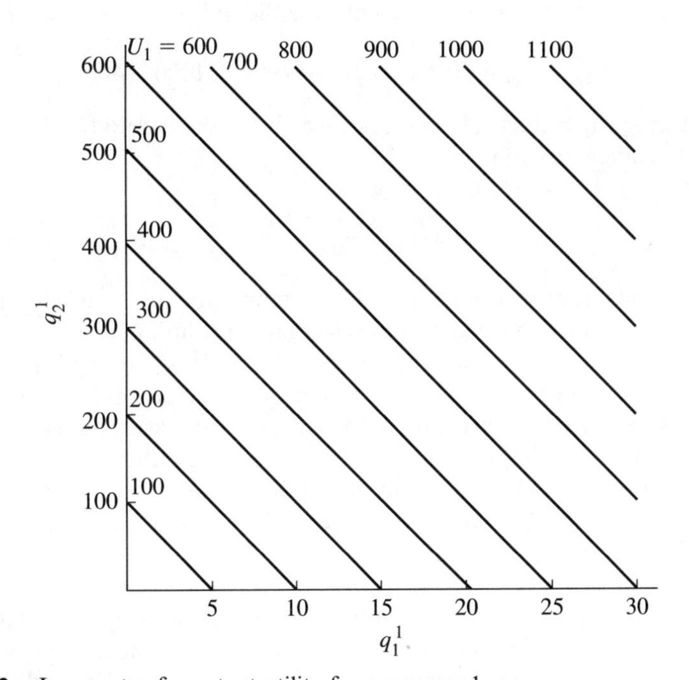

Figure 5.2. Isoquants of constant utility for consumer 1.

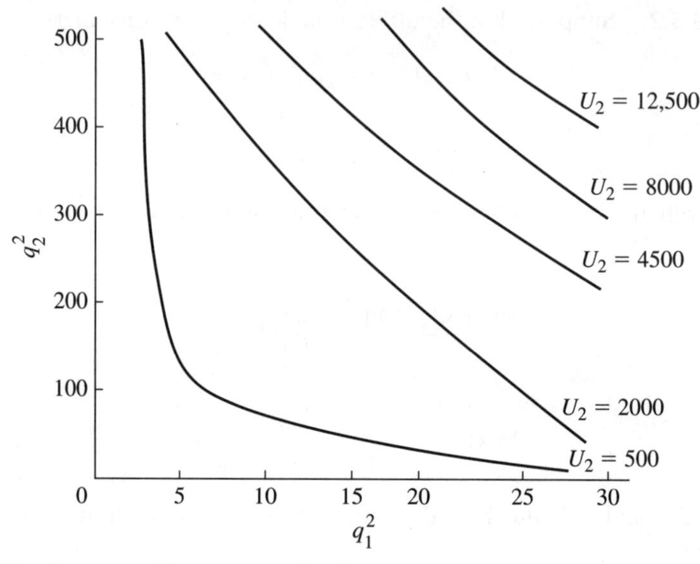

Figure 5.3. Isoquants of constant utility for consumer 2.

Thus from Equation 5.37, we have for the utility of consumer 2,

$$U_2(q_1^2, q_2^2) = 20(q_1^2)^2 \tag{5.41}$$

From Equations 5.36 and 5.40, we have for the utility of consumer 1

$$U_1(q_1^1, q_2^1) = 20(Q_1 - q_1^2) + Q_2 - q_2^2 = 1200 - 40q_1^2 \tag{5.42}$$

By comparing Equations 5.41 and 5.42, we obtain the Pareto frontier or utility possibility curve equation

$$U_2 = \frac{(1200 - U_1)^2}{80} \tag{5.43}$$

This represents the maximum possible combination of consumer levels of satisfaction or utility that yields a Pareto optimal solution.

It is relevant to ask at which values of U_1 and U_2 satisfying Equation 5.43 should equilibrium occur. The answer depends, of course, on the interaction of the firms with consumers in the marketplace, as we will see soon. Hopefully, we can write a social welfare function $W(U_1, U_2)$ to describe the joint utility of consumers 1 and 2 along the utility possibility curve. We will examine some notions of social welfare after we have examined Pareto optimal equilibrium under several competition forms. ■

The relationships between equilibrium in perfect competition and Pareto optimality are of considerable interest. We will first consider the case of two consumers each possessing a single factor input to production and two firms each producing a single product. Thus we assume the production functions of Equations 5.1 and 5.2, the factor inputs to production of Equations 5.3 and 5.4, the utility functions $U_1(q_1^1, q_2^1)$ and $U_2(q_1^2, q_2^2)$, and the produced commodities of Equations 5.28 and 5.29.

The long-term problem of firm 1 is to maximize profits, as given by

$$\Pi_1 = p_1 Q_1 - w_1 x_1^1 - w_2 x_2^1 \tag{5.44}$$

where the production equation of firm 1 is the relation

$$Q_1 = f_1(x_1^1, x_2^1) \tag{5.45}$$

The long-term problem of firm 2 also is to maximize profits, as given by

$$\Pi_2 = p_2 Q_2 - w_1 x_1^2 - w_2 x_2^2 \tag{5.46}$$

where the production equation of firm 2 is the relation

$$Q_2 = f_2(x_1^2, x_2^2) \tag{5.47}$$

The variables in these relations are those that we have been using in our efforts thus far. The terms w_1 and w_2 are the wages paid by the consumers who furnish factor inputs x_1^1 and x_2^1 to firm 1, and x_1^2 and x_2^2 to firm 2. Here, p_1 and p_2 are the price paid by the firms for the two commodities they produce, Q_1 and Q_2.

The problem of consumer 1 is to maximize the utility

$$U_1 = U_1(q_1^1, q_2^1, X_1) \tag{5.48}$$

This is subject to a constraint on the income of consumer 1, as given by

$$I_1 = w_1 X_1 + \Pi_1 s_1^1 + \Pi_2 s_2^1 = p_1 q_1^1 + p_2 q_2^1 \tag{5.49}$$

The problem of consumer 2 also is to maximize utility, as given by

$$U_2 = U_2(q_1^2, q_2^2, X_2) \tag{5.50}$$

This is subject to a constraint on the income of consumer 2, as given by

$$I_2 = w_2 X_2 + \Pi_1 s_1^2 + \Pi_2 s_2^2 = p_1 q_1^2 + p_2 q_2^2 \tag{5.51}$$

Again we use the familiar symbols: q_1^1 and q_2^1 for the commodities purchased by consumer 1; q_1^2 and q_2^2 are the commodities purchased by consumer 2. We assume that the nonsatiation axiom applies and consequently that the consumer spends the entire income from labor as well as from fractional ownership of the firms.

We recognize that the problem we have formulated is a simple case of the general equilibrium problem of Table 4.1. As with the general problem, we have factor market clearing equations that here become

$$X_1 = x_1^1 + x_1^2 \tag{5.52}$$

$$X_2 = x_2^1 + x_2^2 \tag{5.53}$$

and the commodity market clearing equations

$$Q_1 = q_1^1 + q_1^2 \tag{5.54}$$

$$Q_2 = q_2^1 + q_2^2 \tag{5.55}$$

The sum of the fractional ownership by consumers of the firms must be equal to unity. So we have

$$s_1^1 + s_1^2 = 1 \tag{5.56}$$

$$s_2^1 + s_2^2 = 1 \tag{5.57}$$

We will now determine equilibrium requirements under perfect competition conditions. Then we will compare these results to the conditions for Pareto optimality. The necessary conditions for optimality of firm 1 are straightforwardly given by

$$\frac{\partial \Pi_1}{\partial x_1^1} = 0 = p_1 \frac{\partial f_1(x_1^1, x_2^1)}{\partial x_1^1} - w_1 \tag{5.58}$$

$$\frac{\partial \Pi_1}{\partial x_2^1} = 0 = p_1 \frac{\partial f_1(x_1^1, x_2^1)}{\partial x_2^1} - w_2 \tag{5.59}$$

Similarly, the necessary conditions for optimality of firm 2 are given by

$$\frac{\partial \Pi_2}{\partial x_1^2} = 0 = p_2 \frac{\partial f_2(x_1^2, x_2^2)}{\partial x_1^2} - w_1 \tag{5.60}$$

$$\frac{\partial \Pi_2}{\partial x_2^2} = 0 = p_2 \frac{\partial f_2(x_1^2, x_2^2)}{\partial x_2^2} - w_2 \tag{5.61}$$

By combining these four relations, we obtain Equation 5.15, which is the necessary requirement for Pareto optimality of the two firms. Although we will not show it here, this conclusion applies to the general case of F firms using M factor inputs and N commodity inputs. Thus perfect competition conditions ensure Pareto optimality of the production of F firms. Solving Equations 5.58 through 5.61 results in a specific production, whereas solving Equation 5.22 results in a Pareto optimal relationship between Q_1 and Q_2.

To maximize the utility of consumer 1 subject to the budget constraint, we form the Lagrangian

$$L_1 = U_1(q_1^1, q_2^1, X_1) + \lambda_1(w_1 X_1 + \Pi_1 s_1^1 + \Pi_2 s_2^1 - p_1 q_1^1 - p_2 q_2^1) \tag{5.62}$$

We can obtain the necessary conditions for optimality, under the assumption of zero profits or constant profits, such that we avoid the need to take partial derivatives of Π_1 and Π_2, as

$$\frac{\partial L_1}{\partial q_1^1} = 0 = \frac{\partial U_1(q_1^1, q_2^1, X_1)}{\partial q_1^1} - \lambda_1 p_1 \tag{5.63}$$

$$\frac{\partial L_1}{\partial q_2^1} = 0 = \frac{\partial U_2(q_1^1, q_2^1, X_2)}{\partial q_2^1} - \lambda_1 p_2 \tag{5.64}$$

$$\frac{\partial L_1}{\partial \lambda_1} = 0 = w_1 X_1 + \Pi_1 s_1^1 + \Pi_2 s_2^1 - p_1 q_1^1 - p_2 q_2^1 \tag{5.65}$$

$$\frac{\partial L_1}{\partial X_1} = 0 = \frac{\partial U_1(q_1^1, q_2^1, X_1)}{\partial X_1} + \lambda_1 w_1 \tag{5.66}$$

The necessary conditions for optimality of consumer 2 are obtained from the Lagrangian for consumer 2, which is given by

$$L_2 = U_2(q_1^2, q_2^2, X_2) + \lambda_2(w_2 X_2 + \Pi_1 s_1^2 + \Pi_2 s_2^2 - p_1 q_1^2 - p_2 q_2^2) \tag{5.67}$$

as

$$\frac{\partial L_2}{\partial q_1^2} = 0 = \frac{\partial U_2(q_1^2, q_2^2, X_2)}{\partial q_1^2} - \lambda_2 p_1 \tag{5.68}$$

$$\frac{\partial L_2}{\partial q_2^2} = 0 = \frac{\partial U_2(q_1^2, q_2^2, X_2)}{\partial q_2^2} - \lambda_2 p_2 \tag{5.69}$$

$$\frac{\partial L_2}{\partial \lambda_2} = 0 = w_2 X_2 + \Pi_1 s_1^2 + \Pi_2 s_2^2 - p_1 q_1^2 - p_2 q_2^2 \tag{5.70}$$

$$\frac{\partial L_2}{\partial X_2} = 0 = \frac{\partial U_2(q_1^2, q_2^2, X_2)}{\partial X_2} + \lambda_2 w_2 \tag{5.71}$$

Combining Equations 5.63, 5.64, 5.68, and 5.69 easily results in Equation 5.32. Thus we are led to suspect that the requirements for consumer maximization of utility under perfect competition satisfy the conditions for Pareto optimality.

To see this fully, we need to determine the conditions for Pareto optimality of the consumers, since we now have factor inputs to the utility functions of Equations 5.48 and 5.50. To determine Pareto optimality of consumer 1, we maximize the utility of consumer 1 subject to equality constraints, thereby ensuring that

a. the utility of consumer 2 is U_2,
b. the equation relating the production of firms 1 and 2 is satisfied, and
c. the commodity and factor market clearing equations are satisfied.

Thus the Lagrangian, which we use for determining the conditions for Pareto optimality of consumer 1, is given by

$$\begin{aligned} L_1 = {} & U_1(q_1^1, q_2^1, X_1) + \lambda[(U_2 - U_2(q_1^2, q_2^2, X_2)] + \gamma_1[Q_1 - f_1(x_1^1, x_2^1)] \\ & + \nu_2(X_2 - x_2^1 - x_2^2) + \eta_1(Q_1 - q_1^1 - q_1^2) + \eta_2(Q_2 - q_2^1 - q_2^2) \\ & + \gamma_2[Q_2 - f_2(x_1^2, x_2^2)] + \nu_1[X_1 - x_1^1 - x_1^2] \end{aligned} \tag{5.72}$$

where λ, γ_1, γ_2, ν_1, ν_2, η_1, and η_2 are the Lagrange multipliers. We set

$$\begin{aligned} & \frac{\partial L_1}{\partial q_1^1} = \frac{\partial L_1}{\partial q_2^1} = \frac{\partial L_1}{\partial q_1^2} = \frac{\partial L_1}{\partial q_2^2} = \frac{\partial L_1}{\partial X_1} = \frac{\partial L_1}{\partial X_2} = \frac{\partial L_1}{\partial Q_1} = \frac{\partial L_1}{\partial Q_2} = \frac{\partial L_1}{\partial x_1^1} \\ & = \frac{\partial L_1}{\partial x_1^2} = \frac{\partial L_1}{\partial x_2^1} = \frac{\partial L_1}{\partial x_2^2} = 0 \end{aligned} \tag{5.73}$$

and

$$\frac{\partial L_1}{\partial \lambda} = \frac{\partial L_1}{\partial \gamma_1} = \frac{\partial L_1}{\partial \gamma_2} = \frac{\partial L_1}{\partial \nu_1} = \frac{\partial L_1}{\partial \nu_2} = \frac{\partial L_1}{\partial \eta_1} = \frac{\partial L_1}{\partial \eta_2} = 0 \tag{5.74}$$

We obtain from Equation 5.73, after some obvious combinations of terms,

$$\frac{\partial U_1/\partial q_1^1}{\partial U_1/\partial q_2^1} = \frac{\partial U_2/\partial q_1^2}{\partial U_2/\partial q_2^2} = \frac{\partial f_2/\partial x_1^2}{\partial f_1/\partial x_1^1} = \frac{\partial f_2/\partial x_2^2}{\partial f_1/\partial x_2^1} \tag{5.75}$$

$$\frac{\partial U_2/\partial X_2}{\partial U_2/\partial q_1^2} = -\frac{\partial f_1}{\partial x_2^1}, \quad \frac{\partial U_2/\partial X_2}{\partial U_2/\partial q_2^2} = -\frac{\partial f_2}{\partial x_2^2} \tag{5.76}$$

$$\frac{\partial U_1/\partial X_1}{\partial U_1/\partial q_2^1} = -\frac{\partial f_2}{\partial x_1^2}, \quad \frac{\partial U_1/\partial X_1}{\partial U_1/\partial q_1^1} = -\frac{\partial f_1}{\partial x_1^1} \tag{5.77}$$

Careful inspection of these relations shows that they are just those that follow from the necessary conditions for equilibrium under perfect competition conditions: Equations 5.58 through 5.61, Equations 5.63 through 5.66, and Equations 5.68 through (5.71). Solving these equations results in a set of relations between commodities and factors that is Pareto optimal but does not determine specific optimal values of these commodities and factors. Such determination is accomplished using the necessary conditions for perfect competition equilibrium. We will return to this point when we consider social welfare functions. This result is true not only for the case of two commodities and two firms but also for all other cases in general. Thus we conclude that *satisfaction of the conditions of perfect competition and associated equilibrium attainment ensure Pareto optimality*. Doubtlessly this result has done much to confirm the belief that perfect competition is a desirable economic result. Adam Smith formulated, in the eighteenth century, his invisible guiding hand principle, which asserted that free individual private decisions in a competitive economy will be socially optimum in that these decisions lead to that which is most beneficial in the general interest and serve a desirable societal purpose not part of the original intention. To be sure, perfectly competitive conditions do ensure Pareto optimality.

However, the converse is not true. Perfect competition is not necessary (although it is sufficient) for Pareto optimality, and Pareto optimality can result without any form of competition at all, such as in a socialist or dictatorial government.

To achieve Pareto optimality, we need to satisfy three fundamental relations. These are as follows: (1) the *efficient production relation* of Equation 5.15, which ensures that the marginal rate of substitution among the factors of production is the same for all producers; (2) the *efficient consumption relation* of Equation 5.32, which ensures that each consumer places the same relative economic value (at the margin) on all of the products of production; and (3) the *efficient product mix relations*, which are contained as part of Equations 5.75–5.77. These equations necessarily specify the efficient consumption and efficient product mix relations. In addition, the product mix relations require that the ratio of the economic value of two products is given by the ratio of their marginal costs.

We have from Equations 5.75 through 5.77

$$\frac{\partial U_1/\partial q_1^1}{\partial U_1/\partial q_2^1} = \frac{\partial f_2/\partial x_1^2}{\partial f_1/\partial x_1^1}$$

which is the *production mix relationship* for the case where consumer utility does not depend on input factors to production, and

$$\frac{\partial U_1/\partial X_1}{\partial U_1/\partial q_2^1} = -\frac{\partial f_2}{\partial x_1^2}$$

$$\frac{\partial U_2/\partial X_2}{\partial U_2/\partial q_1^2} = -\frac{\partial f_1}{\partial x_2^1}$$

which are the additional relations that must hold when consumer utility does depend on the factor input to production. By solving these relations for efficient production, consumption, and product mix, we obtain necessary conditions for Pareto optimality. To ensure equity, we should also maximize a (scalar) social welfare function $W(U_1, U_2)$. We will turn our attention to techniques to accomplish this in Section 5.5.

In this section, we have considered Pareto optimality under perfect competition conditions. We demonstrated the very useful result that equilibrium under perfect competition conditions ensures Pareto optimality. We did not examine the sufficiency conditions for Pareto optimality, which require that the second-order conditions for each consumer and firm delineated in the previous chapters are satisfied. If some of these conditions are not satisfied, Paretian optimality is not ensured. For example, suppose that one consumer is satiated. Then commodities may be taken from this consumer and given to others. The satiated consumer will suffer no loss of utility by this redistribution, whereas other consumers will increase their utility. Also, we require that there be no externalities, either in production or in consumption. These externalities may be beneficial, such as the effect of a firm's labor-training program on other firms and consumers. They may be detrimental, such as pollution associated with an otherwise good product of a firm. In many cases these externalities will exist. Also there will be many cases in which imperfect competition conditions exist.

Four basic assumptions are necessary for a freely competitive market to exist. When these conditions or assumptions do not exist, a freely competitive economy will *not* be optimum in the sense of being Pareto optimal or maximizing social welfare. In these cases, some form of regulations or other government intervention is potentially desirable. The four conditions that ensure perfect competition are the following

1. *No increasing returns to scale for any firms.* If increasing returns to scale exist along the production frontier, they must be exhausted before the supply–demand equilibrium is reached. We have seen in Chapters 2 and 4 that increasing returns to scale lead to monopolistic production and, therefore, to a noncompetitive economy in which firms will be price setters and not price takers.
2. *No technological external effects.* This is needed to ensure relevant prices and market efficiency. There are at least three forms of external effects: one person's consumption affects another person's utility, one firm's production affects another firm's utility, and one firm's production affects another firm's production.
3. *No ill-effects due to lack of certainty and perfect information.* All firms and consumers must have perfect knowledge of their own utility and production functions, and perfect awareness of market conditions.
4. *Correct ownership of all production factors.* The distribution of ownership of the factors of production enables each consumer to purchase the commodity bundle that corresponds to social welfare maximization.

The first three conditions are required for Pareto optimality to exist. The fourth condition is needed, as we have seen, to maximize social welfare. We will now turn our attention to a detailed consideration of these, and other, factors that alter or prevent perfect competition.

5.3 EXTERNAL EFFECTS AND IMPERFECT COMPETITION: PUBLIC GOODS

Our assumption of perfect competition requires that there exist no external effects in production or distribution. For example, we assume that the cost of production for a given firm does not depend on the production level of any other firm. Also, we assume that the utility of one consumer does not depend on the utility of other consumers. When there are externalities present, the conditions for Pareto optimality will generally not be satisfied. Also, imperfect competition will generally make Pareto optimal allocation of resources impossible. In this section, we will examine the effect of external effects and imperfect competition on Pareto optimality. We will also consider regulations on the economy that will result in Pareto optimal allocation.

A very important type of commodity or good is one that can be shared by all members of the public rather than consumed by individual members. The transportation system is one example of a *public good*, as shared commodities are generally called. Here the notion of a *public good* is very broadly defined such that there is no need to consider a *public bad.* In our initial analysis of public goods, we assume two consumers: a single bundle of ordinary goods and a single bundle of public goods. We assume the utility functions

$$U_1 = U_1(\mathbf{q}^1, \mathbf{g}) \tag{5.78}$$

$$U_2 = U_2(\mathbf{q}^2, \mathbf{g}) \tag{5.79}$$

where the total consumption of the ordinary good is given by

$$Q = \mathbf{q}^1 + \mathbf{q}^2 \tag{5.80}$$

and where $\mathbf{g}$ is the public good. To maximize the utility of consumer 1 subject to the constraint that the utility of the second consumer be fixed at some amount U_2, we form the Lagrangian

$$L = U_1(\mathbf{q}^1, \mathbf{g}) + \lambda[U_2 - U_2(\mathbf{q}^2, \mathbf{g})] + \gamma^T(\mathbf{Q} - \mathbf{q}^1 - \mathbf{q}^2) \tag{5.81}$$

The necessary conditions for optimality are obtained from the Lagrangian as

$$\frac{\partial L}{\partial \mathbf{q}^1} = \mathbf{0} = \frac{\partial U_1(\mathbf{q}^1, \mathbf{g})}{\partial \mathbf{q}^1} - \gamma \tag{5.82}$$

$$\frac{\partial L}{\partial \mathbf{q}^2} = \mathbf{0} = -\lambda \frac{\partial U_2(\mathbf{q}^2, \mathbf{g})}{\partial \mathbf{q}^2} - \gamma \tag{5.83}$$

$$\frac{\partial L}{\partial \mathbf{g}} = \mathbf{0} = \frac{\partial U_1(\mathbf{q}^1, \mathbf{g})}{\partial \mathbf{g}} - \lambda \frac{\partial U_2(\mathbf{q}^2, \mathbf{g})}{\partial \mathbf{g}} \tag{5.84}$$

$$\frac{\partial L}{\partial \lambda} = 0 = Y_2 - U_2(\mathbf{q}^2, \mathbf{g}) \tag{5.85}$$

$$\frac{\partial L}{\partial \gamma} = \mathbf{0} = \mathbf{Q} - \mathbf{q}^1 - \mathbf{q}^2 \tag{5.86}$$

We combine Equations 5.82 and 5.83 to eliminate the Lagrange multiplier γ and then obtain

$$\frac{\partial U_1(\mathbf{q}^1, \mathbf{g})}{\partial \mathbf{q}^1} = -\lambda \frac{\partial U_2(\mathbf{q}^2, \mathbf{g})}{\partial \mathbf{q}^2} \tag{5.87}$$

Taking any two components of the commodity bundle of ordinary goods in Equation 5.87, we can obtain the same results as in Equations 5.32 through 5.34:

$$\frac{\mathrm{MU}^1_{q^1_i}(\mathbf{q}^1, \mathbf{g})}{\mathrm{MU}^1_{q^1_j}(\mathbf{q}^1, \mathbf{g})} = \frac{\mathrm{MU}^2_{q^2_i}(\mathbf{q}^2, \mathbf{g})}{\mathrm{MU}^2_{q^2_j}(\mathbf{q}^2, \mathbf{g})} \tag{5.88}$$

Thus we see that the marginal rate of commodity substitution is the same for the ordinary goods components, just as it was in our earlier analysis in Section 5.2. The efficient consumption relation, as it applies to ordinary goods, is not affected by the presence of public goods.

A similar operation using any two components of the public good bundle in Equation 5.84 leads us to the relation

$$\frac{\partial U_1(\mathbf{q}^1, \mathbf{g})/\partial g_i}{\partial U_1(\mathbf{q}^1, \mathbf{g})/\partial g_j} = \frac{\partial U_2(\mathbf{q}^2, \mathbf{g})/\partial g_i}{\partial U_2(\mathbf{q}^2, \mathbf{g})/\partial g_j} \tag{5.89}$$

or the equivalent statement that

$$\frac{\mathrm{MU}^1_{g_i}(\mathbf{q}^1, \mathbf{g})}{\mathrm{MU}^1_{g_j}(\mathbf{q}^1, \mathbf{g})} = \frac{\mathrm{MU}^2_{g_i}(\mathbf{q}^2, \mathbf{g})}{\mathrm{MU}^2_{g_j}(\mathbf{q}^2, \mathbf{g})} \tag{5.90}$$

Thus we see that the marginal rate of public good substitution is the same for the two consumers under conditions of Pareto optimality.

However, when we calculate the ratio of one component from the ordinary good relation of Equation 5.87 and one from the public good relation of Equation 5.84, we obtain the relation

$$\frac{\partial U_1(\mathbf{q}^1, \mathbf{g})/\partial q^1_i}{\partial U_1(\mathbf{q}^1, \mathbf{g})/\partial g_j} = \frac{\partial U_2(\mathbf{q}^2, \mathbf{g})/\partial q^2_i}{\partial U_2(\mathbf{q}^2, \mathbf{g})/\partial g_j} \tag{5.91}$$

We note that these results are different from those obtained for Pareto optimality of two ordinary goods in Equations 5.32 through 5.34. Since the ratio of marginal utilities of a consumer for two commodities is, under conditions of utility maximization, the ratio of their prices, we see that the conditions for Pareto optimality when both ordinary and public goods are considered will not correspond to utility maximization with these goods. This means that we must distinguish between private and public goods in any realistic economic systems analysis and assessment.

Example 5.3: Suppose that consumer utilities are represented by

$$U_i = q^i_1 q^i_2 + q_3 - 0.5(q_3)^2, \; i = 1, 2 \tag{5.92}$$

Here q_1^i and q_2^i are ordinary goods and q_3 is a public good. The optimum commodity bundle to maximize utility subject to a budget constraint is easily determined, assuming for the moment that each consumer can demand a different quantity of the public good and that each demand can be satisfied, as the expressions

$$q_1^i = \begin{cases} 0, & p_3 \geq I_i \\ \dfrac{(I_i - p_3)p_2}{2p_1p_2 - (p_3)^2}, & \dfrac{2p_1p_2}{p_3} \geq I_i \geq p_3 \\ \dfrac{I_i}{2p_1}, & I_i \geq \dfrac{2p_1p_2}{p_3} \end{cases}$$

$$q_2^i = \begin{cases} 0, & p_3 \geq I_i \\ \dfrac{(I_i - p_3)p_1}{2p_1p_2 - (p_3)^2}, & \dfrac{2p_1p_2}{p_3} \geq I_i \geq p_3 \\ \dfrac{I_i}{2p_2}, & I_i \geq \dfrac{2p_1p_2}{p_3} \end{cases}$$

$$q_3^i = \begin{cases} \dfrac{I_i}{p_3}, & p_3 \geq I_i \\ \dfrac{2p_1p_2 - I_ip_3}{2p_1p_2 - (p_3)^2}, & \dfrac{2p_1p_2}{p_3} \geq I_i \geq p_3 \\ 0, & I_i \geq \dfrac{2p_1p_2}{p_3} \end{cases} \tag{5.93}$$

It can be straightforwardly shown that the sufficiency condition for a maximum of U_i, that $\partial^2 U_i/\partial q^2$ is negative definite along $I_i = \mathbf{p}^{\mathrm{T}}\mathbf{q}$, requires $2p_1p_2 > p_3^2$. For this particular utility function, we see that unless the income of all consumers is greater than p_3, the demand by each consumer for the public good is different and the entire income of each consumer is spent on it. This is a contradiction since the public good can exist only in a given quantity for all consumers. If all consumers have incomes greater than the relation $2p_1p_2/p_3$, then the demand for the public good is the same for all consumers, and Pareto optimality becomes possible. This occurs in this example only because the demand for the public good vanishes for a sufficiently large consumer income. In the range $2p_1p_2/p_3 \geq I_i \geq p_3$, the consumers also demand differing quantities of the public good. Again, this is an impossible demand to fulfill, as only one quantity of a given public good can be made available.

The marginal utilities for this example are given by

$$\mathrm{MU}_1^i = q_2^i$$

$$\mathrm{MU}_2^i = q_1^i$$

$$\mathrm{MU}_3^i = 1 - q_3^i$$

The required ratios for Pareto optimality, for ordinary goods 1 and 2, are given by

$$\frac{\mathrm{MU}_1^1}{\mathrm{MU}_2^1} = \frac{q_2^1}{q_1^1} = \frac{\mathrm{MU}_1^2}{\mathrm{MU}_2^2} = \frac{q_2^2}{q_1^2} \tag{5.94}$$

Between ordinary good 1 and public good 3, we obtain

$$\frac{\mathrm{MU}_3^1}{\mathrm{MU}_1^1} = \frac{1 - q_3^1}{q_2^1} = -\frac{\mathrm{MU}_3^2}{\mathrm{MU}_1^2} = -\frac{1 - q_3^2}{q_2^2} \tag{5.95}$$

Thus we will certainly satisfy Equation 5.94 with the quantities of the ordinary goods demanded in Equation 5.93 for all $I_i > p_3$. The ratio of marginal utilities is then just p_1/p_2, the ratio of prices for consumer optimality. Equation 5.95 will generally not be satisfied unless $I_i = I_2$ or $q_3^i = 1$, when it is routinely satisfied. In this particular example, $q_3 = 1$ for all $I_i = p_3$. So if all consumers have an income equal to precisely this amount, we satisfy Equation 5.95. In this situation, where $p_3 = 1$, each consumer demands all public goods ($I_i < 2p_1p_2$) or all private goods ($I_i > 2p_1p_2$). If the incomes are equal, but not necessarily equal to p_3, then the consumer demands for like goods, public or private, is the same, and it is possible to achieve Pareto optimality with $p_3 \leq I_i \leq 2p_1p_2/p_3$ and with consumers demanding both private and public goods. We can also argue that conditions for Pareto optimality are satisfied for $I_i > 2p_1p_2/p_3$. This satisfaction is somewhat moot as there is no demand at all for the public good in this case. ■

A somewhat unrealistic feature of the results obtained thus far in this section is that there is no mechanism to alter the production of commodities in accordance with consumer demand. There is no difficulty in obtaining Paretian optimality of the firms when there are public goods present, so we do not need to consider this case. Profit maximization, under conditions of perfect competition, will result in Pareto optimality of the firm. We will introduce a production function into the relations determining the behavior of the consumer to allow for a more realistic determination of the commodities demanded and the primary factors delivered from consumers in terms of the utility functions of consumers in a society. We will restrict our discussion to the case of two consumers, since the extension to the case of an arbitrary number of consumers is straightforward. For n consumers, the consumer will maximize personal utility subject to the constraint that the utility of other consumers is unchanged. The resulting equations are solved simultaneously to determine Pareto optimal behavior.

We assume that there exists a disaggregation of a produced vector of commodities, $\mathbf{Q}$, into two components $\mathbf{q}^1$ and $\mathbf{q}^2$. There is a single shared public goods bundle **g**. The two consumers possess primary factors $\mathbf{r}^1$ and $\mathbf{r}^2$ and these combine to form the factor input to production **x**. Here we use $\mathbf{r}_1$ and $\mathbf{r}_2$ as the factor inputs from consumers in keeping with the notion used in Section 4.3. We assume a single firm, which may well represent an aggregation of a number of firms, with production function $\varphi(\mathbf{Q}, \mathbf{g}, \mathbf{x}) = 0$. The problem of consumer 1 is to maximize personal utility subject to the constraint that the utility of consumer 2 remains unchanged and constraints on market clearing equations and the production function. We have the utility functions

$$U_1 = U_1(\mathbf{q}^1, \mathbf{g}, \mathbf{r}^1) \tag{5.96}$$

$$U_2 = U_2(\mathbf{q}^2, \mathbf{g}, \mathbf{r}^2) \tag{5.97}$$

Also, we have the production equation

$$\varphi(\mathbf{Q}, \mathbf{g}, \mathbf{x}) = 0 \tag{5.98}$$

The market clearing (distribution) equations or, more precisely, the Pareto optimal production frontier distribution constraints are given by

$$\mathbf{Q} = \mathbf{q}^1 + \mathbf{q}^2 \tag{5.99}$$

$$\mathbf{x} = \mathbf{r}^1 + \mathbf{r}^2 \tag{5.100}$$

The Lagrangian for consumer 1 is

$$\begin{aligned} L = {} & U_1(\mathbf{q}^1, \mathbf{g}, \mathbf{r}^1) + \lambda[U_2 - U_2(\mathbf{q}^2, \mathbf{g}, r^2)] \\ & + \gamma^{\mathrm{T}}(\mathbf{Q} - \mathbf{q}^1 - \mathbf{q}^2) + \mathbf{v}^{\mathrm{T}}(\mathbf{x} - \mathbf{r}^1 - \mathbf{r}^2) + \theta\varphi(\mathbf{Q}, \mathbf{g}, \mathbf{x}) \end{aligned} \tag{5.101}$$

Here λ, γ, **v**, and **g** are Lagrange multipliers. The necessary conditions for Pareto optimality are obtained from the Lagrangian in the usual way, where we drop arguments of U_i and φ for notational simplicity:

$$\frac{\partial L}{\partial \mathbf{q}^1} = \mathbf{0} = \frac{\partial U_1}{\partial \mathbf{q}^1} - \gamma \tag{5.102}$$

$$\frac{\partial L}{\partial \mathbf{q}^2} = \mathbf{0} = -\lambda \frac{\partial U_2}{\partial \mathbf{q}^2} - \gamma \tag{5.103}$$

$$\frac{\partial L}{\partial \mathbf{g}} = \mathbf{0} = \frac{\partial U_1}{\partial \mathbf{g}} - \lambda \frac{\partial U_2}{\partial \mathbf{g}} + \boldsymbol{\theta} \frac{\partial \varphi}{\partial \mathbf{g}} \tag{5.104}$$

$$\frac{\partial L}{\partial \mathbf{Q}} = \mathbf{0} = \gamma + \theta \frac{\partial \varphi}{\partial \mathbf{Q}} \tag{5.105}$$

$$\frac{\partial L}{\partial \mathbf{x}} = \mathbf{0} = \mathbf{v} + \theta \frac{\partial \varphi}{\partial \mathbf{x}} \tag{5.106}$$

$$\frac{\partial L}{\partial \mathbf{r}^1} = \mathbf{0} = \frac{\partial U_1}{\partial \mathbf{r}^1} - \mathbf{v} \tag{5.107}$$

$$\frac{\partial L}{\partial \mathbf{r}^2} = \mathbf{0} = -\lambda \frac{\partial U_2}{\partial \mathbf{r}^2} - \mathbf{v} \tag{5.108}$$

$$\frac{\partial L}{\partial \lambda} = 0 = U_2 - U_2(\mathbf{q}^2, \mathbf{g}, \mathbf{r}^2) \tag{5.109}$$

$$\frac{\partial L}{\partial \gamma} = \mathbf{0} = \mathbf{Q} - \mathbf{q}^1 - \mathbf{q}^2 \tag{5.110}$$

$$\frac{\partial L}{\partial \mathbf{v}} = \mathbf{0} = \mathbf{x} - \mathbf{r}^1 - \mathbf{r}^2 \tag{5.111}$$

$$\frac{\partial L}{\partial \theta} = 0 = \varphi(\mathbf{Q}, \mathbf{g}, \mathbf{x}) \tag{5.112}$$

We can obtain a very similar set of 11 necessary condition equations for Pareto optimality of consumer 2, which we will not formally write down here, primarily to save space.

We can combine Equations 5.102 and 5.103 to eliminate the Lagrange multiplier γ and then take the ratios of any two scalar components of the resulting vector equation to obtain the same result as previously obtained. This shows that efficient consumption of commodities given by

$$\frac{\partial U_1/\partial q_i^1}{\partial U_1/\partial q_j^1} = \mathrm{MRCS}_{ij}^1 = \frac{\partial U_2/\partial q_i^2}{\partial U_2/\partial q_j^2} = \mathrm{MRCS}_{ij}^2 \tag{5.113}$$

is satisfied, in that the marginal rates of commodity substitution (ordinary goods) are the same for any two consumers, and in general the same for all consumers. In this restrictive sense, we might conclude that the two consumers have the same behavior. By combining Equation 5.102 or (5.103) with Equation 5.105 and taking the ratios of any two scalar components, we obtain

$$\mathrm{MRCS}_{ij}^1 = \mathrm{MRCS}_{ij}^2 = \frac{\partial \varphi/\partial Q_i}{\partial \varphi/\partial Q_j} = \mathrm{MRPT}_{ij} \tag{5.14}$$

This indicates that the marginal rate of (ordinary) commodity substitution for any consumer is equal to the *marginal rate of product transformation* (MRPT) for the firms under conditions of Pareto optimality. Thus the efficient product mix relation is satisfied for (ordinary) commodities, even where there are public goods.

We can combine Equations 5.102, 5.103, and 5.105 to eliminate γ such that we obtain

$$\frac{\partial U_1}{\partial \mathbf{q}^1} = -\lambda \frac{\partial U_2}{\partial \mathbf{q}^2} = -\theta \frac{\partial \varphi}{\partial \mathbf{Q}} \tag{5.115}$$

A specific component of this vector equation is

$$\frac{\partial U_1}{\partial q_i^1} = -\lambda \frac{\partial U_2}{\partial q_i^2} = -\theta \frac{\partial \varphi}{\partial Q_i}$$

This can be used to obtain Equation 5.114. A specific component of Equation 5.104 is

$$\frac{\partial U_1}{\partial g_j} - \lambda \frac{\partial U_2}{\partial g_j} + \theta \frac{\partial \varphi}{\partial g_j} = 0$$

We may rearrange the two preceding equations to eliminate the Lagrange multiplier such that we obtain

$$\frac{\partial U_1/\partial g_j}{\partial U_1/\partial q_i^1} + \frac{\partial U_2/\partial g_j}{\partial U_2/\partial q_i^2} = \frac{\partial \varphi/\partial g_j}{\partial \varphi/\partial Q_i} \tag{5.116}$$

This relation shows that the sum of the marginal rates of substitution of the public good j for the ordinary commodity i for the two consumers is equal to the marginal rate of product transformation from producing ordinary commodity Q_i to producing public good g_j. Unlike the requirement for marginal rates of commodity substitution, the rates of substitution of public goods for ordinary commodities need not be equal. Consumers do not control producers and there is generally no way in which they can force the producers to adjust their production so that Equation 5.116 is satisfied and such that consumer Pareto optimality will result.

Two other useful results, which can be obtained from Equations 5.102 through 5.112, are

$$\frac{\partial U_1/\partial q_i^1}{\partial U_1/\partial r_j^1} + \frac{\partial U_2/\partial q_i^2}{\partial U_2/\partial r_j^2} = \frac{\partial \varphi/\partial Q_i}{\partial \varphi/\partial x_j} \tag{5.117}$$

and

$$\frac{\partial U_1/\partial g_i}{\partial U_1/\partial r_j^1} + \frac{\partial U_2/\partial g_i}{\partial U_2/\partial r_j^2} = \frac{\partial \varphi/\partial g_i}{\partial \varphi/\partial x_j} \tag{5.118}$$

These results indicate that the marginal rate of substitution of a factor input from consumers for an ordinary good must, for each consumer, equal the marginal rate of productivity of the factor input to production of the ordinary good. Also, these results show that the sum of the marginal rates of substitution of a primary factor input for a public good is, for all consumers, equal to the marginal productivity of the factor input to production of the public good.

Example 5.4: Suppose there is one ordinary good, one public good, and one factor input and that consumer utility, for $i = 1,2$, is

$$U_i = U_i(q^i, g, r^i) = q^i g + 0.5a_i(r_e^i - r^i)^2 \tag{5.119}$$

Here we assume that q^i, g, and r^i are each greater than zero and that r^i is also less than the endowed value r^i_{e}. The firm's production will be assumed to be

$$bQ + cg = x = r^1 + r^2$$

Here a_i, b, and c are assumed to be known.

The set of necessary conditions for Pareto optimality for consumer 1, from Equations 5.102 through 5.112, are given by a set of 11 equations of the form

$$\frac{\partial U_1}{\partial q^1} - \gamma = 0 = g - \gamma$$

$$-\lambda \frac{\partial U_2}{\partial q^2} - \gamma = 0 = -\lambda g - \gamma$$

$$\frac{\partial U_1}{\partial g} - \lambda \frac{\partial U_2}{\partial g} + \theta \frac{\partial \varphi}{\partial g} = 0 = q^1 - \lambda q^2 + \theta c$$

$$\gamma + \theta \frac{\partial \varphi}{\partial Q} = 0 = \gamma + \theta b$$

$$v + \theta \frac{\partial \varphi}{\partial x} = 0 = v - \theta$$

$$\frac{\partial U_1}{\partial r^1} - v = 0 = a_1(r^1 - r^1_{\mathrm{e}}) - v$$

$$-\lambda \frac{\partial U_2}{\partial r^2} - v = 0 = -\lambda a_2(r^2 - r^2_{\mathrm{e}}) - v$$

$$U_2 - q^2 g - 0.5a_2(r^2_{\mathrm{e}} - r^2)^2 = 0$$

$$Q - q^1 - q^2 = 0$$

$$x - r^1 - r^2 = 0$$

$$bQ + cg - x = 0$$

By solving these 11 equations in 11 unknowns, we can obtain the desired operating conditions for Pareto optimality of consumer 1 in terms of the assumed fixed utility of consumer 2, U_2. We can formulate a similar set of 11 equations in 11 unknowns and can obtain the desired operating conditions for Pareto optimality of consumer 2 in terms of an assumed fixed utility of consumer 1, U_1. We obtain the following additional equations:

$$\frac{\partial U_2}{\partial q^2} - \overline{\gamma} = 0 = g - \overline{\gamma}$$

$$-\bar{\lambda}\frac{\partial U_1}{\partial q^1} - \bar{\gamma} = 0 = -\bar{\lambda}g - \bar{\gamma}$$

$$\frac{\partial U_2}{\partial g} - \bar{\lambda}\frac{\partial U_1}{\partial g} + \bar{\theta}\frac{\partial \varphi}{\partial g} = 0 = q^2 - \bar{\lambda}q^1 + \bar{\theta}c$$

$$\bar{\gamma} + \bar{\theta}\frac{\partial \varphi}{\partial Q} = 0 = \bar{\gamma} + \bar{\theta}b$$

$$\bar{v} + \bar{\theta}\frac{\partial \varphi}{\partial x} = 0 = \bar{v} - \bar{\theta}$$

$$\frac{\partial U_2}{\partial r^2} - \bar{v} = 0 = a_2(r^2 - r_{\mathrm{e}}^2) - \bar{v}$$

$$-\bar{\lambda}\frac{\partial U_1}{\partial r^1} - \bar{\gamma} = 0 = -\bar{\lambda}a_1(r^1 - r_{\mathrm{e}}^1) - \bar{v}$$

$$U_1 - q^1 g - 0.5a_1(r_{\mathrm{e}}^1 - r^1)^2 = 0$$

Here we use the overbars to denote the Lagrange multipliers for the problem of consumer 2. These equations easily show that the corresponding Lagrange multipliers for the two problems are the same. We then have the following 11 equations:

$$g - \gamma = 0$$

$$q^1 + q^2 + \theta c = 0$$

$$\gamma + \theta b = 0$$

$$v - \theta = 0$$

$$a_1(r^1 - r_{\mathrm{e}}^1) - v = 0$$

$$a_2(r^2 - r_{\mathrm{e}}^2) - v = 0$$

$$Q - q^1 - q^2 = 0$$

$$x - r^1 - r^2 = 0$$

$$bQ + cg - x = 0$$

$$U_1 - q^1 g - 0.5a_1(r_{\mathrm{e}}^1 - r^1)^2 = 0$$

$$U_2 - q^2 g - 0.5a_2(r_{\mathrm{e}}^2 - r^2)^2 = 0$$

Solving these 11 relations for the Pareto optimal equilibrium, we obtain

$$\theta = \frac{(a_1 r_e^1 + a_2 r_e^2) + a_1 a_2}{2a_1 a_2 bc + a_1 + a_2} = v$$

$$r^1 = \frac{\theta}{a_1} + r_e^1$$

$$r^2 = \frac{\theta}{a_2} + r_e^2$$

$$Q = -\theta c$$

$$g = -\theta b = \gamma$$

These represent most of the Pareto optimum solution components for this example and the ones of most interest. We note that we cannot obtain explicit solutions for q^1 and q^2 here. From utility optimality of each consumer, we can assume a wage–income–expenditure relation that may be somewhat unrealistic only in the sense that consumers do not usually pay directly for public goods. This relation is

$$w_i r^i = I^i = p_0 q^i + \frac{p_p g}{2}$$

where p_0 is the price of the ordinary good and p_p the price of the public good. We arbitrarily assume that each consumer pays half the cost of the public good. These can be used to determine q^i. If p_0 and p_p are not given, they can be obtained from economic optimization of the firm and simultaneous solution of the resulting equations together with the equations for Pareto consumer optimality.

The marginal substitution rates are of considerable interest in this example. From Equation 5.116 we obtain

$$\frac{q^1}{g} = \frac{q^2}{g} = \frac{c}{b} = \frac{Q}{g}$$

Thus we see, again, that we must impose a requirement in the firm's production equation to ensure Pareto optimality. From Equation 5.117 we obtain

$$\frac{g}{a_1(r^1 - r_e^1)} = \frac{g}{a_2(r^2 - r_e^2)} = -b$$

Also, from Equation 5.118 we have

$$\frac{q^1}{a_1(r^1 - r_e^1)} = \frac{q^2}{a_2(r^2 - r_e^2)} = -c$$

These three equations are not independent and can, of course, be obtained directly from the optimum solutions obtained previously. ■

We have seen that the presence of public goods makes attainment of Pareto optimality difficult owing to requirements that are imposed on the production characteristics of the firm. It may require, for example, that firms sublimate production of automobiles by production of mass transit systems when the economic incentives to the firm for such a transition are not present. Public goods are generally financed by public agencies from taxation and not paid for directly by consumers. Ultimately the consumer must pay for a public good, but this payment will be through a, perhaps complicated, tax structure and not by such a simple scheme as dividing the cost of the public good by the number of consumers. We now focus on other externalities, such as taxes, and imperfections associated with production.

5.4 EXTERNAL EFFECTS AND IMPERFECT COMPETITION: NONINDEPENDENT PRODUCTION AND CONSUMPTION

Imperfections in the production process such as monopoly, oligopoly, monopsony, oligopsony, and the like generally prevent the attainment of Pareto optimality. This also occurs when consumer utility functions are nonindependent. We will first derive conditions of Pareto optimality for the consumer for a case where there are no external or imperfect competition effects. We assume two consumers, two commodity bundles and factor inputs, and a single production equation. This is precisely the problem posed in Equation 5.101 if we simply force the public good bundle equal to zero. Equations 5.113 and 5.114 result for the perfect competition case. We know that the ratio of marginal utilities is just the ratio of prices for the commodities. Thus we have from Equations 5.113 and 5.114 the efficient commodity consumption and efficient product mix relations

$$\frac{\partial U_1/\partial q_i^1}{\partial U_1/\partial q_j^1} + \frac{\partial U_2/\partial q_i^2}{\partial U_2/\partial q_j^2} = \frac{\partial \varphi/\partial Q_i}{\partial \varphi/\partial Q_j} = \frac{p_i}{p_j} \tag{5.120}$$

Also, we have from Equations 5.106 through 5.108 the efficient factor supply and efficient factor mix relations

$$\frac{\partial U_1/\partial r_i^1}{\partial U_1/\partial r_j^1} + \frac{\partial U_2/\partial r_i^2}{\partial U_2/\partial r_j^2} = \frac{\partial \varphi/\partial x_i}{\partial \varphi/\partial x_j} = \frac{w_i}{w_j} \tag{5.121}$$

From Equations 5.102, 5.103, and 5.105 through 5.108, we obtain the efficient product and factor mix relations

$$\frac{\partial U_1/\partial r_i^1}{\partial U_1/\partial q_j^1} + \frac{\partial U_2/\partial r_i^2}{\partial U_2/\partial q_j^2} = \frac{\partial \varphi/\partial x_i}{\partial \varphi/\partial Q_j} \tag{5.122}$$

One condition for profit maximization of the firm is obtained by setting the derivative of the profit with respect to factor inputs equal to zero. From this we obtain the relation that the marginal factor productivity of production is just the ratio of wages for the factor to the price of the good produced. Also, we can take the derivative of profit with respect to the quantity produced and obtain the result that the price of the product is just the marginal cost of producing the product. Generally those firms operating under imperfect competition will not obey these rules derived for the perfect competition case. We have shown that when firms and consumers operate under the rules of perfect competition, and there are ordinary goods only, Pareto optimality will result. We should expect that, unless a firm operates under perfectly competitive conditions, we cannot obtain Pareto optimality. However, as we have previously indicated, Pareto optimality is certainly possible even though firms and consumers do not operate under conditions of perfect competition. In this sense, perfect competition is a sufficient, but not necessary, condition for Pareto optimality.

Often the utility of one consumer will depend on the utility, and hence the level of consumption, of other consumers. One consumer's utility may increase when neighbor Smith buys a new Mercedes, whereas that of another consumer may decrease dramatically because of this same phenomenon. Under these circumstances, to show the dependence of their utilities on the consumption of the other consumer, the utility functions of two consumers may be written as

$$U_1 = U_1(\mathbf{q}^1, \mathbf{q}^2) \tag{5.123}$$

$$U_2 = U_2(\mathbf{q}^1, \mathbf{q}^2) \tag{5.124}$$

where the total bundle is constrained by

$$\mathbf{q}^1 + \mathbf{q}^2 = \mathbf{Q} \tag{5.125}$$

To maximize the utility of consumer 1 subject to an equality constraint on the utility of consumer 2 and the size of the commodity bundle, we define the Lagrangian

$$L = U_1(\mathbf{q}^1, \mathbf{q}^2) + \lambda[\underline{U}_2 - U_2(\mathbf{q}^1, \mathbf{q}^2)] + \gamma^T(\mathbf{Q} - \mathbf{q}^1 - \mathbf{q}^2) \tag{5.126}$$

The necessary conditions for maximization of the utility of consumer 1 are given by

$$\frac{\partial L}{\partial \mathbf{q}^1} = \mathbf{0} = \frac{\partial U_1}{\partial \mathbf{q}^1} - \lambda \frac{\partial U_2}{\partial \mathbf{q}^1} - \gamma \tag{5.127}$$

$$\frac{\partial L}{\partial \mathbf{q}^2} = \mathbf{0} = \frac{\partial U_1}{\partial \mathbf{q}^2} - \lambda \frac{\partial U_2}{\partial \mathbf{q}^2} - \gamma \tag{5.128}$$

$$\frac{\partial L}{\partial \gamma} = \mathbf{0} = \mathbf{Q} - \mathbf{q}^1 - \mathbf{q}^2 \tag{5.129}$$

$$\frac{\partial L}{\partial \lambda} = 0 = \underline{U}_2 - U_2(\mathbf{q}^1, \mathbf{q}^2) \tag{5.130}$$

We eliminate the Lagrange multiplier γ by combining Equations 5.127 and 5.128 to obtain

$$\frac{\partial U_1}{\partial \mathbf{q}^1} - \frac{\partial U_1}{\partial \mathbf{q}^2} = \lambda\left(\frac{\partial U_2}{\partial \mathbf{q}^1} - \frac{\partial U_2}{\partial \mathbf{q}^2}\right) \tag{5.131}$$

We find two scalar components of this vector equation and form the ratio of these equations to eliminate λ. We obtain

$$\frac{\partial U_1/\partial q_i^1 - \partial U_1/\partial q_i^2}{\partial U_1/\partial q_j^1 - \partial U_1/\partial q_j^2} = \frac{\partial U_2/\partial q_i^1 - \partial U_2/\partial q_i^2}{\partial U_2/\partial q_j^1 - \partial U_2/\partial q_j^2} \tag{5.132}$$

as one of the necessary conditions for Pareto optimality. Equation 5.132 is not the same as Equation 5.120, which is the equation we obtain for perfect competition. Thus we see that perfect competition, which results in Equation 5.120, will not result in Pareto optimality when the consumers have nonindependent utility functions.

Example 5.5: Suppose that utility functions of two consumers are given by

$$U_1 = U_1(\mathbf{q}^1, \mathbf{q}^2) = (q_1^1 - aq_1^2)(q_2^1 - bq_2^2) \tag{5.133}$$

$$U_2 = U_2(\mathbf{q}^2) = q_1^2 q_2^2 \tag{5.134}$$

such that consumer l's utility is reduced, for fixed q_1^1 and q_2^1, as the consumption of consumer 2 increases.

To maximize the utility of consumer 2, we instruct consumer 2 to use income I^2 to purchase the commodity bundle

$$q_1^2 = \frac{I^2}{2p_1} \tag{5.135}$$

$$q_2^2 = \frac{I^2}{2p_2} \tag{5.136}$$

The optimum commodity bundle for consumer 1, who suffers from a *keeping up with the Jones's* phenomenon (for fixed purchases by consumer 2), is given by

$$q_1^1 = \frac{p_1 aq_1^2 - p_2 bq_2^2 + I^1}{2p_1} \tag{5.137}$$

$$q_2^1 = \frac{p_2 bq_2^2 - p_1 aq_1^2 + I^1}{2p_2} \tag{5.138}$$

When we substitute the values of the commodity bundle actually purchased by consumer 2 into the two preceding relations, we obtain

$$q_1^1 = \frac{0.5I^2(a-b) + I^1}{2p_1} \tag{5.139}$$

$$q_2^1 = \frac{0.5I^2(b-a) + I^1}{2p_2} \tag{5.140}$$

These expressions are independent of the results obtained for consumer 2 only if $a = b$, or $I^2 = 0$, which is degenerate. This is a very special case that occurs because consumer l's utility is symmetrically reduced with that of consumer 2 for $a = b$.

The requirement for Pareto optimality (Eq. 5.132) for this example becomes

$$\frac{(q_2^1 - bq_2^2)(1-a)}{(q_1^1 - aq_1^2)(1-b)} = \frac{q_2^2}{q_1^2} \tag{5.141}$$

If we substitute the results of utility optimization with an income constraint, Equations 5.135, 5.136, 5.139, and 5.140, into Equation 5.141, we obtain $a = b$ as the requirement under which individual maximization of utility under a budget constraint leads to Pareto optimality. For example, with $q_1^2 = 50$ and $q_2^2 = 25$, we have $U_2 = 1250$. This is the optimum utility of consumer 2 under the perfect competition utility maximization. If we assume $Q_1 = 112.5$ and $Q_2 = 43.75$, the same total consumption as under individual utility optimization, then the necessary conditions for Pareto optimality of consumer 1 (Eqs. 5.127–5.130) become

$$0 = q_2^1 - \gamma_1$$

$$0 = q_1^1 - 0.5q_1^2 - \gamma_2$$

$$0 = 0.5q_2^1 - \lambda q_2^2 - \gamma_1$$

$$0 = -\lambda q_1^2 - \lambda_2$$

$$112.5 = Q_1 = q_1^1 + q_1^2$$

$$43.75 = Q_2 = q_2^1 + q_2^2$$

$$1250 = U_2 = q_1^2 q_2^2$$

Solving these equations leads to the following conditions for Pareto optimality: $q_1^1 = 66.21$, $q_2^1 = 16.75$, $q_1^2 = 46.29$, and $q_2^2 = 27.00$. Under both Pareto optimality and individual utility maximization, we have $U_2 = 1250$. Under Pareto optimality we have $U_1 = 721.31$, whereas under individual utility maximization we have $U_1 = 703.13$. As expected, we can get a better result for one consumer (consumer 1) under Pareto optimality by readjusting the distribution of commodities to the other consumer (consumer 2) while maintaining constant utility for consumer 2. This occurs because the utility functions are specified such that Pareto optimality requirements do not correspond to the requirements for maximization of each individual's utility.

An imposed or negotiated Pareto optimal solution results from aggregation of the utility functions of Equations 5.133 and 5.134 into a single utility function given by

$$U = (q_1^1 - 0.5q_1^2)q_2^1 + q_1^2 q_2^2 \tag{5.142}$$

We maximize this subject to the two equality constraints on income. The Lagrangian for this problem is given by

$$L = (q_1^1 - 0.5q_1^2)q_2^1 + q_1^2 q_2^2 + \lambda_1(100 - q_1^1 - 2q_2^1) + \lambda_2(100 - q_1^2 - 2q_2^2)$$

Following the usual optimization approach, we obtain the optimum purchased commodities as $q_1^1 = 60$, $q_2^1 = 20$, $q_1^2 = 40$, and $q_2^2 = 30$. The consumer utilities corresponding to these purchased commodities are $U_1 = 800$ and $U_2 = 1200$. The sum of the utility functions is greater than that examined thus far for any other case. Unfortunately, however, the utility of consumer 2 has been reduced. If there were some way to redistribute the gain in utility of consumer 1 to enhance the satisfaction of consumer 2, then, since the total utility is higher, there would have been some net societal gain. We will not pursue this point further in this example. However, some of our later discussions in this chapter concerning taxes and social welfare functions will address distribution problems such as this. ■

We will conclude our efforts in this section by illustrating how taxes and subsidies can be used to obtain Pareto optimality under external disturbance conditions when each economic unit maximizes its own objective function. We could consider a consumer tax and/or subsidy, but will instead focus on the behavior of two firms, each producing a single product and subject to external effects possibly including taxation and subsidy. We assume production cost functions for the two firms, $\mathrm{PC}_1(q_1, q_2)$ and $\mathrm{PC}_2(q_1, q_2)$, such that the profit of the firms is

$$\Pi_1 = p_1 q_1 - \mathrm{PC}_1(q_1, q_2) \tag{5.143}$$

$$\Pi_2 = p_2 q_2 - \mathrm{PC}_2(q_1, q_2) \tag{5.144}$$

If each firm maximizes profits by setting its own production level, the necessary conditions for perfectly competitive optimal behavior are

$$\frac{\partial \Pi_1}{\partial q_1} = 0 = p_1 - \frac{\partial \mathrm{PC}_1(q_1, q_2)}{\partial q_1} \tag{5.145}$$

and

$$\frac{\partial \Pi_2}{\partial q_2} = 0 = p_2 - \frac{\partial \mathrm{PC}_2(q_1, q_2)}{\partial q_2} \tag{5.146}$$

A form of negotiated or imposed Pareto optimality may be obtained by forming an aggregate profit function[1] given by

$$\Pi = \Pi_1 + \Pi_2 = p_1 q_1 + p_2 q_2 - \mathrm{PC}_1(q_1, q_2) - \mathrm{PC}_2(q_1, q_2) \tag{5.147}$$

and then adjusting q_1 and q_2 for optimality. In this case, we obtain the results

$$\frac{\partial \Pi}{\partial q_1} = p_1 - \frac{\partial \mathrm{PC}_1(q_1, q_2)}{\partial q_1} - \frac{\partial \mathrm{PC}_2(q_1, q_2)}{\partial q_1} = 0 \tag{5.148}$$

and

$$\frac{\partial \Pi}{\partial q_2} = p_2 - \frac{\partial \mathrm{PC}_1(q_1, q_2)}{\partial q_2} - \frac{\partial \mathrm{PC}_2(q_1, q_2)}{\partial q_2} = 0 \tag{5.149}$$

The necessary conditions obtained for perfect competition optimality (Eqs. 5.145 and 5.146) will generally be different from those obtained for the imposed Pareto optimality (Eqs. 5.148 and 5.149). We can compensate for the differences by adding a tax or subsidy to the production cost function for each firm. We will assume a unit tax or subsidy, although an *ad valorem* tax or subsidy is certainly possible. Thus we have the modified cost functions including the tax or the subsidy, $\mathrm{PC}_1(q_1, q_2) + q_1 \mathrm{TS}_1$ and $\mathrm{PC}_2(q_1, q_2) + q_2 \mathrm{TS}_2$, where TS is positive for a tax and negative for a subsidy. The tax or subsidy is generally determined such that when the firms individually optimize their profits, they obtain the production levels corresponding to Pareto optimal conditions. The profit functions with the tax and/or subsidy are given by

$$\Pi_1 = p_1 q_1 - \mathrm{PC}_1(q_1, q_2) - q_1 \mathrm{TS}_1 \tag{5.150}$$

$$\Pi_2 = p_2 q_2 - \mathrm{PC}_2(q_1, q_2) - q_2 \mathrm{TS}_2 \tag{5.151}$$

The new necessary conditions for perfectly competitive optimality are

$$\frac{\partial \Pi_1}{\partial q_1} = p_1 - \mathrm{TS}_1 - \frac{\partial \mathrm{PC}_1(q_1, q_2)}{\partial q_1} = 0 \tag{5.152}$$

$$\frac{\partial \Pi_1}{\partial q_2} = p_2 - \mathrm{TS}_2 - \frac{\partial \mathrm{PC}_2(q_1, q_2)}{\partial q_2} = 0 \tag{5.153}$$

We wish to obtain the same production levels from the results of Equations 5.152 and 5.153, and Equations 5.148 and 5.149. Thus we see that the taxes and/or subsidies are obtained by solving

$$\mathrm{TS}_1 = \frac{\partial \mathrm{PC}_2(q_1, q_2)}{\partial q_1} \tag{5.154}$$

[1]This is *not* the general result we obtain by maximizing Π_1 with an equality constraint on Π_2 and maximizing Π_2 with an equality constraint on Π_1.

$$\mathrm{TS}_2 = \frac{\partial \mathrm{PC}_1(q_1, q_2)}{\partial q_2} \tag{5.155}$$

evaluated at the Pareto optimal q_1 and q_2.

Total profits under conditions of Pareto optimality will generally be greater than profits under perfect competition conditions when market imperfections exist. These profits can be distributed to the firms such that their profits return to the original level as obtained from the production levels in Equations 5.145 and 5.146. There will be a surplus or social dividend, since Pareto optimal profits are greater than perfect competition profits under market imperfections. This social dividend should be used in some beneficial way. This process and results can be best illustrated with an example.

Example 5.6: We consider the production cost functions

$$\mathrm{PC}_1(q_1, q_2) = 10q_1 + q_1^2 + q_2 \tag{5.156}$$

$$\mathrm{PC}_2(q_1, q_2) = 5q_2 + q_2^2 + 4q_1 \tag{5.157}$$

This indicates that each firm experiences diseconomies due to the presence of the other firm. We assume that $p_1 = 20$ and $p_2 = 25$. From Equations 5.145 and 5.146 we obtain

$$p_1 = \mathrm{MC}_1 = 20 = 10 + 2q_1$$

$$p_2 = \mathrm{MC}_2 = 25 = 5 + 2q_2$$

Now, under individual profit optimization we obtain $q_1 = 5$ and $q_2 = 10$. The profits of the firms are $\Pi_1 = 15$ and $\Pi_2 = 80$, and the total profit is 95.

Under imposed Pareto optimality, we have the necessary conditions of Equations 5.148 and 5.149, which result in

$$20 - 10 - 2q_1 - 4 = 0$$

$$25 - 1 - 5 - 2q_2 = 0$$

This leads to the optimal results $q_1 = 3$ and $q_2 = 9.5$. The profits of the firms are now $\Pi_1 = 11.5$ and $\Pi_2 = 87.75$, and the total profit is 99.25, some 4.25 units greater than the earlier profit.

From Equations 5.154 and 5.155, we compute the unit taxes to be added to the production costs of the firms as $\mathrm{TS}_1 = 4$ and $\mathrm{TS}_2 = 1$. We now determine the profits of the firms with the tax imposed from Equations 5.150 and 5.151 as $\Pi_1 = -0.50$ and $\Pi_2 = 78.25$. Thus we must make a lump sum subsidy of 15.5 units to firm 1 to bring it back to the original operating-level profit of $\Pi_1 = 15$, and a subsidy of 1.75 units to firm 2 to bring it back to the original operating-level profit of $\Pi_2 = 80$. But where does the total subsidy of 17.25 units come from? The government has collected taxes totaling $\mathrm{TS}_1 q_1 + \mathrm{TS}_2 q_2 = 21.5$ units. A total of 17.25 units of this tax must be returned to the firms to bring them back to their original profit levels, and 4.25 units represent social dividends and can be used for other purposes.

We should not delude ourselves with the thought that we have just found the Pareto optimal solution. All we can really find is a *Pareto frontier* or a production possibility set. There are two optimization problems to determine Pareto optimality. The first problem is to maximize the profit of firm 1 subject to the constraint that the profit of firm 2 is constant. The second problem is to maximize the profit of firm 2 subject to the constraint that the profit of firm 1 is constant. We will now illustrate the complete Paretian solution to this problem.

For the first problem we define the Lagrangian

$$L_1 = 10q_1 - q_1^2 - q_2 + \lambda(\Pi_2 - 20q_2 + q_2^2 + 4q_1) \tag{5.158}$$

and then obtain the necessary conditions for optimality as

$$\frac{\partial L_1}{\partial q_1} = 10 - 2q_1 + 4\lambda = 0 \tag{5.159}$$

$$\frac{\partial L_1}{\partial q_2} = -1 - 20\lambda + 2\lambda q_2 = 0 \tag{5.160}$$

$$\frac{\partial L_1}{\partial \lambda} = \Pi_2 - 20q_2 + q_2^2 + 4q_1 \tag{5.161}$$

For the second problem we define the Lagrangian

$$L_2 = 20q_2 - q_2^2 - 4q_1 + \gamma(\Pi_1 - 10q_1 + q_1^2 + q_2) \tag{5.162}$$

Then, we obtain the necessary conditions for optimality of firm 2 as

$$\frac{\partial L_2}{\partial q_1} = -4 - 10\gamma + 2\gamma q_1 = 0 \tag{5.163}$$

$$\frac{\partial L_2}{\partial q_2} = 20 - 2q_2 + \gamma = 0 \tag{5.164}$$

$$\frac{\partial L_2}{\partial \gamma} = \Pi_1 - 10q_1 + q_1^2 + q_2 = 0 \tag{5.165}$$

By comparing Equations 5.159, 5.160, 5.163, and 5.164, we easily see that the relationship between the Lagrange multipliers is $\lambda\gamma = 1$. Thus Equation 5.159 is equivalent to Equation 5.163, and Equation 5.160 is equivalent to Equation 5.164. Equations (5.159)–(5.161) and (5.164) represent four equations in five unknowns. We obtain from these equations

$$q_1 = 5 + 2\lambda$$

$$q_2 = 10 + \frac{1}{2\lambda}$$

$$\Pi_1 = 15 - 4\lambda^2 - \frac{1}{2\lambda}$$

$$\Pi_2 = 80 - 8\lambda - \frac{1}{4\lambda^2}$$

$$\Pi = \Pi_1 + \Pi_2 = 95 - 4\lambda^2 - 8\lambda - \frac{1}{2\lambda} - \frac{1}{4\lambda^2}$$

These five equations represent solutions for all negative values of λ that yield all positive solutions. For $\lambda = -1$ we obtain the negotiated or imposed Pareto optimal solution just discussed. For $\lambda = -1$ the total profit is increased, and if the taxes and subsidies are designed properly, then total profit is increased and no firm should complain if the gain of one firm is used to compensate the loss of another by adopting this imposed solution.

In this example we have seen how taxes and subsidies can be used to entice firms to production levels that enhance social welfare. The results of this example would have been far more dramatic had we allowed firm l's production to add to the cost of firm 2's production and firm 2's production to be deducted from the cost of firm l's production. ■

In this section and in Section 5.3 we examined departures from perfect competition. We saw that this will generally create a situation such that non-Paretian allocation of resources will result if firms and consumers optimize according to the rules of perfect competition. We saw how taxes and subsidies can be used to obtain Paretian results from firms (and consumers) that follow the rules of perfect competition.

An interesting question is, what if some constraint prohibits fulfillment of one or more conditions of Pareto optimality? Is there any benefit in adjusting the remaining conditions for Pareto optimality? The answer is, it depends. A sensitivity analysis of pertinent conditions would appear to be the best approach to providing an answer to this question. The theory of the second best can be used to show that a solution with three equations different from the Pareto optimal equations may be better than a solution with two equations that are different from the ones that would result in true Pareto optimality.

Thus we have seen here and in our previous discussions that the notion of Pareto optimality is a very fundamental one and enters in many ways in economic systems analysis and assessment efforts. It is strongly related to other approaches as we have seen in our discussions. We need to strive for Pareto optimality such that there will be no possible allocation of effort or reallocation of organizational resources that will increase product differentiation without increasing costs, and no reallocation that will decrease costs and which will not also decrease differentiation. Generally, an organization should select a single perspective for competitive advantage, that is to be a low cost producer or a high differentiation producer, and then adopt a competitive strategy that is supportive of this. This will insure that we deal with issues involving cost, value, and competition in information and knowledge intensive systems, organizations, and enterprises in an efficient and effective manner.

5.5 WELFARE MAXIMIZATION AND SOCIAL CHOICE

As we have seen, the attainment of Pareto optimality results in situations in which it is not possible to make anyone better-off without making someone else worse-off. Thus Pareto optimal solutions are *efficient*. There is no guidance concerning equity in that any shift along a Pareto optimal frontier is bound to make someone better-off at the expense of making others worse-off. Thus it should be easy to get agreement on making all distributions Pareto efficient, but there is every reason to expect difficulty in getting agreement concerning which of the many Pareto efficient solutions to implement.

To resolve this difficulty, we pose a scalar *social welfare function* that expresses a social or normative utility of a, perhaps large, set of individual utility functions. Since increasing the utility of any individual while holding that of all other individuals constant must result in an increase in social utility, we see that the social welfare function must be convex. A social welfare function maximum must, of course, be Pareto efficient. The converse is also true.

We can verify the statement that if $\mathbf{X}$ maximizes social welfare, it is Pareto efficient by showing the negative statement is false. If $\mathbf{X}$ is *not* Pareto efficient, then there must exist some other resource allocation $\overline{\mathbf{X}}$ such that $U_j(\overline{\mathbf{x}^j}) \geq U_j(\mathbf{x}^j)$ for all j with strict inequality holding for at least one j. But if this is the case, then it is not possible for $\mathbf{x}^j$ and the associated $\mathbf{X}$ to maximize social welfare. Demonstrating that every Pareto efficient solution maximizes a (not necessarily *the*) social welfare function is relatively simple as well. All we need to do is assume that $\mathbf{X}$ is Pareto efficient and that individual utilities $U_j(\mathbf{x}^j)$ are continuous, concave, and monotonic in $\mathbf{x}^j$. Then for the linear social welfare function

$$W[U_1(\mathbf{x}^1), U_2(\mathbf{x}^2), \ldots, U_N(\mathbf{x}^N)] = \sum_{j=1}^{N} w_j U_j(\mathbf{x}^j) \tag{5.166}$$

the particular choice of weights w_j such that resource allocation $\mathbf{X}$ maximizes W subject to resource constraints $\sum_{j=1}^{N} \mathbf{x}^j = \mathbf{X} \leq \mathbf{R}$ is given by $w_j = 1/\lambda_j$, where λ_j is the jth consumer's marginal utility of income and $\mathbf{R}$ is the resource constraint. This is equivalent to

$$w_j^{-1} = \lambda_j = \frac{\partial U_j(\mathbf{p}, \mathbf{I}^j = \mathbf{p}^T \mathbf{x}^j)}{\partial \mathbf{I}^j} \tag{5.167}$$

This result (see Problems 26 and 27), that the weights w_j for the linear social welfare function are the reciprocals of the marginal utility of income, has a very interesting economic interpretation. If a person has a large income, then that person's marginal utility for additional income will typically be smaller than that for a person with a lower income. The weight corresponding to those with large incomes will be larger than that for those with smaller incomes. We must here remember that this is an implicit social welfare function associated with a perfect competition result involving Pareto efficiency. This shows that a perfectly competitive market will give efficient allocations. Nothing at all is said here about equitable allocations except perhaps the statement that under perfect competition conditions, those with larger incomes will have greater weight in determining the social welfare function than those with lower incomes.

The type of social welfare function needed for normative economics is a *cardinal* welfare or utility function. It is often difficult to establish group cardinal welfare functions, as they involve interpersonal comparisons of values. It would be preferable to use only ordinal utility functions, which provide numerical representations that allow preference comparisons to be made for each individual. Generally, this cannot be done. Let us now elaborate a little on this point.

We can use ordinal utilities to determine the commodity bundle that would be purchased by a consumer with a fixed income. For example, we can find the commodity bundle that maximizes

$$U_i(\mathbf{x}^i) = \sum_{j=1}^{M} a_j^i \ln x_j^i$$

subject to the income constraint

$$I^i = \mathbf{p}^{\mathrm{T}}\mathbf{x}^i = \sum_{j=1}^{M} p_j x_j^i$$

for a known price vector. The utility function here need only be an ordinal utility function. Such an ordinal utility function would not be sufficient to allow us to determine by how much the change from commodity bundle A to commodity bundle B would be more or less preferred over the change from commodity bundle C to commodity bundle D. This is generally needed if we have to cope with uncertainty or the chance associated with the outcomes that will result from particular allocations. Also, one needs to compare the relative pleasure that two consumers receive from changes in resource allocations.

It is this latter case that is of concern to us here. If we could work with ordinal preferences only, then it should be possible for us to combine the ordinal preferences of three consumers to obtain a social choice among alternatives. Suppose that there are three possible social distributions of resources $\mathbf{X}^A$, $\mathbf{X}^B$, and $\mathbf{X}^C$. Also suppose that the preferences of the three individuals are as follows:

Person	Preference structure
1	$\mathbf{X}^A > \mathbf{X}^B > \mathbf{X}^C$
2	$\mathbf{X}^B > \mathbf{X}^C > \mathbf{X}^A$
3	$\mathbf{X}^C > \mathbf{X}^A > \mathbf{X}^B$

What is the preferred equitable ordering of society's resources in this case? There is simply no answer to this question. We surely cannot use voting, for two people have preferences $\mathbf{X}^A < \mathbf{X}^B$, two have preferences $\mathbf{X}^B > \mathbf{X}^C$, and two have preferences $\mathbf{X}^C > \mathbf{X}^A$. If society were transitive for these ordinal preferences, we would require that if $\mathbf{X}^A > \mathbf{X}^B$ and if $\mathbf{X}^B > \mathbf{X}^C$, then $\mathbf{X}^A > \mathbf{X}^C$. This result does not imply in any sense that society is intransitive; it only says that we may well obtain intransitive group results for transitive group members if we insist on using ordinal preference or ordinal utility functions.

In 1951, Kenneth Arrow demonstrated this result in a very elegant way. He assumed that individual and social preference orderings will satisfy two very reasonable and weak axioms:

Axiom 1. Completeness or Total Ordering.

For all alternative pairs $\mathbf{X}$ and $\mathbf{Y}$, either $\mathbf{X}\mathfrak{R}\mathbf{Y}$ or $\mathbf{Y}\mathfrak{R}\mathbf{X}$, where $\mathfrak{R}$ may stand for strong preference $>$, weak preference $\geq$, or indifference $\sim$.

Axiom 2. Transitivity.

For all $\mathbf{X}$, $\mathbf{Y}$, and $\mathbf{Z}$, if $\mathbf{X}\mathfrak{R}\mathbf{Y}$ and $\mathbf{Y}\mathfrak{R}\,\mathbf{Z}$, then $\mathbf{X}\mathfrak{R}\mathbf{Z}$.

Next, Arrow postulated five very reasonable conditions:

1. *Complexity and free choice*. There are at least three alternatives under consideration and all possible orderings are allowed.
2. *Positive association of individual and social values*. If an individual increases his or her preference for an alternative, then society increases its preference for that alternative.
3. *Independence of irrelevant alternatives*. Introduction of a *dummy* alternative will not influence individual or societal choices, that is, ordinal preference orderings, for the alternatives originally considered.
4. *Sovereignty of individual citizens*. The social preference ordering cannot be imposed by society such that $\mathbf{X}^A > \mathbf{X}^B$, even if $\mathbf{X}^B > \mathbf{X}^A$ by all members of society. This is simply a weak version of Pareto efficiency that says it must be possible to reach the Pareto frontier.

5. *Nondictatorship*. There must not be an individual such that if he or she has preference $\mathbf{X}^{\mathrm{A}} > \mathbf{X}^{\mathrm{B}}$, then society prefers $\mathbf{X}^{\mathrm{A}} > \mathbf{X}^{\mathrm{B}}$ no matter what the individual preference orderings of society are.

Then Arrow showed, by a rather complex argument, that there exists no social welfare function that is guaranteed to satisfy the two axioms and five conditions. This is known as the *impossibility theorem* of Arrow. It shows us that all voting methods based on ordinal rankings may be irrational. This is not guaranteed, however. Indeed, it can be shown that if the alternatives can be arranged such that all individual preferences among alternatives are single peaked, then majority rule is transitive. But to *require* that individual preferences be single peaked, such that there exists a common point set that represents alternatives and where preferences decrease with increasing distance from the most preferred alternative, results in a violation of the free-choice condition.

Condition 3 is the most vulnerable of Arrow's conditions. It in fact involves two parts. The first part requires the irrelevance of *dummy* alternatives, and that, by itself, is entirely reasonable. The second part requires that this irrelevance be true when we have knowledge only of ordinal preferences or orderings. Many believe that this requirement is unreasonable for cases in which we have only ordinal preference relations. If we allow cardinal utilities or preferences, then we can obtain a possibility theorem using Arrow's axioms and conditions. To allow this requires that we should obtain revealed preference intensities or cardinal utilities for individuals and combine them into a scalar choice function. To fully pursue this topic would, sadly, take us far away on a lengthy journey that would be somewhat removed from our present goals. Suffice it to say that we advocate taking a cardinalist perspective with respect to individual utility and social choice functions.

5.6 SUMMARY

Our goal in this chapter has been to extend the concepts concerning consumers, firms, and market equilibriums to cases where perfectly competitive economic assumptions do not apply. We introduced the important concept of Pareto optimality and showed some very simple results from multiple objective optimization. We indicated that Pareto optimal surfaces are efficient but not necessarily equitable and introduced the concept of a social welfare function that could, in principle, ensure equity. Finally we indicated that ordinal preferences are generally insufficient to result in meaningful social choice and that cardinal preference intensities must be used. We will extend these concepts of normative economics to cost–benefit and cost–effectiveness analyses in Chapter 6.

PROBLEMS

1. Derive the Pareto production curve or Pareto frontier in Example 5.1.

2. In a two-consumer, two-producer economy, the utility and production functions are given by

$$u_1 = u_1(q_1^1, q_2^1)$$

$$u_2 = u_2(q_1^2, q_2^2)$$

$$q_1 = f_1(K_1, L_1)$$

$$q_2 = f_2(K_2, L_2)$$

where

$$q_1 = q_1^1 + q_1^2$$

$$q_2 = q_2^1 + q_2^2$$

are the quantities produced. Capital and labor are constrained by

$$K = K_1 + K_2$$

$$L = L_1 + L_2$$

Show that the efficient production, efficient consumption, and efficient product mix equations are given by

$$\frac{\partial q_1/\partial L_1}{\partial q_1/\partial K_1} = \frac{\partial q_2/\partial L_2}{\partial q_2/\partial K_2}$$

$$\frac{\partial u_1/\partial q_1^1}{\partial u_1/\partial q_2^1} = \frac{\partial u_2/\partial q_1^2}{\partial u_2/\partial q_2^2}$$

$$\frac{\partial u_1/\partial q_1^1}{\partial u_1/\partial q_2^1} = \frac{\partial q_2/\partial K_2}{\partial q_1/\partial K_1}$$

where

$$q_i = f_i(K_i, L_i)$$

3. What are the relations for Problem 2 that ensure that the utility of each consumer is maximized subject to fixed income and known prices? Does the efficient consumption relation follow?

4. What are the relations for Problem 2 that ensure maximum profit on each firm, assuming fixed wages for capital and labor? Does the efficient production relation follow?

5. Two consumers in a closed system have the following utility functions:

$$U_1(q_1^1, q_2^1) = a \ln q_1^1 + (1-a) \ln q_2^1$$

$$U_2(q_1^2, q_2^2) = \min(q_1^2, q_2^2)$$

The initial endowment for consumer 1 is one unit of commodity 1; the initial endowment for consumer 2 is one unit of commodity 2. What is the Paretian efficient redistribution of commodities? What are the market clearing prices that should exist for equivalence to the economic theory of the consumer?

6. Repeat Problem 5 with $U_2(q_1^2, q_2^2) = \max(q_1^2, q_2^2)$.

7. In an economy of 50 consumers and 3 commodities, consumer 6 has a Cobb–Douglas utility function

$$U_6(q_1^6, q_2^6) = 0.5\ln q_1^6 + 0.5\ln q_2^6 + \ln q_3^6$$

Consumer 6's commodity bundle under a certain Paretian efficient allocation is $q_1^6 = 20, q_2^6 = 5, q_3^6 = 8$. What are the perfectly competitive equilibrium prices of these commodities? By extrapolating on the implications of this example, can you establish a related method to determine parameters for a utility function of fixed form from the knowledge of the prices and the size of the commodity bundle? What is the role of Pareto optimality in this?

8. Suppose that there are two firms and two consumers in an economy whose behaviors are described by

$$Q_1 = f_1(x_1^1, x_2^1)$$

$$Q_2 = f_2(x_1^2, x_2^2)$$

$$U_1 = U_1(q_1^1, q_2^1, x_1)$$

$$U_2 = U_2(q_1^2, q_2^2, x_2)$$

where

$$Q_1 = q_1^1 + q_2^1$$

$$Q_2 = q_2^1 + q_2^2$$

$$X_1 = x_1^1 + x_1^2$$

$$X_2 = x_2^1 + x_2^2$$

a. What are the conditions for Pareto optimality for simultaneous consumption and production?

b. How do these compare with Paretian optimality conditions for production only, and for consumption only?

c. What are the Paretian production and distribution frontiers for the case where

$$X_1 = 10$$

$$X_2 = 100$$

$$U_i = q_1^i q_2^i, \quad i = 1, 2$$

$$Q_i = x_1^i x_2^i, \quad i = 1, 2$$

d. How do the results in part (c) compare with the general economic equilibrium results that you obtain for the specifications of part (c)?

9. Suppose that the production functions of two firms are given by

$$q_1 = \gamma K_1^{\alpha} L_1^{1-\alpha}$$

$$q_2 = \Delta K_2^{\beta} L_2^{1-\beta}$$

Initially one-half of the total capital and labor is associated with each firm. Can capital and labor be reallocated to enhance production?

10. Suppose that in Problem 9 there is a single consumer with utility function $U(q_1, q_2) = q_1 q_2$. How should capital and labor be allocated to maximize utility?

11. Suppose that a person may divide his or her labor L between two activities q_1 and q_2. The person's utility function is $U(q_1, q_2) = q_1 q_2$ and the production equations are $q_1 = aL_1^{\alpha}$ and $q_2 = bL_2^{\beta}$. What is the optimum division of labor among the two production processes? How does the efficient product mix relation apply here?

12. What are the relevant efficient production, consumption, and production mix equations for a model such as that in Problem 2 where there are three consumers and three firms?

13. The efficient production frontier in a certain economic system is given by

$$Q_1^2 + Q_2^2 = 41$$

There are three consumers with identical utility functions given by

$$U_i = q_1^i + q_2^i, \quad i = 1, 2, 3$$

The actual consumption of the three consumers at present is given in the table below.

Consumer	Consumption	
	q_1^i	q_2^i
1	1	0
2	1	3
3	2	2

What is the Pareto optimal reallocation of the commodities?

14. Investigate the Pareto optimal allocation of n commodities among m consumers with utility functions

$$U_i = \sum_{j=1}^{n} \alpha_j^i \ln q_j^i, \quad i = 1, 2, \ldots, m$$

for an economy in which the production frontier is given by

$$\mathbf{Q}^{\mathrm{T}}\mathbf{Q} = \sum_{j=1}^{n} Q_j^2 = R^2$$

15. Derive Equations 5.120 through 5.122.

16. Determine the conditions for Pareto optimality of the two firms whose profit equations are 5.143 and 5.144. Use this result to determine the efficient production frontier in general terms and for the specific production functions used in Example 5.6.

17. Suppose that there are two firms that produce products q_1 and q_2. The production cost equations are

$$\mathrm{PC}_1 = 10q_1^2 + 50q_1 - 2.5q_2^2$$

$$\mathrm{PC}_2 = 5q_2^2 + 40q_2 + 2.5q_1^2$$

The price of each product is \$10.

a. What are the requirements for the profit maximization of each firm?

b. What are the requirements for Pareto optimality?

c. How can an *imposed* Pareto optimality to maximize the sum of the profits of the two firms be obtained?

18. In this example, we consider the case where Paretian efficient consumption does not exist owing to presence of a single monopolistic firm. Assume that the consumer demand function is $p = D(q)$ and that the production cost function is $\mathrm{PC} = c(q)$. Generally the price charged by the monopolist is too high and the produced quantity too low for Pareto optimality of consumers to exist. Show that a unit subsidy added to the profit of the monopolist may encourage a production level and subsidy that are those of perfect competition and that therefore satisfy (consumer) Pareto optimality. What is the amount of this subsidy?

19. With respect to the issue posed in Problem 18, suppose that the consumer demand and production cost function are given by $p = 3 - q$ and $\mathrm{PC} = 0.5q^2$. What are the imperfect competition price, the quantity produced, and the profit? What are the Pareto optimal price, the quantity produced, and the profit? What are the amounts of the subsidy per unit product and the lump sum tax on consumers?

20. The production cost functions for two firms are $\mathrm{PC}_1 = 40q_1 + 8q_1^2 - 6q_1q_2$ and $\mathrm{PC}_2 = 150q_2 + 20q_2^2$. The market prices are $p_1 = 3$ and $p_2 = 5$. Determine the optimum profit for each firm. What is the optimum profit for the firms if the sum of the profits of the individual firms is maximized? What are the taxes and subsidies that will result in this profit?

21. Suppose that the utility of consumer i can be described by

$$U_i(\mathbf{q}^i) = \prod_{j=1}^{n} q_j^i - \sum_{j=1}^{m} 0.5(\mathbf{q}^j)^{\mathrm{T}}\mathbf{R}^j\mathbf{q}^j$$

where there are m consumers and n commodities. What are the requirements for Pareto optimality? How can they be imposed?

22. Investigate the hypothesis that efficiency and equity can be disaggregated in the sense that we can maximize efficiency by a correct allocation of factors (capital and labor) between production inputs. Then, as a separate problem, we can determine a unique allocation of these produced goods among people such as to ensure equity. You may assume, to simplify your efforts, that an efficient production frontier $F(q) = 0$ has been determined and is assumed to be known. As a particular vehicle for your discussion, you may wish to use the utility functions for two consumers

$$U_1 = q_1^1 q_2^2$$

$$U_2 = q_1^2 q_2^2$$

the social welfare function

$$W = U_1 U_2$$

and the resource constraint

$$q_1 + q_2 \leq 10$$

23. Did your response in Problem 22 depend on the knowledge of the production frontier? In other words, is the efficient production frontier a function of the social welfare function?

24. Suppose that the utility function of two individuals in a perfectly competitive closed economy is

$$U_1 = x_1^1 x_2^2, \quad U_2 = x_1^2 x_2^2$$

and that societal utility is

$$W = k_1 U_1 + k_2 U_2 + \Xi k_1 k_2 U_1 U_2$$

where

$$k_1 + k_2 + \Xi k_1 k_2 = 1$$

The production function for the two goods is

$$x_1 = K_1^{2/3} L_1^{1/3}, \quad x_2 = K_2^{1/3} L_2^{2/3}$$

Suppose that person 1 is initially endowed with one-fourth of the total labor. What portion of the capital should be owned by consumer 1 if we wish to maximize social welfare? How is this influenced by Ξ? Provide an interpretation to this.

25. Consider the two-household social welfare function

$$W = W[U_1(q_1^1, q_2^1, x_1), U_2(q_1^2, q_2^2, x_2)]$$

where U_i is the utility of the ith household. The production characteristic for the two firms is given by

$$\varphi(Q_1, Q_2, X_1, X_2) = 0$$

where

$$Q_1 = q_1^1 + q_1^2, \ Q_2 = q_2^1 + q_2^2$$

a. Show that the optimal solution to maximize social welfare is Pareto optimal.

b. Suppose that

$$W = aU_1 + (1-a)U_2$$

$$U_1 = q_1^1 q_2^1 + (1-x_1)^2$$

$$V_2 = q_1^2 + q_2^2 + (1-x_2)^2$$

$$Q_1 + 2Q_2 = X_1 + 2X_2$$

Find the optimum distribution of commodities as a function of a. How could a be specified?

c. Suppose that the factors are owned by households and that each of the two households maximizes its utility U_i subject to a budget constraint $p_1 q_1^i + p_2 q_2^i = w_i x_i = I^i$, $i = 1,2$. Find the economic equilibrium conditions and contrast and compare them with your results in part (b).

26. Verify Equation 5.167 for a perfectly competitive economy.

27. Suppose that two consumers have the utility functions

$$U_1(\mathbf{p}, I^1 = \mathbf{p}^{\mathrm{T}}\mathbf{x}^1) = \ln I_1 - a_1^1 \ln p_1 - (1-a_1)\ln p_2$$

$$U_2(\mathbf{p}, I^2 = \mathbf{p}^{\mathrm{T}}\mathbf{x}^2) = b\ln I^2 - a_2 \ln p_1 - (1-a_2)\ln p_2$$

and that the initial allocation of commodities is $\mathbf{r}^1 = \mathbf{r}^2 = [1 \ 1]^{\mathrm{T}}$

a. What are the market equilibrium prices?

(a) Reformulate this problem in terms of conventional utility functions and incomes.

b. What are the resulting weights for the social welfare function of Equation 5.166?

28. Suppose that there are 50 consumers and 3 goods in a simple economy. Consumer 1 has a Cobb–Douglas utility function $U_1(\mathbf{x}^1) = x_1^1 x_2^1 x_3^1$ and one Pareto efficient allocation is $\mathbf{x}^1 = [5, 2, 2]^{\mathrm{T}}$. What are the perfectly competitive prices? How do these change with changing the Pareto efficient allocation and what does this say with respect to the income of consumer 1? Suppose that the utility of consumer 1 is given by $U_i(\mathbf{x}^i) = x_1^i x_2^i x_3^i$ and that the social welfare function is $\sum_{i=1}^{50} w_i U_i(\mathbf{x}^i)$. What is the equitable distribution of commodities? How would this differ from the distribution under perfect competition?

29. Suppose that the production function in an economy is linear and given by

$$\mathbf{a}^{\mathrm{T}}\mathbf{X} = b$$

and that the social welfare function is

$$W(\mathbf{X}) = \sum_{j=1}^{M} c_j \ln X_j$$

What is the optimum set of commodities **X** to maximize social welfare?

30. Suppose that there are three alternatives to choose from, and a group of three people elects to *vote* to determine the preference ordering among alternatives. Each individual is transitive and the preferences of the individuals in the group are such that one person prefers a to b to c, one prefers b to c to a, and one prefers c to a to b. Majority rule decides. Does the majority prefer a to b or b to c or a to c? How could an interpersonal comparison of preferences using a cardinal scale over the interval 0 to 1 resolve fundamental difficulties encountered here?

31. In attempting to rank alternatives, we might assign points in accordance with preferences. For example, our first choice might get five votes, our second choice four votes, etc. Suppose that in ranking five alternatives, two people assign votes as follows:

Alternative	Voter 1	Voter 2	Total points
A	5	3	8
B	2	5	7
C	4	2	6
D	1	4	5
E	3	1	4

Now suppose that alternatives C and E are, for some reason, eliminated. Show that the new voting is such that B is assigned a higher priority than A and preferences for the first two alternatives are reversed. This results from the nonindependence of irrelevant alternatives, one of Arrow's conditions for the existence of an *ordinal* social welfare function. Show that if the preferences had been scaled in a cardinal fashion over the interval 0 to 1, we would not have this problem. What potential difficulties exist in obtaining this interpersonal comparison of values?

BIBLIOGRAPHY AND REFERENCES

Three references that provide excellent and detailed coverage of welfare economics are as follows:

Layard PRG, and Walters AA. Microeconomic theory. New York: McGraw-Hill; 1978.

Mishan EJ. Introduction to normative economics. New York: Oxford University Press; 1981.

Ng YK. Welfare economics: towards a more complete analysis. New York: Macmillan; 2004.

The above references provide much historical perspective and many references to original journal sources. The following reference is of interest with respect to a survey of modern theories of social choice and systems engineering and management science applications

Sage AP. Behavioral and organizational considerations in the design of information systems and processes for planning and decision support. *IEEE Trans Syst Man Cybern* 1981;SMC-11(9): 640–678.

A brief excellent philosophical text concerning social welfare is

MacKay AF. Arrow's theorem: the paradox of social choice. New Haven Connecticut: Yale University Press; 1984.

The celebrated works of Arrow, which we have mentioned in Section 5.5, are discussed in the aforementioned texts and in

Sage AP. Methodology for large scale systems. New York: McGraw-Hill; 1977.

CHAPTER 6

COST–BENEFIT AND COST–EFFECTIVENESS ANALYSES AND ASSESSMENTS

6.1 INTRODUCTION

The broad goals of cost–benefit analysis and cost–benefit assessment are to provide procedures for the estimation and evaluation of the benefits and the costs associated with alternative courses of action, including their analysis and assessment. In many cases it will not be possible to obtain a completely economic evaluation of the benefits of proposed courses of action. In this case the word *benefit* is replaced by the term *effectiveness*, and we determine a cost–effectiveness analysis and assessment rather than a strictly economic analysis and assessment of the net benefits associated with alternative courses of action.

We may view economics as a descriptive or as a normative science. The often retrospective study of the decisions people and organizations make with respect to the employment of scarce resources for production and consumption is a descriptive study. The study of the decisions that people and organizations *should* make with respect to resource allocations is a normative study. Our studies of the behavior of firms and consumers in Chapters 2 to 4 are basically studies of the normative behavior of individual firms and consumers. Often descriptive behavior will be approximately the same as normative behavior, at least in a substantive or "as if" fashion. Of course, normative implies value judgments. Here "normative" would have to imply "assuming that firms and consumers wish to maximize their profit and utility, respectively." Some related comments may be found in Section 1.8.

In this chapter we restate the fundamental assumptions and requirements for rational (unaided) economic behavior. As we shall see, these requirements are somewhat difficult, especially with respect to their information demands, for consumers and firms to meet, in an unaided, descriptive, or positive sense.

Economic Systems Analysis and Assessment,
by Andrew P. Sage and William B. Rouse

They are especially difficult to meet when there are groups, either of individuals or of firms, involved.

In Chapter 5 we introduced concepts from welfare or normative economics. In these situations, there is invariably more than one person or firm present. Thus questions of equity in the distribution of allocations will necessarily arise, unless some very restrictive assumptions are imposed. These assumptions will ensure that all of the requirements for the existence of perfect competition are met. When there is more than one person's judgment involved, questions generally arise concerning how to combine the judgments of the individuals involved. There are a large number of ways to accomplish this combination, and five are particularly important here. We illustrate these for the case where there are two alternatives under consideration.

1. ***Unanimity.*** Alternative 1 is superior to alternative 2 if each member of society individually judges alternative 1 superior to alternative 2.
2. ***Majority Rule.*** Alternative 1 is superior to alternative 2 if the majority of the members of society prefer alternative 1 over alternative 2.
3. ***Pareto Superiority.*** Alternative 1 is Pareto superior to alternative 2 if at least one person judges alternative 1 superior to alternative 2 and no one judges alternative 2 superior to alternative 1.
4. ***Potential Pareto Superiority.*** Alternative 1 is potentially Pareto superior to alternative 2 if those who "gain" by the choice of alterative 1 over alternative 2 *could* compensate those who "lose," so that if compensation were paid, the final result would be that no one would be worse off than if alternative 2 had been selected.
5. ***Social Welfare Superiority.*** Alternative 1 is superior to alternative 2 in terms of social welfare if the cardinal utility function representing social welfare is larger for alternative 1 than it is for alternative 2.

A brief discussion of each of these criteria is in order. Obviously unanimity is very desirable, but it will often not exist. Majority rule may lead to significant group intransitivity, as we have seen in Chapter 5, even when the preference structure of each individual in the group is transitive. For example, if there are three people with preferences among three alternatives $a > b > c$, $b > c > a$, and $c > a > b >$, then two out of the three people express the preferences $a > b$, $b > c$, and $c > a$ and there is a major intransitivity. This can be, in principle, completely resolved by using a cardinal preference scale such that a scalar social welfare function results. There are considerable pragmatic difficulties, however, in constructing a cardinal scalar social welfare function.

Pareto superiority is only a slight weakening of the unanimity conditions in that individuals are now allowed to be indifferent or to judge alternatives as incomparable. It is doubtful that Pareto superiority will exist among many alternatives in realistic situations. Thus we will be left with a Pareto-efficient frontier with no way of determining a single Pareto-superior alternative. The concept of potential Pareto superiority is then of major importance. It is the potential Pareto-superiority criterion that forms the basic criterion for cost–benefit analysis. It is not especially different from the social welfare superiority criterion for cases where a strictly economic analysis can be performed. A key concept contained in the potential Pareto-superiority criterion, which is not explicitly contained in the scalar social welfare criterion, is that of gainers *potentially* compensating losers. This implies a precise measurement of costs and benefits, which will often be very difficult to accomplish. It will require a cardinal preference scale.

In Section 6.2 we introduce some necessary preliminary concepts. Included among these are various subjects from engineering economics, including net present worth (NPW), rate of return, depreciation, inflation, opportunity costs, portfolios of projects, and discount rates. Then we turn our attention to a more formal definition of cost–benefit analysis. Following this, we discuss the role of consumer and producer surplus, shadow pricing, and valuation of unmarketed goods in cost–benefit analysis. Finally, we present a brief discussion of cost–effectiveness analysis. The generic principles and concepts of cost–benefit and cost–effectiveness analyses are equally applicable in the private and public sectors. Our discussions focus on the usage of these important concepts in both sectors.

6.2 THE TIME VALUE OF MONEY

Generally, it is not fully meaningful to simply add monetary values when these exist at different points in time. Few of us would be just as happy with $10,000 received five years from now as we are to have the $10,000 now. There are several reasons for this. Normally we value present consumption more than we do future consumption. Also, opportunities may be lost due to not having capital at an earlier point in time. Consequently, we are generally willing to pay a premium for present consumption, or the opportunity for consumption, which is a measure of the worth of money over time. This premium is known as the discount rate or interest rate. This interest rate will vary with the supply and demand for money and with the risk that is associated with the venture. As we have noted, we will not specifically deal with risk and uncertainty considerations here, despite their great importance. The interest rate will also have to include the inflation rate.

6.2.1 Present and Future Worth

When we deposit an amount P_0 into a savings account at year 0 and the yearly interest earned on the investment is i times the principal, then we expect to earn iP_0 in interest in one year, and to have available at the end of the first year the principal plus interest, or $F_1 = P_0 + iP_0 = P_0(1 + i)$. If this amount is invested for the next year, we will have $F_2 = P_1(1 + i) = F_1(1 + i) = P_0(1 + i)^2$ at the end of the second year. At the end of N years we will have accumulated

$$F_N = P_0(1+i)^N \tag{6.1}$$

In a similar way, the present value P_0 of a future amount F_N is

$$P_0 = \frac{F_N}{(1+i)^N} \tag{6.2}$$

Often amounts are invested over a number of years. As we have noted, it would not be generally meaningful to simply add monies that occur at different points in time. But it is generally meaningful to add the present values of amounts A_n that occur at different points in time. The relation

$$P_{0,n} = \frac{A_n}{(1+i)^n} \tag{6.3}$$

expresses the worth, really the discounted worth, at time 0 of an amount A, at time n, when the interest rate is i. We simply add these amounts for all n under consideration and obtain

$$P_0 = \sum_{n=0}^{N} P_{0,n} = \sum_{n=0}^{N} \frac{A_n}{(1+i)^n}$$

There is no need for the interest to remain constant from year to year. Should the interest rate from year k to year $k+1$ be i_k, then we have

$$P_{0,n} = \frac{A_n}{\Pi_{k=0}^{n-1}(1+i_k)} \tag{6.4}$$

$$P_0 = \sum_{n=0}^{N} P_{0,n} \tag{6.5}$$

This expression represents the present worth of amounts A_n invested for n years, where the interest rate varies from year to year.

In some cases the annual amounts of a benefit or a cost are constant. Then, if the discount rate is constant over the period under consideration, relatively simple closed-form expressions for present worth result. Many ways can be used to obtain these. One simple way is to note that the present worth, at time 0, of an amount 1 invested forever on an annual basis beginning at year 1 is

$$\frac{1}{1+i} + \frac{1}{(1+i)^2} + \cdots + \frac{1}{(1+i)^n} + \cdots = \frac{1}{i}$$

This follows in a simple manner from

$$\frac{a}{1-a} = a + a^2 + a^3 + \cdots, \quad a < 1$$

where $a = (1 + i)^{-1}$. The present value at time N of an amount i invested annually from time $N + 1$ is also i^{-1}, and the present value at time 0 of this is $i^{-1}(1 + i)^{-N}$. We subtract this from the expression i^{-1} and obtain the expression for the present worth, at time 0, of an amount 1 invested annually for N years. We have therefore determined that the present (discounted) worth P_0 of a constant amount A invested at each of the N periods from $n = 1$ to N is given by

$$P = A\frac{(1 + i)^N - 1}{i(1 + i)^N} \tag{6.6}$$

We may easily calculate the future worth, at the Nth period, of annual payments from this relation as

$$F = P(1 + i)^N = \frac{A}{i}\left[(1 + i)^N - 1\right] \tag{6.7}$$

We now drop the subscripts on P and F for convenience. Relations such as these are especially useful for calculating the amount of annuity payments that an initial principal investment will purchase, or for calculating constant-amount mortgage payments. We must, however, be sure that we understand the precise meaning of the calculation, as we will soon see.

Example 6.1: Traditional home mortgage payments are adjusted such that the monthly payments A are constant. For monthly compounding of interest and monthly payments for N years to retire principal amount P, at annual interest i, we have from Equation 6.7

$$A = \frac{(i/12)(1 + i/12)^{12N}}{(1 + i/12)^{12N} - 1}P$$

We may easily obtain a table showing the ratio A/P, which represents the fraction of the principal paid off monthly, and $12NA/P$, which is the ratio of total payments of principal and interest over the life of the mortgage to the principal.

Among other things, there is generally great difficulty in coping with fixed monthly payments and a fixed-size mortgage as interest rates increase. For example, if we could afford to pay 1.7% of the mortgage principal per month, we could pay off a loan in about 6.25 years if the interest rate were 8%, whereas it will take 20 years at 20% interest. There is no way that the monthly payment will be less than 1.67% of the principal if the interest rate is 20%. The monthly payments required to service the interest are $0.20/12 = 0.01667$ of the principal. Of course, these sorts of facts have led to great difficulty in home purchase in recent times.

There are extensive tables available of the form presented here. With the advent of the handheld stored program calculator, and with inexpensive microcomputers, there seems to be little need for them at this time, however. ■

Other factors such as inflation and depreciation (or appreciation) need to be considered when evaluating the present or future worth of a project. For the most part these can be considered just as if they were interest. For example, the worth at year $n+1$ of an investment at year n of P_n that is subject to interest i, inflation r, and depreciation d is given by

$$P_{n+1,n} = P_{n,n}(1+i_n)(1-r_n)(1-d_n) \tag{6.8}$$

Here i_n is the interest in year n, r_n the inflation in year n, and d_n the depreciation in year n. If we assume that this is augmented by an amount A_{n+1} in year $n+1$, then the worth of the investment at the start of the next period is

$$P_{n+1,n+1} = P_{n+1,n} + A_{n+1} \tag{6.9}$$

or

$$P_{n+1,n+1} = P_{n,n}(1+i_n)(1-r_n)(1-d_n) + A_{n+1} \tag{6.10}$$

We can solve this difference equation with arbitrary values of the parameters i_n, r_n, d_n, and A_{n+1} to yield the future worth of any given investment path and annual investment over time.

In the simplest case the initial investment $P_{n,n}$ is zero and the annual investment and rates are constant. Then it is a simple matter for us to show that the result of an investment over N years is given by the future amount at year N:

$$F = P_{N,N} = \frac{\left[(1+I)^N - 1\right]A}{I} \tag{6.11}$$

Here

$$I = i - r - d + rd - ir - id - ird$$

If the percentage rates are all quite small, little error results from using the approximation obtained by dropping the products of small terms. We have

$$I \approx i - r - d$$

such that we simply use an effective interest rate that is the true interest rate less inflation and depreciation.

6.2.2 Economic Appraisal Methods

There are several existing methods that we can use to yield a single number that is reflective of the economic value of a project. The crudest of these do not consider the time value of money at all. The *payback period*, for example, is the time required from the start of a project before total cash flow becomes positive. Generally, projects involve initial outflows of capital followed by returns on the investment. As can be easily shown, the payback period is a rather naive criterion to use in evaluating projects.

Another common method of project evaluation is called the *internal rate of return* (IRR). It is that interest rate that will result in economic benefits from the project being equal to economic costs, assuming that *all* cash flows can be invested at the IRR. The assumption that it is possible to invest cash flows *from the project* at the IRR will often be incorrect. Thus the number that results from the IRR calculation may give an unfortunate impression of the actual return on investment (ROI). It tends to favor short-term projects that yield benefits quickly, as contrasted with projects that yield long-term benefits only, because of this reinvestment at the constant IRR assumption.

The relation from which the IRR may be calculated is easily obtained. If we assume that a project, of duration N years, has benefits B_n in year n and costs C_N, then the present value of the benefits and costs are, assuming that the interest rate is constant,

$$\mathrm{PB}_0 = \sum_{n=0}^{N} \frac{B_n}{(1+i)^n} \tag{6.12}$$

$$\mathrm{PC}_0 = \sum_{n=0}^{N} \frac{C_n}{(1+i)^n} \tag{6.13}$$

The NPW of the project is given by

$$\mathrm{NPW} = \mathrm{PB}_0 - \mathrm{PC}_0 = \sum_{n=0}^{N} \frac{B_n - C_n}{(1+i)^n} \tag{6.14}$$

To obtain the IRR we set the NPW equal to zero and solve the resulting Nth-order algebraic equation for the IRR.

Three other criteria result from these relations. We have just defined the NPW criterion as the relation from which the IRR is obtained. We note, however, that the interest rate used in the NPW calculation is the actual interest rate, whereas the interest rate obtained from the IRR criterion is a fictitious rate that assumes reinvestment possibilities at the IRR. The *benefit–cost ratio* (BCR) is just the ratio of benefits to cost and is given by

$$\mathrm{BCR} = \frac{\mathrm{PB}_0}{\mathrm{PC}_0} = \frac{\sum_{n=0}^{N} B_n/(1+i)^n}{\sum_{n=0}^{N} C_n/(1+i)^n}$$

The ROI is just the ratio of the net present value to the net present costs and is given by

$$\mathrm{ROI} = \frac{\mathrm{PB}_0 + \mathrm{PC}_0}{\mathrm{PC}_0} = \mathrm{BCR} - 1$$

Sometimes the ROI criterion is called the net benefit to cost ratio (NBCR) or the NPW to cost ratio.

It is of considerable interest to contrast and compare these criteria so as to enable a determination of the criterion, or criteria, most appropriate in

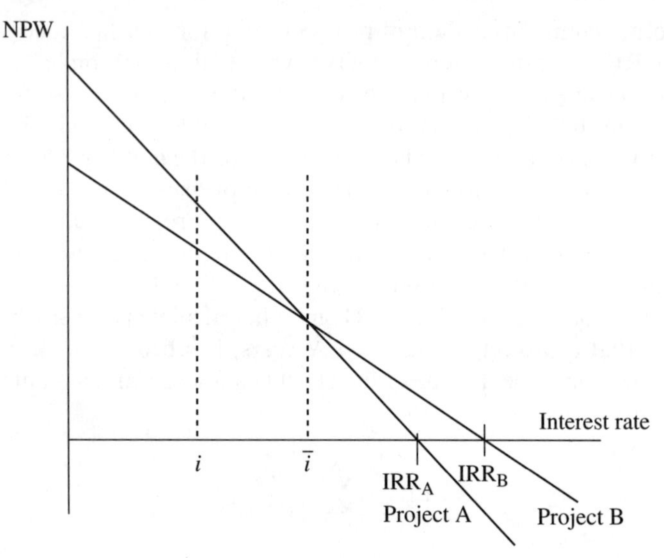

Figure 6.1. New Present Worth and Internal Rate of Return for Two Simple Projects.

particular circumstances. We will first consider the selection of a single project from several. An immediate problem that arises is that NPW and IRR may lead to conflicting results. The BCR criterion may result in a different ranking also. Unless costs are constrained, there is no real reason to use a BCR criterion, however.

Two projects may have the NPW versus interest rate curves shown in Fig. 6.1. Here we see that if the actual interest rate i is less than interest rate $\bar{i}$, the rate at which the NPW of the two investments are the same, we prefer investment A over investment B if we use the NPW criterion. If the interest rate i is greater than $\bar{i}$, we prefer investment B. However, if the actual interest rate is greater than IRR_B, we would prefer not to invest at all, if this is possible, since the NPW of each investment is negative.

The IRR of project A is IRR_A and that of project B is IRR_B. Since IRR_B is greater than IRR_A, we would prefer project B to project A by the simplest IRR criterion. Some analysts would calculate a differential IRR by assuming one project as a base project and calculate a differential IRR (ΔIRR) for switching from the first project to the second. If, for some reason, we are able to select project A as the base project, then the ΔIRR curve is obtained from $\Delta NPW = NPW_B - NPW_A$, as shown in Fig. 6.2. The ΔIRR of the differential investment is i. If i is greater than the actual interest rate i, or possibly a minimum attractive rate of return (MARR), then we should select project B. This is opposite to the conclusion obtained before. So perhaps we should have computed $\Delta NPW = NPW_A - NPW_B$, but there is no logic to suggest this. Let us examine these two criteria in greater detail such that we obtain a greater appreciation for what they actually measure.

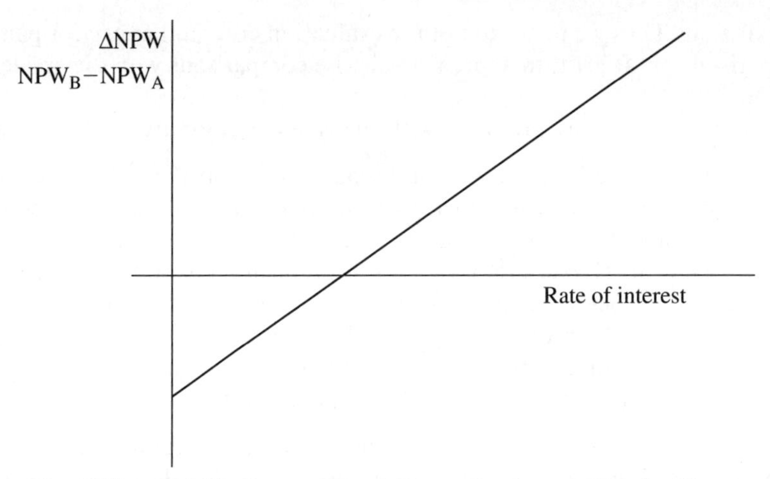

Figure 6.2. Differential Net Present Worth Curve: Net Present Worth of Investment B Minus That of Investment A.

6.2.3 Comparison of Net Present Worth and Internal Rate of Return Criteria

To set the stage for a comparison of the NPW and IRR as investment rules, it is of value to discuss desirable properties for an investment rule.

The following are four essential properties or assumptions that any rational investment rule or criterion should satisfy:

1. All cash flows should be considered.
2. The cash flows should be discounted at the opportunity cost of capital (OCC) or some specified discount rate.
3. The rule should select, from a set of mutually exclusive projects, the one that maximizes benefit or wealth broadly defined.
4. One should valuate one project independently of all others.

Property 1 is trivially desirable, since we cannot obtain a valid measure of the worth of an investment unless we consider all the cash flows that constitute that investment. The OCC is the interest rate that can be obtained on an investment, and we will obtain an unrealistic picture of the worth of an investment unless we consider this and discount the investment at this or some other specified (social) discount rate. Wealth or benefit maximization is clearly the scalar performance criterion we should use, as this, broadly defined, represents economic effectiveness. Finally, it is desirable that we evaluate a project independently of all others, perhaps by comparing it to an absolute standard. This assures us that economic evaluation will be relatively simple to accomplish. For n projects we need only accomplish n evaluations, as contrasted with a much larger number that would result if the evaluations were not independent. Also, there can exist transitivity difficulties in making selective pairwise alternative comparisons. Thus we prefer to not use evaluation criteria that call for pairwise comparisons or, at least, to approach pairwise comparisons with the greatest of caution.

Some of the many problems with the IRR criterion are the following:

A. The IRR criterion does not discount money at the OCC. There is an implicit assumption inherent in the IRR criterion that the time value of money is the IRR, since all cash flows are discounted at that rate. Thus the IRR criterion violates the implicit reinvestment rate assumption property 2 that money should be discounted at the OCC.

B. Because of problem A, the IRR criterion does not allow one to consider time-varying opportunity costs of capital. Since this often occurs, this is a rather significant potential flaw.

As a consequence of violating rational rule 2, due to either or both problems A and B, use of the IRR as a decision criterion will often lead to choice of an inferior investment. To illustrate a simple example of this, consider two investment alternatives that yield the NPW versus discount rate curves shown in Fig. 6.3. The IRR criterion will select project 2 as being better than project 1, whereas in reality project 1 is better at the OCC shown in Fig. 6.3. The reason why the IRR criterion errs here, and in other cases as well, is that the unstated implicit assumption is made that all cash flows can be reinvested at a rate equal to the IRR. This implicit assumption defies common sense and logic. Thus use of the IRR as a criterion does not assure selection of the project that maximizes wealth.

C. Consequently, the IRR criterion violates rule 3. This violation was apparently first discovered by the noted economist, statistician, and decision theorist Savage, together with coauthor Lorie, in 1955 in their famous paper "Three problems in capital rationing."

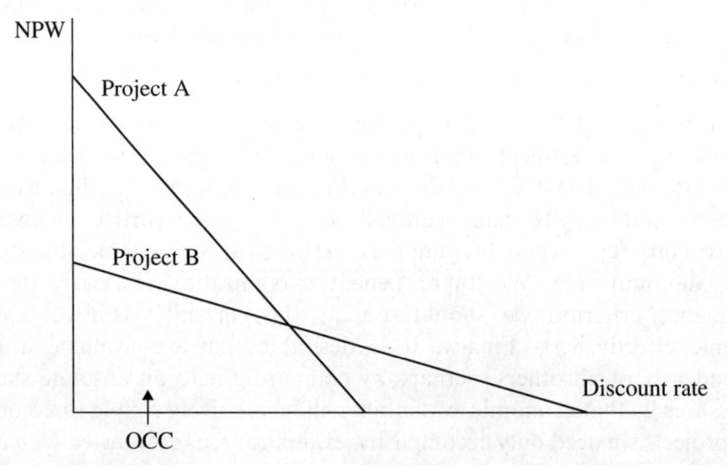

Figure 6.3. Net Present Worth Versus Interest Rate, Discount Rate, or Opportunity Cost of Capital.

One way in which some people propose to avoid this fundamental flaw with the IRR criterion is to use the IRR criterion on the cash flow differences between two investment alternatives. But the difficulties with the IRR do not end by using these differential calculations. Calculations using the incremental rate of return using the algorithms often posed may give precisely the wrong answer. The following steps are often recommended to accomplish the differential IRR calculations:

a. For two alternatives, the incremental rate of return (ΔROR) on the difference between the alternatives is computed. This is compared with a threshold of acceptability called the minimum attractive rate of return (MARR):
If $\Delta\text{ROR} \geq \text{MARR}$, the higher cost alternative is chosen.
If $\Delta\text{ROR} < \text{MARR}$, the lower cost alternative is chosen.

b. For three or more alternatives the typical procedure used is a bit more complicated. An incremental pairwise analysis is generally used in which the computed incremental IRR is compared to the MARR for essentially all differences between pairwise comparisons of projects. An increment is said to be economically desirable by the IRR criterion whenever $\Delta\text{IRR} \geq \text{MARR}$.

The typical steps in the incremental rate of return analysis for more than two alternatives are the following:

1. Compute the rate of return for each alternative and reject any alternatives whose IRR is less than the MARR.
2. Rank the remaining alternatives in their order of increasing present worth of cost. If a higher cost alternative has an IRR greater than that of a lower cost alternative, then the lower cost alternative may be immediately rejected according to the IRR criterion.
3. Consider only those alternatives not rejected in steps 1 and 2. Compute the incremental IRR (ΔIRR) on the differences between the two lowest cost alternatives by subtracting the alternative with the higher present value of cost from that with the lower present value of cost. If the $\Delta\text{IRR} \geq \text{MARR}$, the increment is desirable according to this rule. The higher cost alternative should be retained for further consideration and the lower cost alternative rejected. The opposite judgment is made when $\Delta\text{IRR} < \text{MARR}$.
4. Take the preferred alternative from step 3, consider the next higher cost alternative, and proceed with another two alternative pairwise comparisons.
5. This procedure is continued until all alternatives have been examined and the best of the multiple alternatives has been identified according to this particular criterion.

In all of this analysis, it is implicitly assumed that all positive differential increments of investments—all investments with ΔIRRs that exceed the

MARR—are desirable. In every situation, the best project will invariably depend on the assumed discount rate. *There cannot, in general, be any "best project" that is independent of the OCC.* Thus our "desirable property 2" is indeed desirable. Even if these complexities and difficulties were not present, there are still other problems with the IRR criterion:

D. Use of the IRR necessitates confounding all projects that are under consideration to select the best project, even if the projects are mutually exclusive. Thus if there are N projects, there will need to be $N!$ pairwise comparisons of projects. We cannot, or at least should not, use the simplified approach of eliminating an alternative once it is shown to be inferior to another. This approach is generally not applicable, as its use presumes transitivity of the IRR as an evaluation rule, and transitivity is not guaranteed. If there are 8 mutually exclusive projects to consider, we might need to make $8! = 40{,}320$ differential pairwise comparisons to select the best single project. Thus the amount of computational labor required to select a single project from eight can be enormous. This would be bad by itself, but there are even more significant deficiencies in the IRR as a criterion. When one considers portfolio selection, the problem becomes truly unmanageable, for one must evaluate $2^{N!}$ portfolios for N projects.

E. Even the addition of independent projects alters project selection when using the IRR as a criterion. This and problem D violate property 4, the value additivity property that is desirable of an investment criterion.

The conclusion that the IRR is only of any value when one has a very "simple" investment, such as an initial cost followed by a constant steady stream of benefits, appears correct and unavoidable. But let us continue with our discussion of the virtues of this IRR criterion. Another problem is the following:

F. The IRR rule can, and generally will, result in multiple rates of return. None of these may make sense because of the violation of rule 2. Furthermore, these multiple rates of return will typically violate rule or property 4.

Example 6.2: A classic example[1] in which this violation occurs is the oil well pump problem. An oil company wants to know whether to install a new, higher-speed pump on a well already in operation. With the existing pump, the cash flow will be \$10 million in years 1 and 2, and the oil well will be fully depleted. A new pump costs \$1.6 million, but will allow extraction of the \$20 million worth of oil in year 1. The basic analysis using the IRR criterion is always a differential or incremental analysis, and the following are incremental cash flows representing the difference between the new pump

[1]Due to Lorie and Savage (1955).

and the existing pump. For the cash flow due to the new pump, we have the following:

Year	Cash flow
0	−1,600,000
1	−10,000,000
2	10,000,000

The reason for the negative cash flow in year 2 is that all of the oil is extricated in year 1 with the new "superpump." With the old pump, there would be some oil to extract in year 2. The project has two IRRs, 25% and 400%. Suppose the OCC = MARR = 10%. We have an abundance of IRRs (although perhaps not oil) now, and each exceeds the OCC or MARR. Should the project be accepted, since both of these IRR rates are very good? The NPW of this investment is easily shown to be *negative*, about −773,554, and the oil company should not buy the pump, even though the IRR approach would suggest it.

It is actually possible, however, to use a modified form of IRR correctly. There is a real physical and conceptual problem associated with the IRR criterion. There *cannot* be more than one rate of return, although incorrect thought may lead to a formal computation of several rates of return. What we should note is that the cash the oil company gives to the project earns at the IRR, but the cash the project gives to the company earns at the OCC. There is simply no way in general that the cash payout at year 1 can be invested at the IRR. The firm invests $1.6 million initially in the project and this is worth, at the end of the first year (in millions of dollars, or M dollars), 1.6 (1 + IRR).

This returns a gain at year 1 at the IRR, since there is an investment in the project. The amount paid to the firm from the oil extracted exceeds the investment in the project. This money is invested back in the project to the extent of the discounted initial investment. This money, over the next period, earns at the OCC. So we have (in M dollars) $10(1 + i) - 1.6(1 + \text{IRR})(1 + i)$ as the amount the firm has earned at the end of the second period. The firm then pays $10M at the end of the second interest period, and so we have for the net worth at that time $F = -10 + 10(1 + i) + 1.6(1 + \text{IRR})(1 + i)$. The $\text{OCC} = i = 10\%$, and we can easily compute the IRR and obtain, for $F = 0$, IRR = −43.18%. The IRR is *negative* as well, as it should be. The project should be rejected. We note that this correct analysis using a modified IRR gives the same conclusion. However, this is a very cumbersome analysis. There are several nonlinear logic steps involved, since one must make a decision at each point concerning whether to use OCC or IRR as the interest rate. The basic IRR approach is much simpler, easy to present, easy to understand, and *wrong*. We do not advocate it, except in very simple cases, such as the one noted earlier. For complex situations, the calculations needed to obtain meaningful results using the IRR criterion are very tedious to perform and not necessarily easy to understand. ■

There have been a number of efforts to salvage the IRR concept by developing a taxonomy of investments and then developing IRR procedures for each resulting category. It is perhaps easiest to distinguish between what we might regard as *simple conventional* and *nonsimple nonconventional* investments. A simple conventional investment is one in which there is an initial flow of capital into the project in years $0, 1, \ldots, M$ and an outflow of cash from the project in later years $M+1$, $M+\cdots$, N. A nonsimple nonconventional investment is one that is not simple or conventional—an investment has one or more positive cash inflows into the investment *after* a positive cash outflow from the project. For example, the cash flow from Example 6.1 represents a conventional investment, whereas that in Example 6.2 represents a nonconventional investment. A conventional investment is one whose cumulative net worth from 0 to a point in time T necessarily improves after some point in time where the cash flows first become (and remain) positive. In this sense, a conventional investment is necessarily a *pure* investment. A nonconventional investment may be a pure investment, but it may also be a *mixed investment* in the sense that the net worth from 0 to a point in time T may well not be monotonically increasing in T for sufficiently large T; it may become negative for sufficiently large T. A mixed project investment is one that has unrecovered investment balances such that the firm "loans" money to the project and overrecovered investment balances such that the firm "borrows" money from the investment. At a *valid* IRR, a pure investment is one such that the firm never "borrows" from the project. It is to obtain this valid IRR that we corrected the conventional IRR calculations in Example 6.2, which represents a nonsimple mixed investment strategy. When investments are nonsimple and mixed, we need to use two interest rates: an IRR, perhaps more properly called return from the project on invested capital, and a return on capital loaned to the firm from the project that can only earn at the OCC.

It is easy to examine the cash flow from a project to determine whether we are dealing with a simple or nonsimple investment. If a sequence of positive returns to the firm follows a sequence of negative returns (or investments), we have a simple investment. A simple investment is necessarily pure, and there can exist only one positive IRR.

If we have a nonsimple investment, we can calculate the project investment balances as a function of time using sequences of an IRR that is not yet known and the OCC discount rate. The value of IRR that would cause this balance to go to zero would be determined. If this computed IRR is between the values of interest that produce positive NPW at that point in time, we can generally assume that the project is returning money to the firm at that point in time. After this, the project is loaning money to the firm, and the OCC should be used to compute balances at the next period. A general statement of an appropriate algorithm is available, and contained in the work by Bussey (1978), but we will not present it here.

We have presented a number of discussions here to indicate that the *IRR is the interest rate earned only on the unrecovered balances of an investment* such that a precise zero balance of costs and benefits occurs at the assumed end of the project. As we will indicate by means of an example, the IRR is not just a rate of interest on the positive investments into the project. When there are positive recovered investment balances, or intermediate cash flows to the firm from the projects, as typically there will be when we compare two project alternatives, we must use a *correct* interpretation of the ΔIRR calculations if we are to obtain consistent results. Generally this is complex. Even when we have initial simple pure investments, we can only determine whether or not a project is acceptable in terms of whether or not IRR > MARR. We *cannot* ever justify using IRR as a figure of merit to use in ranking projects. To do this we must use the corrected ΔIRR concept, which generally results in a nonsimple mixed investment project and a tedious evaluation effort.

Example 6.3: As a final illustration of the IRR and NPW criteria as decision rules, let us briefly consider the cash flow from three very simple investments:

Year	Investment A, $	Investment B, $	Investment C, $
0	−100.01	−100	−100
1	50	0	0
2	50	0	0
3	50	168.72	180

Each of these investments represents a simple pure investment. Suppose that we assume that OCC = 12%. Then we easily calculate the NPW and IRR and obtain the following:

	NPV, $	IRR, %
A	20.08	23.35
B	20.09	19.05
C	28.12	21.645

Clearly investment *C* is the superior investment, as it has the higher NPW. It does not have the highest IRR, however. For all practical purposes, from the viewpoint of initial cost and NPW, investments A and B are equivalent, although investment B has a 1¢ higher NPW and a 1¢ lower initial cost.

If we use the IRR criterion, investment A is the clear winner. In reality, it is the worst loser. We note, incidentally, that if we could purchase investments A, B, and C, the total NPW is just the sum of the NPWs for the three investments, that is, $68.30. This indicates the simplicity of the NPW calculation. But what about the IRR for the combined three investments? It turns out to be IRR(A + B + C) = 21.084% and is obtained from the solution of a nonlinear algebraic equation. There is no way that the three individual IRRs can be combined to yield the total IRR.

If we apply rule 2 for use of the IRR criterion, we note that higher cost alternative A has an IRR greater than the rate of return of *all* lower cost alternatives; so immediately we may pick A as the winner. Of course, it is the logical loser. The incremental flows are given by the following:

Year	Investment A–B, $	Investment A–C, $
0	−0.01	−0.01
1	50	50
2	50	50
3	−118.72	130

Each of these represents a nonsimple mixed investment. We calculate the NPW and ΔIRR and obtain the following:

Project	NPW, $	ΔIRR, %
A–B	−0.01	12.01, ∞
A–C	−8.03	18.82, ∞

If we assume that MARR = OCC = 12%, we see that we should purchase the differential projects (i.e., the higher cost alternative A) in both cases. Even though the ΔIRR criterion says purchase A compared to B, there is virtually no difference between them; the ΔIRR criterion says there is a 12% difference in internal return (whatever that is). Now, the ΔIRR criterion convinces us to purchase project A compared to project C. This differential comparison might even suggest that B is better than C. We see this if we look at the difference between the ΔIRRs for A–B and A–C: A is much more to be preferred to C than it is to B.

But the differential NPW criterion, which we never have to use because of the value additivity property of NPW, clearly shows that the differential projects all have negative worth at the OCC. Once again the use of IRR leads to results that defy logic. Project A is just not better than B or C except at a very large OCC, as a simple sketch of NPW versus i will show.

Again we can formally make the IRR yield the correct answer by correcting for the fundamental conceptual error, the constant reinvestment rate assumption, implicit in our use of the IRR criterion thus far in this example. If we assume that all cash disbursements from the investment to us can only be reinvested at the OCC and that cash balance is obtained at year 2, we obtain $-100.01(1+\mathrm{RII})^3 + 50(1.12)^2 + 50(1.12) + 50 = 0$ as the relation to solve for the return on initial investment (RII). Here project A returns 19.048%, project B returns 19.05%, and project C returns 21.645%. We see that we can rank the investments using this correct, but cumbersome, interpretation of the RII and obtain the correct and proper preference ranking C > B > A.

This conclusion can also be obtained from a corrected interpretation of the incremental cash flows that we have just obtained. Initially the differential projects A–B and A–C result in a cash inflow to the project that earns at the ΔIRR. The returns of 50 for years 1 and 2 represent cash outflows to the firm from the project that can only be invested at the OCC. Thus we have for cash flow balance at year 3

$$0 = -118.72 + 50(1+i) + 50(1+i)^2 - 0.01(1+\text{IRR})(1+i)^2$$

for differential investment project A–B. For differential project A–C we get

$$0 = -130 + 50(1+i) + 50(1+i)^2 - 0.01(1+\text{IRR})(1+i)^2$$

We could use another cost than the OCC for i here, but the most useful interpretation of the MARR seems to be that it should be the OCC. Using $i = \text{OCC} = 0.12$, we obtain negative numbers for both IRRs. This indicates that both B and C are better than project A. Project C can easily be shown to be better than B, and so it is the clear winner. We would get these same results in a somewhat more meaningful way by determining the corrected IRRs for project difference C–A.

So while we *can* get correct answers from a correct interpretation of the IRR criterion, it is very complicated, and it is simply not true that we can generally determine the IRR in a meaningful and correct way without knowledge of a discount rate for return of borrowed capital.

Another potentially fundamental concern arises when we leave the small decision-type problems of selecting pressure relief valve A or B to major impact problems in which we are trying to aid in the decision, say, concerning whether to build a nuclear power plant or a coal-fired power plant. This concern arises because the incremental differences between the two major, perhaps fundamentally different, projects may make no sense at all except perhaps in a life-cycle cost sense. How do we difference the system effectiveness attributes for the two decision alternatives just suggested? How does one trade off the differential IRR when there exist noncommensurate attributes associated with the projects? If we accept the premise that the solution to an economic systems analysis problem is really just the solution of a mathematical equation, then the arguments posed here are not cogent. Of course they are, for it is always necessary that the mathematics we use replicate reality. If this is not done, major troubles can ensue. ■

In this section, we have examined some salient attributes of the IRR criterion. Of four desired conditions for a rational economic decision criterion, the IRR criterion is shown to satisfy only one of them, that is, we can consider all cash flows. The NPW criterion can, however, accommodate all four criteria. In particular, it obeys value additivity; it correctly discounts at the OCC or social discount rate; and finally, it results in maximization of profit or worth. Furthermore, it is compatible with most forms of microeconomic and welfare economics used today as a basis for planning and decision making in industry and government. Thus it is the criterion of choice to use in comparing individual projects and, as we will see, for portfolios of projects.

Our arguments thus far in this section show that the IRR can often be expected to be a very unreliable criterion, despite the fact that it is, in practice, often used. There is nothing in our argument thus far, however, that assures us that the NPW criterion is the criterion of choice. It is possible to pose a set of very reasonable axioms, or properties of an investment decision, that only the NPW criterion satisfies. We will now do this.

We wish to specify an investment criterion for investments over the interval (0, N). We will require that the investment criterion be

a. ***Complete:*** Given any two sequences of cash flows **x** and **y**, we can always say that **x** is not preferred to $\mathbf{y}[\mathbf{y}\Re\mathbf{x}$ or $\mathbf{y} \geq \mathbf{x}]$ or **y** is not preferred to **x** $[\mathbf{x}\Re\mathbf{y}$ or $\mathbf{x} \geq \mathbf{y}]$.
b. ***Transitive:*** With three cash flow vectors **x**, **y**, and **z**, if $\mathbf{x} \geq \mathbf{y}$ and $\mathbf{y} \geq \mathbf{z}$, then transitivity requires $\mathbf{x} \geq \mathbf{z}$.

Next we state five axioms or desirable properties that should be possessed by a preference ordering of cash flow vectors.

Axiom 1. Continuity.

If **x** is preferred to **y** $[\mathbf{x} > \mathbf{y}]$, then for sufficiently small $\varepsilon > \mathbf{0}$, cash flow vector $\mathbf{x} - \varepsilon$ is also preferred to **y** $[\mathbf{x} - \varepsilon > \mathbf{y}]$. This continuity axiom is very reasonable in that it assures us that the preference criterion and associated preferences will not be schizophrenic for small arbitrary changes in the return of the investments.

Axiom 2. Dominance.

If cash flow **x** is at least as large as **y** at every period in (0, N) and strictly greater in at least one period, then $\mathbf{x} > \mathbf{y}$. This dominance criterion is equivalent to greed—more is preferred to less—or consumer satiation. All it says is that if $x_i \geq y_i$ for all i and $x_i > y_i$ for at least one i, then $\mathbf{x} \geq \mathbf{y}$ and $\mathbf{x} > \mathbf{y}$.

Axiom 3. Time Value of Money.

If two investments **x** and **y** are identical except that an incremental cash flow obtained by investment **x** at some period n does not result until period $n+1$ for investment **y**, then **x** is preferred to **y**. This investment axiom is equivalent to impatience. We would prefer to have an incremental amount of money now than at some future time.

Axiom 4. Consistency at the Margin.

We prefer cash flow **x** to **y** if and only if the differential cash flow $\mathbf{x} - \mathbf{y} = [x_0 - y_0, x_1 - y_1, \ldots, x_N - y_N]^{\mathrm{T}}$ is preferred to a cash flow of $\mathbf{0} = [0, 0, \ldots, 0]^{\mathrm{T}}$.

Axiom 5. Consistency in Time.

If we shift all investment returns by some number of periods, the preference order is unchanged.

Surely, it would be difficult to argue that any of these axioms are unreasonable. If only Axioms 1–4 are satisfied, then it is possible to show[2] the following:

Theorem A.

The only preferences that satisfy Axioms 1–4 are those given by the NPW criterion in which interest or discount rates, which may vary from period to period, are positive.

This theorem just says that we should value investment **x** according to an NPW criterion, or decision rule, in which we obtain the present value of the investment component at the nth period x_n by using the standard discounting relation

$$\mathrm{NPW}(x_n) = x_n \prod_{j=1}^{n} \frac{1}{1 + i_{j-1}} \tag{6.15}$$

and then add together these present values over all periods to obtain

$$\mathrm{NPW}(\mathbf{x}) = \sum_{n=0}^{N} \mathrm{NPW}(x_n) \tag{6.16}$$

or

$$\mathrm{NPW}(\mathbf{x}) = \sum_{n=0}^{N} \left(x_n \prod_{j=1}^{n} \frac{1}{1 + i_{j-1}} \right) \tag{6.17}$$

Here i_{j-1} is the interest received in the time interval from period $j-1$ to present j.

When we also impose Axiom 5, we require further that the interest or discount rate be constant throughout the investment time interval $(0, N)$. Thus we have the following:

Theorem B.

The only preferences that satisfy Axioms 1–5 are those given by the NPW criterion in which the interest or discount rates are constant and positive.

If we accept these axioms as reasonable, then we can only accept the NPW criterion, or one that would yield the same preference ordering, as truly reasonable. Since the IRR criterion does not generally yield the same preferences as the NPW criterion, it must be rejected as an inferior criterion.

[2]See Williams and Nassar (1966).

6.2.4 Benefit–Cost Ratio and Portfolio Analysis

In this section we discuss calculation of the BCR. Also, we indicate why and where it is such an important criterion. To do this we need to introduce some concepts from portfolio analysis that concern decisions with a constrained budget. Generally BCRs are always defined by discounted benefits and discounted costs. We compute the BCR from Equations 6.12 and 6.13:

$$\mathrm{BCR} = \frac{\mathrm{PB}_0}{\mathrm{PC}_0} = \frac{\sum_{n=0}^{N} B_n/(1+i)^n}{\sum_{n=0}^{N} C_n/(1+i)^n} \tag{6.18}$$

As is easily seen, we do not have to discount the benefits and costs to the present time to determine the BCR. Any time will do. Nor do we need to do this to obtain the ROI as BCR/(1 + BCR). It is convenient, however, to discount benefits and costs to the present time, for then we can easily obtain the NPW from $\mathrm{PB}_0 - \mathrm{PC}_0$.

It is easily seen that ranking projects by BCR will generally lead to results different from those obtained by the use of NPW.

Example 6.4: Suppose that we have six projects, each of which involves initial costs at year 0 and benefits that flow over a five-year period. Suppose that the discounted present value of benefits, initial costs, and BCR and NPW are given by the following:

Project	PB_0	PC_0	BCR	NPW
A	10	5	2	5
B	20	12	1.67	8
C	5	2	2.50	3
D	10	5.56	1.80	4.44
E	5	2.22	2.25	2.78
F	10	5.88	1.70	4.12

If we rank the projects by NPW, we have

$$B > A > D > F > C > E$$

However, if we rank them by BCR, we obtain

$$C > E > A > D > F > B$$

We see that B is the best alternative from the viewpoint of NPW and the worst alternative from the point of view of BCR.

Actually, alternative B is both best and worst. If we can only pick a single alternative, and have as much as 12 units of money to spend, *and have nothing that we can do with the money not spent*, then B is indeed our best alternative. NPW is, in this sense, an effectiveness criterion. But if we value projects from the point of view of efficiency only, then clearly BCR is the criterion of choice and project C is the most efficient project. It returns the greatest fraction (2.5 times) of its cost. The next most efficient project E returns 2.25 times its costs; B is not at all an efficient project here.

If we had 12 units of money to spend *and* could purchase as many units of the same project as we wished, we would surely wish to purchase a number of projects of type C. In this particular case, we can spend all of our money on type C projects. Six of them will cost 12 units of money and will return 30 units, discounted to the present. The most effective, and efficient, use of our 12 units of money would be to purchase 6 units of project C.

Generally we would not expect that we could use all of our money on type C projects and would have some money left over. This suggests that we need to form a number of portfolios of projects, each portfolio costing 12 units, and select from that portfolio the one that has the greatest NPW *or* BCR. Obviously the NPW and BCR criteria are the same if we constrain PC_0.

The precise complete formulation and solution of the above-posed problem is an exercise in integer programming that we will not develop here. An important subcase of this problem occurs when we can incorporate each project in the portfolio at most a single time. Clearly we wish to incorporate the most efficient projects in the portfolio, but we cannot exceed the constraint of 12 units cost. Thus we rank-order the projects according to decreasing BCR and include as many as we can until we exceed the cost constraint. Again we will have problems because of the integral nature of the projects. Often simple heuristics can be posed that will allow easy identification of several candidate best portfolios, and we would then select the one with the highest NPW.

Here, for example, relatively efficient portfolios are the following:

Portfolio	PB_0	PC_0	NPW
C, E, A	20	9.22	10.78
C, E, D	20	9.78	10.22
C, E, F	20	10.10	9.9

and the {C, E, A} portfolio is clearly the best. The NPW from this portfolio is 10.78, to which we should add the unspent 2.78, from the available funding of 12, to get 13.56. ■

6.2.5 The Discount Rate

In our discussions thus far in this chapter, we have assumed that there is a market rate of interest, which we have called the discount rate or opportunity cost of capital. However, a perfect capital market will generally not exist, and there is no single interest rate. Several related approaches to determining a discount rate are possible: marginal interest rates, marginal time preference rates (MTPRs) for individuals, corporate discount rates, MARRs, government borrowing rates, and the *social* opportunity costs of capital.

Corporate discount rates include a premium for risk, a markup for corporate taxes on incomes above a certain level, and a profit to be returned to shareholders. For example, if government bonds return 10% and the corporation expects a 4% premium for risk to be reasonable, then the corporation will have to obtain almost 28% ROI to provide a 14% return to stockholders. This is the case for "large" businesses. Small business corporations can elect to pass all earnings directly to shareholders, who must pay ordinary income taxes on these earnings rather than capital gains taxes, which are generally lower than taxes on ordinary income.

Market interest rates for government and corporate bonds vary, for example, with the perceived risk that the lender of capital takes on an investment. We can compute a market value of outputs, perhaps as a function of industry, and the inputs to the project, all measured in dollars and discounted in time. Then if a return of $112 results from an input investment of $100, we would say that the market interest rate, which would be the realized MARR, is 12%. This would need to be essentially doubled by corporations because of taxes they must pay. In the last paragraph of this section we describe a normative approach to determining a discount rate. Here we describe a descriptive approach.

The individual MTPR or personal discount rate is the rate at which an individual is willing to trade off present personal consumption for future personal consumption. Related to this is the concept of a personal social discount rate. The personal discount rate represents personal deferred consumption preferences. The social discount rate of an individual represents that individual's preferences in terms of societal behavior. Many modifications to rates such as these can be made. One might believe that individuals will set too high a personal social discount rate to enjoy consumption today at the expense of future generations. Thus one might argue, as does the Pigouvian[3] discount rate, for a social discount rate that is lower than the personal social discount rate of present-day individuals.

Doing this will, of course, enhance the NPW of projects with very long-term payoffs. But it seems reasonable to infer a *minimum* social discount rate on the market rate for the private or public organization potentially undertaking the project. The reasoning behind this is relatively straightforward. Projects must be paid for. If future costs and benefits are discounted at too low a rate, an unfortunately erroneous inference of ability and willingness to pay for the project in the future is obtained. But development projects must be paid for in one way or another. If too low a discount rate is assumed, then the project will not be really paid for when higher existing market rates extract their toll. So the burden of paying for the project will fall on groups that are perhaps not even identified in a cost–benefit analysis.

It would seem better for arguments to be made that important attributes of projects have been omitted and that the benefits of a project are greater than those identified by an analysis than to distort the analysis through use of an

[3]Named after Cecil Pigou, a noted nineteenth-century economist.

unreal discount rate. Thus the argument that the social discount rate should represent the opportunity costs of the project—the returns that could be obtained from these funds if used on some other project—is a very potent one. In other words, resources must be used in the most productive way. To not acknowledge this is to forget, in effect, that there are always resource constraints on the use of capital. The task should be to define *productive* in an appropriate way and not to artificially lower discount rates such that projects are justified that cannot pay their own way and which thereby cast the burden of payment in an undefined and unexplored manner, perhaps even on the unborn generations the artificially low discount rate is supposed to project. This seems to be wishful thinking in a most maladroit form.

Most of the discussion in this section applies to cost–benefit analysis and discount rates in the public sector. In the private sector, appropriate definition and use of a discount rate seems much less ambiguous and much more straightforward than in the public sector. In fact, it is the effect on the private sector of capital allocations in the public sector that acts to complicate resource allocation concerns in the private sector. At least three related questions arise:

1. To what extent does a public sector decision to implement a project result in a transfer of funds from the private to the public sector?
2. To what extent does implementation of a project in the public sector result in a reduction of private sector resource allocations for similar projects?
3. If there are foregone private sector investments due to a public sector project implementation, what would be the NPW of the foregone private sector projects, and how would this compare with the NPW of the public sector project?

The answer to these questions is clearly project dependent. We might expect to get considerably different answers, for example, for national defense projects and energy projects. Although these questions are both interesting and important, we will not answer them here.

So, where does this leave us with respect to the selection of a discount rate? One of at least three judgment situations typically exists relative to the choice of a discount rate. First, the rate may be fixed by corporate or government policy, and there is then little that can be done except to use the established rates. Second, the decision maker may have strong personal feelings about the appropriate rate. This is especially the case in private sector efforts. One role of the economic systems analyst and assessor is to work with clients in solving problems. This can include consultations with the decision maker concerning the appropriate rate to use, but, ultimately, the decision maker's wishes must prevail.

In the third situation in which the economic systems analyst and assessor is free to work with the client in the specification of a discount rate, there will generally be no substitute for a sensitivity analysis that will allow determination of the critical discount rate at which decisions switch. Often these critical discount rates are such that arguments about the actual discount rate to use in a problem are futile, in that any reasonable rate will produce the same best decision. A reexamination of the graphs of NPW versus interest rates that have been produced in this section cannot help but reinforce this conclusion. An important use of sensitivity analysis is to allow pertinent parameters to vary over the imprecise values identified by the client.

6.3 IDENTIFICATION OF COSTS AND BENEFITS

Identification and quantification of the benefits and costs of possible alternative courses of action, or projects, or decisions, is usually a difficult task. It is generally not as difficult, perhaps, as formulation of the issue and identification of the alternatives themselves, but it is still not an easy task.

Here we use the word *benefits* to mean the possible effects of a project, both positive benefits and negative benefits, or disbenefits. We must first identify benefits and then we should quantify them by assigning a value to them. Many benefits (and disbenefits) will be intangible and will occur to differing groups or individuals in differing amounts. Problems with intangibles may be especially difficult in the public sector, where associated agencies are designed primarily to deliver services or public goods, rather than products for individual consumption. A major goal, however, of a private sector organization is profit maximization, and it is relatively easy to measure profit as a benefit. The benefits of a public service, such as a school, or a public good, such as a subway system, are much more difficult to define and identify because they are intangible or indivisible (or both). The very political environment of many public sector efforts further complicates measurements—and this means that factors other than variables associated with efficiency, economy, and equity need to be measured.

One valuation philosophy that we might adopt is based on two premises:

Premise A. The value of a project to an individual is equal to the fully informed willingness of the individual to pay for the project.

Premise B. The social value of a project is the sum of the values of the project to the individual members of society.

The fundamental conclusion of microeconomics, under perfect competition conditions, ensures us that these two premises should, under ideal conditions, serve as good guideposts for cost–benefit analysis. As we have seen, when faced with a given price for a good, the rational individual seeking to maximize satisfaction will purchase a number of units of that good so that, at the margin, the individual's willingness to pay for that good is precisely equal to its price.

There is much that is contained in these two premises, and in this fundamental conclusion of the microeconomic theory of the consumer, that needs to be more fully explained. Willingness to pay seems like a reasonable concept, but it is a meaningful measure of value only if it is based on a sound knowledge of all the true benefits of a project. Premise A also implies that the consumer knows best what is good for them, or consumer sovereignty, and that the existing distribution of income is necessarily the most equitable. Postulate B assumes additivity and the equal weighting of individual values. Under these conditions, society is better off if one individual gains a value of $2 + \varepsilon$ from a project while another individual gains nothing than it is if each individual gains one unit of value.

Thus, there are information-related measurement difficulties associated with these premises, and there are potential equity problems as well. Measurement problems arise from the difficulty of assigning comparable values that reflect the value of a complex project. Generally, decomposition of the benefit of a project into attributes and assigning number values to these attributes will result in a more reliable indication of the value of a project than unaided wholistic judgment. A number of studies in behavioral psychology have shown this. We will look at ways to accomplish this decomposition later.

Example 6.5: The benefits associated with a project can be due to an increased value of the project outputs compared with some basic do-nothing alternative, or to reduced costs. A change can result, as we have indicated, in positive or negative benefits. For a manufacturing project, for example, we might develop the attribute tree of benefits shown in Fig. 6.4. For cost–benefit analysis, we need to quantify these values in strictly economic terms. In cost–effectiveness analysis, we do not need to necessarily associate economic value with benefits.

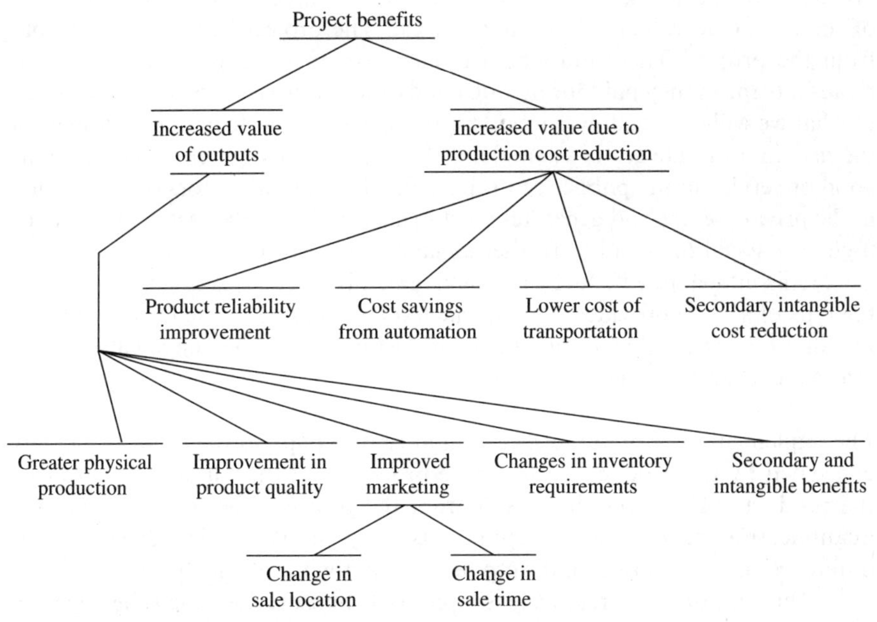

Figure 6.4. Hypothetical Attribute Tree of Project Benefits. ■

Example 6.6: A balance sheet or profile of quantified costs and benefits for the projects may be tabulated after they have been identified. This has great value in indicating the identified and quantified costs and benefits of alternative projects, including secondary and intangible costs and benefits. Sometimes it will be possible to quantify secondary and intangible effects. In many instances, the presence of a large number of important secondary and intangible effects should serve as an indication that a cost–effectiveness analysis might be more suitable than a cost–benefit analysis, which, traditionally, converts all costs and benefits into monetary terms. For example, we could show a balance sheet of benefits for one hypothetical project alternative associated with implementation of a new transportation system within a city. A similar balance sheet can be displayed for costs. Balance sheets of this sort have existed at least since the time of Benjamin Franklin, who called them "decision balance sheets." de Neufville and Stafford (1971) have presented a useful impact incidence matrix that is an alternate "sheet" and can also be used for these purposes.

In practice, it will not be often that a complex project can be displayed in a simple single-page profile, such as we have shown here. Much information is contained in a very well-constructed balance sheet. We can display benefits and costs, discounted to the present using an acceptable discount rate. We can illustrate these impacts spatially. They can be shown for various income groups, for various city services, perhaps private industry groups affected by the system, and in different regions of the community, and at the state (or national) level.

The net economic value added due to a project is its monetary benefits less its monetary costs. The "costs" of a project represent the external inputs to the project. If the project has merit, there are internal values added to it in the form of the economic value of labor and capital. The project benefit is the output from the project. There must be a money flow in the reverse direction that represents the money paid for private goods and services, or their value in terms of what we will call shadow prices. This revenue payment from the customers of the firm in a private sector example, or from use (broadly defined) of a public good or service in the public sector, pays for the internal factors of production in the private sector and generates a net benefit in the public sector that may be used as a social dividend to further enhance the welfare of various publics.

This model may be further disaggregated by sectors, according to region, the type of service provided, the income level of workers in the region, etc. It is this disaggregation that results in the need for the transfer payments, due to internal exchanges within these sectors. ■

The implementation of any project will alter the supply of "inputs," which are consumed by the production or service process, and the supply of "outputs," the products or services that result from the project. The identification and quantification of the benefits and costs of projects involve exploring the difference between inputs and outputs with and without the "project."

This notion is a relatively simple and comfortable one when we are evaluating operational-level projects such as the possible introduction of word processing equipment into an office to replace electric typewriters. However, strategic-level concerns, especially in the face of changing environments, involve changed situations that might prevail if a project is not implemented. Many projects are considered for implementation because of contingencies that will result, if they materialize, in increased costs and/or reduced benefits if one or more new projects are not implemented. In situations such as this, the "no-project" or do-nothing alternative must be described as evolving over time and it must include benefit reductions, or cost increases. Useful descriptions of situations such as these often require complex judgments.

The development of alternative future scenarios is generally of value in situations such as these. It would take us somewhat away from our primary objectives to describe scenario construction in detail; however, the essential steps are easily stated. Definitive discussions of scenario construction are found in Porter *et al.* (1980). These involve the fundamental steps of the systems process. The formulation step involves identification of the needs, constraints, and alterables that may influence environmental change. Potential impacts from these alterables are determined in the analysis step, and these may be evaluated in an interpretation effort if desired. The result of this is a set of possible future scenarios that may or may not occur. The probabilities of each scenario occurring may be obtained. Proposed projects, including the do-nothing alternative, are then embedded into these scenarios and evaluated.

Usually a financial accounting of the various alternative projects results in the information that is needed to identify and quantify economic benefits and costs, perhaps in the format of a balance sheet as illustrated in Fig. 6.4. Adjustments to the financial accounting data are often needed. Sometimes we will find that the financial accounting will neglect some benefits and costs, particularly those of a secondary and/or intangible nature, that are necessary for a useful analysis of the economic benefits and costs of a project. Also, it will often be necessary to adjust or revalue financial data to reflect the fact that market prices either do not exist or do not reflect true economic value even if they do exist. We will now identify some types of economic information that may be missing in financial data; then we will discuss the important topic of "shadow" pricing to obtain true economic value from market prices.

Among information that reflects economic benefits and costs and that may be omitted, or need exclusion, from financial data are the following:

1. ***Transfer Payments*** Often there are intersectoral flows of benefits and costs that do not reflect the production of goods or services. Suppose, for example, a before-project state in which one unemployed person receives \$100 per week in unemployment benefits. This \$100 comes from the taxes of another person (or group). Suppose that the after-project state is such that the unemployed person is employed by the public sector at a weekly salary of \$200, delivers benefits to society of \$150, and loses the unemployment benefits. The group's taxes are increased to \$200 to cover this salary. The before-project costs and benefits are each \$100 per week. The costs represent consumption

foregone by the group, and the benefits represent the consumption enjoyed by the unemployed person. The with-project costs are \$200, which represent the taxes paid by the group. The benefits to society are the \$200 economic value of consumption of the one formerly unemployed person and the \$150 value to society of the product of this person's labor. The net value to society of the project is the increase in benefits less the increase in costs, or (\$200 + \$150 − \$100) − (\$200 − \$100) = \$150. We see that the implementation of the project has resulted in the employment of one unemployed person, and this has resulted in an increase in the aggregate societal value of this *unemployed* person's labor. We note also that there is zero *aggregate* social cost associated with employing an unemployed person. In a similar way there are zero costs associated with using otherwise unemployed capital or land on a project. Of course, there may be a *transfer* of costs associated with using otherwise unemployed labor, capital, or land.

Payment of interest by a project transfers this amount of purchasing power from the project to the lender of the money. So we should treat a loan principal investment as a real economic cost, a cost that occurs at a point in time when this loan principal is spent. We should not be concerned, however, with the interest involved in financing the investment. In a similar way, depreciation should not be considered in economic valuation as the economic cost of using an investment is the initial investment less the (discounted) terminal value of the investment. There are other transfer payments, such as taxes and subsidies, that are also improperly treated as costs or benefits.

2. ***Sunk Costs*** Economic sunk costs are those that have occurred prior to the time at which a project decision is made. They represent money already spent and are costs that cannot be avoided (any more), regardless of the wisdom or judgment once used in making this resource allocation. It is a very common failure of people to not ignore sunk costs. Sometimes this is due to realistic concerns, but more generally it is not. Often a person who makes a partial investment in a project will discover that the project is not working well at all. Rather than suffer the immediate self-admitted criticism of poor judgment, or criticism by others for poor judgment, the person will continue the project investment. Whether the "embarrassment deferral" made possible by strategies such as this is worth the cost is determinable using cost–effectiveness analysis, or decision analysis. We would simply consider "embarrassment deferral" as an attribute and evaluate the extent to which it has value, or utility, relative to other attributes.
3. ***Secondary Effects and Externalities*** Often some of the impacts of a project are such that they do not produce an immediate benefit or cost within the immediate environment of the project, as narrowly defined over a restricted planning horizon. However, if they produce an effect

on other entities, then they should be considered as part of an economic cost–benefit analysis. These secondary effects and externalities, which are often very difficult to identify and quantify, should certainly be considered. Often, there is a presumption that secondary effects and externalities are harmful. This is not necessarily so. While inflation, pollution, and traffic congestion may be "bads," education and learning are "goods." Any of these may be secondary effects and externalities that result from the implementation of a project. Often the implementation of a project will have a beneficial, or detrimental, multiplier effect on other projects. To the extent possible, all of these factors must be considered.

4. ***Contingencies and Risk*** Often, perhaps more often than not, a project is implemented without precise knowledge of the environment that will exist over the entire planning horizon for the project. Consequently, various contingency plans are considered to account, to the extent possible, for risks should these materialize. Generally the costs and benefits of these, such as contingencies to account for anticipated product price rises, should be included. Changes in interest rates represent, however, a contingency that should generally not be included. The key point in this is that projects are evaluated *prior* to being implemented, and it is the expected project costs and benefits that are estimated. It is reasonable that these be used to judge the success of projects and not the actual outcomes, which may differ from the expected outcome. In other words, cost–benefit analysis is an approach to enhance judgment quality. There is no assurance that good outcomes necessarily follow from good judgments. As a case in point, we exercise good judgment if we pay $100 for a "project" that will return $1000 with probability 0.9 and $0 with probability 0.1. But even with this "good judgment," we may lose our $100.

To be sure, there are many cases in which market prices exist but where they are *not* reflective of the economic value of a product or service. A perfectly competitive economy, as we have discussed throughout this book, is one such that the price of everything precisely represents the value the last unit of that product or service contributes to production and consumption. Under conditions of perfect competition, there will generally exist an economic equilibrium. In this economic equilibrium, the "best" use of all productive units, yielding economic efficiency, will be achieved. There will be no alternate use of the resources of labor, capital, and land that would result in more efficient production and in greater total satisfaction in consumption. But markets are imperfect, and so prices will not, in all cases, perfectly reflect value. This inadequacy of market prices as a true value measurement is a strong reason why we should make a separate determination of value, and not just equate value to price, in all instances except those of private individual investments, where market price is what the individual is concerned with.

In those cases where the production of goods or services increases with the price of the good or service remaining constant, the social benefit of increased production is just the fixed price times the change in production quantity. However, when there is a substantial change in price, we must use an alternate approach. We need to determine a quantity called the compensating variation, which, we will show, is essentially equivalent to the concepts of consumer surplus and producer surplus. Our earlier discussions concerning consumer and producer surplus in Chapter 4 are of interest here.

The effects on the welfare of individuals that result from changes in the prices of products and services are measured by the consumer surplus. People obtain similar effects from changes in the wages of the factor services that they supply. These changes are denoted as producer surplus. A producer surplus also results to producers who are able to sell goods at a price higher than the smallest price they are willing to accept for these products and services.

We have indicated that willingness to pay is equivalent to the approximate area under a demand curve. A useful way to interpret a demand curve follows from the willingness to pay concept. Let $D(q)$ be the demand curve for a specified time interval. Then if the price is constant at p_0 the consumer will purchase q_0 over that interval; if the price is p_1 the consumer will purchase q_1 for that interval; and if the price is p_2 the consumer will purchase q_2. The demand curve does not mean that, strictly speaking, if the price is initially at p_0 the consumer will instantly purchase q_0; it says that over the time horizon for which the demand curve was constructed, the consumer will purchase q_0 if the price is constant at p_o. What this might also mean is that if the price drops to p_1 during the time interval for which the demand curve is valid, the consumer will purchase an additional $q_1 - q_0$ times the fraction of the total demand horizon time that the price is p_1. If it drops again to p_2 the consumer might purchase an additional $q_2 - q_1$ times the fraction of the total demand horizon time that the price is p_1. While this interpretation needs to be verified against whatever construct was used to derive it, and modified to give it precision for the true dynamic case, it is easy to see that it is this interpretation on which the consumer surplus concepts are based. We have commented before on the dynamic interpretation of demand curves in Chapter 1.

To derive a measure of the value of a price decrease, it is convenient to pose the question in the following way: What is the maximum amount of money one would be willing to pay to buy as much as they want of the product or service at the decreased price rather than at the initial price? The answer should be that amount of money that results in the same level of utility at the lower price as at the higher price. This amount of money is called the *compensating variation*. To determine a consumer's compensating variation, we must either know the consumer utility function or measure the compensating variation, perhaps through measurement of the utility function.

The following are two approximations that are generally valid and will eliminate the need to measure the compensating variation:

a. The smaller the price change, the closer the consumer surplus is to the compensating variation.
b. The smaller the income portion of the consumer spent on the product or service, the closer the consumer surplus is to the compensating variation.

Some discussions in Chapter 4 contain additional commentary relative to consumer surplus and these approximations. If the price changes are small, then, as we have previously noted, we do not really need the consumer surplus concept. In advanced economies, almost every product or service purchased, with housing as one possible exception, is a small proportion of total consumer expenditures. Thus consumer surplus is a close measure of the value of a price change. This measure is the compensating variation.

What we are noting here is that when implementation of a project results in lowered prices to consumers, this effect must be considered in determining the value of the project, as consumers would be willing to pay more for some of the quantity of the product or service consumed than they pay with the project. We have shown that the compensating variation for an individual is the value measure to the consumer that results from the project. Also, we have indicated that this compensating variation is in practical circumstances equivalent to the consumer surplus. If the social welfare gain is set equal to the sum of the individual welfare gains, then the aggregation of the compensating variation over all people gives a measure of social welfare. A government agency may wish to assign a higher value for consumer surplus accruing to poor people than it does to rich people. Also, the government might wish to encourage consumption rather than savings and may wish to assign a higher weight to consumer surplus components that result in consumption as contrasted with consumer surplus components that result in investments. Squire and van der Tak (1975) expanded considerably on the concept of "weight" in cost–benefit analysis, and their work should be consulted for further details.

Shadow prices may also be used as a measure of project value. As we have seen, the dual variables of linear programming, the Lagrange multipliers, may be interpreted as shadow prices. These represent the amount by which a cost function will change with changes in the marginal unit of a product or service that is consumed. Thus when we maximize social welfare, the Lagrange multiplier takes on particular meaning. Lagrange multipliers, which are shadow prices, are the social values of goods that are created by a project. The need for these shadow prices arises when market prices do not reflect social value. When a value other than market price is used in a cost–benefit analysis, that value is a shadow price. The justification for shadow pricing is that decisions must be made, and decisions imply that valuations have been made. If market values are not available, or appropriate, then other values must be used. These values should reflect social values if they are to improve decisions in a social context.

We have already discussed some approaches to determining "shadow prices" in this section. It is useful to view shadow prices as the dual variables in a linear program. We now consider a simple economy. There are two types of products: final consumption products X and raw materials Y. Society has somehow valued the final goods by associating prices with them, p_i, $i = 1, 2, \ldots, N$. A linear technology, through which the raw materials are transformed into final consumption products,

$$\begin{aligned} X_1 &= A_{11}Y_{11} + A_{12}Y_{21} + \cdots + A_{1M}Y_{M1} \\ X_1 &= A_{21}Y_{12} + A_{22}Y_{22} + \cdots + A_{2M}Y_{M2} \\ &\vdots \\ X_N &= A_{N1}Y_{1N} + A_{N2}Y_{2N} + \cdots + A_{NM}Y_{MN} \end{aligned}$$

or

$$\mathbf{X} = \mathbf{AY}$$

is assumed to exist. Here X_i is the number of units produced of product i and Y_{ji} the amount of raw material j used in the production of product i. A_{ij} are nonnegative production process parameters. The raw material available for use is constrained. The production process must also satisfy the raw material constraints

$$\begin{aligned} Y_{11} + Y_{12} + \cdots + Y_{1N} &\leq \overline{Y_1} \\ Y_{21} + Y_{22} + \cdots + Y_{2N} &\leq \overline{Y_2} \\ &\vdots \\ Y_{M1} + Y_{M2} + \cdots + Y_{MN} &\leq \overline{Y_M} \end{aligned}$$

Our goal here is to maximize the social value of production. This is

$$J = p_1X_1 + p_2X_2 + \cdots + p_NX_N$$

Thus we wish to maximize

$$J = \mathbf{p}^{\mathrm{T}}\mathbf{X} = \sum_{i=1}^{N} p_i X_i$$

subject to the equality constraint

$$X_i = \mathbf{A}_{i^{\mathrm{T}}}\mathbf{Y}_i = \sum_{j=1}^{M} A_{ij}Y_{ji}$$

where $\mathbf{Y}_i$ represents the vector of raw materials going into product X_i. We also have the inequality constraint equation

$$\sum_{i=1}^{N} Y_{ni} = Y \leq \overline{Y}$$

Example 6.7: We consider the performance of two projects involving new methods of extracting raw materials. Project A involves taking one unit of Y_1 and two units of Y_2 out of the present use in the production of final goods and using them instead to increase extraction of Y_3 by three units. The new Y_3 is used in the production of consumer products **X**. Project B uses two units of Y_1 and one of Y_3 to get two more units of Y_2. The shadow prices of Y_1, Y_2, and Y_3 are somehow computed as 1, 2, and 3.

We are interested primarily in the output production of final consumer products **X**. Our interest in raw materials is their effect on production. Raw materials have no value by themselves to us. We determine how projects A and B affect the value of the final output, which is J. We need a way to relate changes in Y to J. This is the role of the shadow price or Lagrange multiplier.

The benefit–cost analysis may be performed using the *known* shadow prices. The do-nothing alternative involves zero benefit and zero cost. Project A results in an increased benefit of 9 and an increased cost of 5, whereas project B results in an increased benefit of 6 and an increased cost of 4. So project B increases J by 2, whereas A increases J by 4 and is the preferred alternative.

Benefit–cost analysis is simple here. We should ask: How did we get the shadow prices? Somehow we determine the value of $\partial J/\partial Y_i$. One answer is through use of the technology matrix $X_i = A_i Y_i$. We determine $\partial J/\partial X$ and $\partial X/\partial Y_i$, and then we multiply these terms. The term $\partial J/\partial X$ is just the given price vector p. Determination of $\partial X/\partial Y_i$ must be made at the optimum operating condition. This involves linear programming solutions to the problem. This may not be a simple task if there are a large number of products in the economy such that determination of the technology production coefficient matrix **A** is difficult. ■

In this section, we have discussed approaches toward the identification and quantification of costs and benefits. The fundamental concept here is willingness to pay. Much of our discussion concerned approaches that could be used to identify the value of costs and benefits when prices did not exist or when they were biased. It is not easy to get people to "willingly" reveal their willingness to pay. The clever use of questionnaires and public participation efforts are sometimes useful toward these ends, in addition to experimental and empirical results.

6.4 THE IDENTIFICATION AND QUANTIFICATION OF EFFECTIVENESS

In Section 6.3 we discussed various approaches toward the identification and quantification of economic costs and benefits associated with alternative action options or projects. Often there will be a variety of reasons why people will be uncomfortable with providing a strict economic measure for benefits. The word *effectiveness* is often used for *benefit* when a strictly economic valuation is not intended. When *effectiveness* is substituted for *benefit* we obtain a cost–effectiveness analysis.

In cost–effectiveness analysis, we desire to rank projects in terms of economic costs and effectiveness. The reason for this is that there are noncommensurate attributes of a project. Certainly we would wish to eliminate dominated or inferior projects—projects that are more expensive and less effective than other projects—from consideration for selection. Beyond this, a cost--effectiveness analysis does not specify which of the several nondominated projects is "best." This can be accomplished if one is willing to trade off cost for effectiveness, so as to obtain a scalar performance index. It can be done by considering cost as one of the attributes in the effectiveness evaluation approach we will now briefly describe. This is but one of the many generally similar approaches in the subject area of decision analysis, an important field, but one that we are unable to explore in any depth here.

The effectiveness of an alternative is the degree to which that alternative is perceived by the decision maker as satisfying identified objectives. The effectiveness assessment approach described here provides an explicit procedure for the translation of qualitative impressions of values, or effectiveness indices, into a quantitative evaluation of alternatives when the impacts of the alternatives are described by multiple attributes. This is accomplished by identifying and organizing the attributes or proposed alternatives into a tree-type hierarchy, attribute tree, or worth structure that is used, together with measures of effectiveness, to compare alternatives as a basis of choice making. Effectiveness assessment is most appropriate when a single approach is desired that will enable us to specify and interpret the effectiveness of the impacts of proposed policies on individuals or groups, and to rank or prioritize programs in terms of effectiveness such as to enable the selection of that policy which has maximum effectiveness for a particular group.

The typical final results or product of the effectiveness evaluation approach we describe is an explicit evaluation of the worth or value of the outcomes or impacts associated with specific proposed policies in terms of the attributes or objectives that have led to the policy proposals or projects. The procedure that we describe also results in an easily communicable picture of the effectiveness, value, or importance that individuals or groups place on different attributes or objectives associated with the impacts of the proposed projects. Typically, also, there results a significant amount of learning by decision makers concerning their own preference structure and the consistency of their evaluations and associated decisions with their preferences. An increased understanding of the decision situation through a careful definition of alternatives, outcomes, and their relationship, and of decision-maker preferences for possible outcomes is another result of the effort. Generally an effectiveness assessment study involves the following major steps:

0. ***Preanalysis: Formulation or Framing of the Issue*** The individual, group, or organization whose effectiveness measures or preferences

among alternative projects are to be assessed is identified. The scope of the effort is determined by the objectives or attributes of the impacts of projects that are important in the problem. The attributes should be restricted to those of the highest degree of importance, and all relevant attributes that can be identified should be included. No attribute should encompass any other attribute, and the attributes or objectives should be "independent" in the sense that the decision maker is willing to trade partial satisfaction of one objective for reduced satisfaction of another objective without regard for the level of satisfaction attained by either. Once the high-level effectiveness attributes or objectives have been established, they must be disaggregated into lower-level attributes. Each of these is further subdivided until the decision maker feels that project effectiveness can be measured. This dividing and subdividing process results in a tree-type hierarchical structure of effectiveness attributes. This preanalysis should be performed in detail using other methods for issue formulation and analysis as we have described in Chapter 1.

1. ***Selection of Appropriate Attributes or Effectiveness Performance Measures*** Some physical characteristic of performance or effectiveness for an alternative is assigned to each lowest-level attribute to measure the degree of attribute or objective satisfaction.
2. ***Definition of the Relationship Between Low-Level Attributes and Physical Attribute Measures*** This relation is established by assigning a worth or effectiveness score w to all possible values of a given attribute measure. The worth score given a particular attribute measure can range from 0 to 1. Zero is the worst score that any alternative project can have on any given project, and 1 is the best score. In determining the worth or effectiveness score of each alternative on all lowest-level attributes, several questions must be answered:
   a. Is the scale of attribute measure values continuous or discrete? Generally it is continuous; occasionally it can be discrete, such as a good or bad outcome.
   b. For a continuous scale, does the attribute measure possess either a logical upper bound or a logical lower bound or both? If this is not the case, the particular attribute in question needs to be redefined.
   c. What values of each attribute measure are identified with worth scores of 0 to 1?
   d. Does the rate of change of worth or effectiveness with respect to attribute measures stay fixed, increase, or decrease?
   e. If the rate of change of worth with respect to the attribute measure changes, does it always decrease (increase) or does it first decrease (increase) and then increase (decrease)?

   The result of doing this is generally either a scale or curve of effectiveness score versus attribute measure, as indicated in Fig. 6.5.

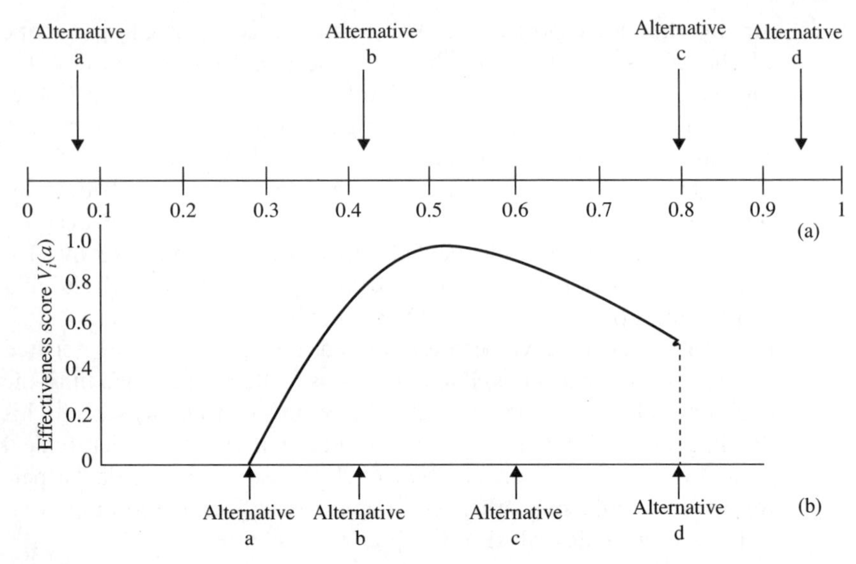

Figure 6.5. Determination of Lowest-Level Attribute Scores: Two Alternate Approaches: (a) Direct Determination of Effectiveness (Worth) Score on One Attribute and (b) Determination of Effectiveness in Terms of an Indirect Measure.

3. ***Establishment of Relative Importance of Each Level of the Attributes*** Some of the attributes within a certain level may be more important than others, and the effectiveness attribute weights are defined to indicate the perceived relative importance of satisfying one attribute or objective with respect to satisfying others. The first step in this process is to rank the subattributes of a particular attribute by relative importance with respect to overall satisfaction. The most important subattribute is assigned a temporary value of 1.0. If the second most important subattribute is three-fifths as important as the first, it is assigned a temporary value of $1.0 \times (3/5) = (3/5)$. If the third is two-thirds as important as the second, it is assigned a temporary value of $(3/5) \times (2/3) = 2/5$. This is accomplished by determining the relative importance of the alternative value scores of 0 and 1 for the two attributes in question. This process is continued until all subattributes have been assigned temporary weights. Then these temporary weights are scaled so that the sum of attribute weights (for each particular attribute) is unity.
4. ***Determination of the Equivalent Weights for Each Lowest-Level Attribute*** Appropriate scaled weights in the hierarchy are multiplied. Because of the sum to one property of the weights at each hierarchical level, the sum of equivalent weights is unity.
5. ***Effectiveness Is Calculated*** The effectiveness of each project is calculated by multiplying the equivalent weights by the individual worth scores and summing to yield an overall effectiveness score. We obtain for the effectiveness of an alternative

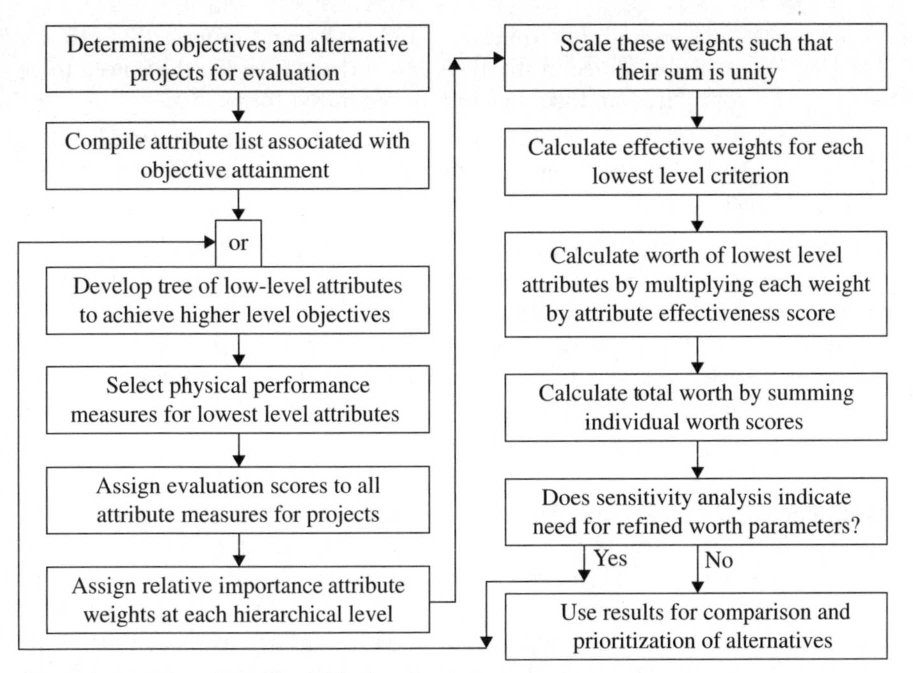

Figure 6.6. Flow Chart for Effectiveness Assessment.

$$E(a) = \sum_{i=1}^{N} \rho_i v_i(a) = \boldsymbol{\rho}^{\mathrm{T}} \mathbf{v}(a)$$

Here $v_i(a)$ are the effectiveness scores of alternatives on the ith lowest-level attribute i and ρ_i are the weights.

6. ***Sensitivity Analysis*** The sensitivity of the effectiveness scores to variation in parameters is determined by a sensitivity analysis in which different values are assigned to the attribute worth scores or effectiveness scores on lowest-level attributes, and the worth scores are recalculated.

7. ***Final Results Are Used for the Comparison, Ranking, and Prioritization of Alternatives According to Effectiveness*** Figure 6.6 presents a flow chart of the activities involved in conducting an effectiveness analysis.

   Among the appropriate conditions for use of this effectiveness analysis approach,

   a. there is a need to evaluate systematically the effectiveness or desirability of many proposed activities, or projects, to assist in choice making;

   b. there is a need to predict the decision behavior of individuals;

   c. there is a need to communicate individual or group values to others;

d. there are multiple objectives and assessment criteria that are not easily quantified in strictly *economic* benefit terms (these need to be considered and arranged in an organized form); and

e. the events that follow from alternative actions are predictable, such that risk is not a dominant part of the effectiveness analysis situation.

Effectiveness assessment can be very useful for the interpretation and evaluation of the results of an analysis effort. To use the approach we need a set of attribute or objective measures deemed to be of importance for issue resolution, information on the relative importance of attributes or objectives, and sufficient knowledge about project alternatives and their outcomes to be able to assign effectiveness scores to the attribute measures that characterize the impacts of each outcome.

Example 6.8: As an example, we consider the purchase of an automobile. We decide to use effectiveness assessment to guide our decision. We assume that we have reduced our choice to a set of three feasible automobiles. Any of the three automobiles would be acceptable if it were the only possible purchase. To illustrate the approach we will initially consider cost as an attribute. Thus we obtain an overall worth assessment of effectiveness and cost. We define the first-level performance attributes as (1) cost, (2) esthetics, and (3) safety. The subdivision of these attributes and the resultant hierarchy are obtained as are the selected attribute measures. The sum of the scaled weights is unity for each subdivision of attributes. These weights are obtained from the decision maker by eliciting the relative importance of the attributes on the difference between the best and worst performing alternatives for the attributes under consideration. The cost objective will be analyzed in detail here.

Cost is one of the three first-level performance attributes that we have identified for consideration in the purchase of an automobile. Cost can be divided into two subattributes: initial cost and maintenance cost. The maintenance cost can be subdivided into two more specific attributes: scheduled maintenance and repairs. Initial cost, scheduled maintenance, and repairs are considered as lowest-level cost attributes. Suppose that initial cost and maintenance cost carry normalized branch weights of 0.6 and 0.4, respectively. The sum of these weights equals 1.0, as required, and these weights may be obtained as follows. The individual evaluating the attributes determined initial cost to be more important than maintenance cost with respect to the difference between best and worst performing alternatives on these attributes. Thus it is reasonable to assign a temporary weight of 1.0 to initial cost (w_1). The individual determined maintenance to be two-thirds as important as initial cost with respect to the difference between best and worst performance on these attributes. Thus we may assign it a temporary value (w_2) of $1.0 \times 2/3 = 2/3$. From these temporary values the scaled weights were obtained.

Here we have combined the cost parameters with effectiveness parameters. The decision maker may not wish to do this. It might be argued here that the scheduled maintenance costs and average repair costs have nuisance value (rather than economic value) only. There might be no difference in fuel economy among the three cars. Thus we could use the worth measure that we have calculated less that due to initial cost as an effectiveness measure. To explore these concerns fully would take us into the primary subject area for yet another textbook—one in multiple objective decision analysis. ■

6.5 SUMMARY

As we have noted, cost–benefit analysis is a method used by systems analysts and assessors to aid decision makers in the evaluation and comparison of proposed alternative plans or projects. Objectives must be identified and alternatives generated and defined carefully prior to initiation of formal analysis efforts. Then the costs as well as the benefits of proposed projects are identified. These costs and benefits are next quantified and expressed in common economic units whenever this is possible. Discounting is used to compare costs and/or benefits at different points in time. Present worth is the discount criterion of choice. Overall performance measures, such as the total costs and benefits, are computed for each alternative. In addition to this quantitative analysis, an account is made of qualitative impacts due to intangibles such as social, esthetic, and environmental effects. Equity considerations are considered to determine the distribution of costs and benefits across various societal groups. The cost–benefit method is based on the principle that a proposed economic condition is superior to the present state if the total benefits of a proposed project exceed the total costs, so that if there are provisions whereby gainers compensate losers, everyone is better off. Distribution or equity issues are addressed in the qualitative part of the analysis. Results are presented to the decision makers, who may use them to select one or more of the proposed project alternatives.

Among the results of a cost–benefit analysis are the following:

1. tables containing a detailed explanation of the economic costs and benefits over time of each alternative project and the present value of costs or benefits of each alternative project (see Section 6.2);
2. computations and comparisons of overall performance measures, in terms of benefits and costs, or effectiveness and costs if a cost-effectiveness determination is desired, for each of the alternative projects (Section 6.3 discusses this topic and presents a balance sheet that is useful in presenting a summary of this information); and
3. an accounting of intangible and secondary (social, environmental, aesthetic) costs and benefits associated with each alternative project.

The following major activities are generally accomplished in a cost–benefit analysis:

1. ***Formulation of the Issue*** This is generally done using techniques specifically suited for issue formulation, such as identification of

objectives to be achieved by projects, some bounding of the issue in terms of constraints and alterables, and generation of alternative projects. The result of this formulation of the issue consists of a number of clearly defined alternatives, the time horizon for the study and its scope, a list of impacted individuals or groups, and perhaps some general knowledge of the impacts of each alternative.

2. ***Identification of Costs (Negative Impacts) and Benefits (Positive Impacts) of Each Alternative*** A list is made of the costs and benefits for each project. Measures for different types of costs and benefits are specified and, if possible, conversion factors derived to express different types of costs or benefits in the same economic units. For example, one of the benefits of a proposed highway project might be the reduced travel time between two cities. To make this comparable to monetary costs, we have to determine how many dollars per time unit are gained by the reduction of the travel time. The determination of such conversion factors can be a sensitive issue, since the worth of various attributes can be totally different for different stakeholders. For example, consider the difficulties involved in transforming additional safety benefits of a proposed project, measured in human lives saved, into monetary benefit units. Further complicating economic benefit evaluation issues are equity considerations. The costs and benefits of a project may be allocated in different amounts to different groups. It is not unusual for one group to pay the costs and for another group to receive the benefits.
3. ***Collection of Data Concerning Costs and Benefits*** Specific information is gathered concerning the economic costs and benefits of each alternative. Much of this may be available from other analysis and assessment efforts involving modeling and optimization.
4. ***Quantitative Analysis of Costs and Benefits*** Quantified costs and benefits are expressed in common economic units as far as possible. Comparisons of projects may also be made with respect to two different quantified units, such as dollars (of cost) and human lives (for benefits). When free-market considerations do not exist, market prices may not reflect the true costs or benefits to society of the projects. In such cases, shadow prices should be computed to replace market prices in the cost–benefit analysis. Economic discounting by means of the present value criterion is used to convert costs and benefits at various times to values at the same point in time. Subsequently, various measures of performance may be computed, for example, the net present value, equal to the total balance of benefits and costs converted to present values, and the BCR, equal to the ratio of total costs and total benefits converted to present values.

Much of the data used in a cost–benefit analysis are based on uncertain assumptions about future conditions. The choice of a discount rate may be highly controversial. Owing to all of these factors, a sensitivity analysis of the quantitative results is generally desired. Results of the sensitivity analysis may give an indication of how the overall performance indices for different alternatives change for different assumptions or controversial choices.

5. ***Analysis of Qualitative Impacts*** The impacts that cannot easily be quantified are assessed for each alternative project. This usually includes intangible and secondary or indirect effects such as social and environmental impacts, legal considerations, safety, esthetic aspects, and equity considerations.
6. ***Communication of Results*** This usually takes the form of a report on both the quantitative and qualitative parts of the study. The report may include a ranking or prioritization of alternative projects or a recommended course of action. It is important that all assumptions made in the study are clearly stated in the report. The report should be especially clear with respect to
   - **a.** the costs and benefits that have been included in the study;
   - **b.** the costs and benefits that have been excluded from the study;
   - **c.** the approaches used to attach instrumental values to the costs and benefits;
   - **d.** the discount rates that have been used; and
   - **e.** relevant constraints and assumptions used to bound the analysis.

Figure 6.7 illustrates typical steps of a cost–benefit analysis. The following are appropriate conditions for the use of a cost–benefit approach:

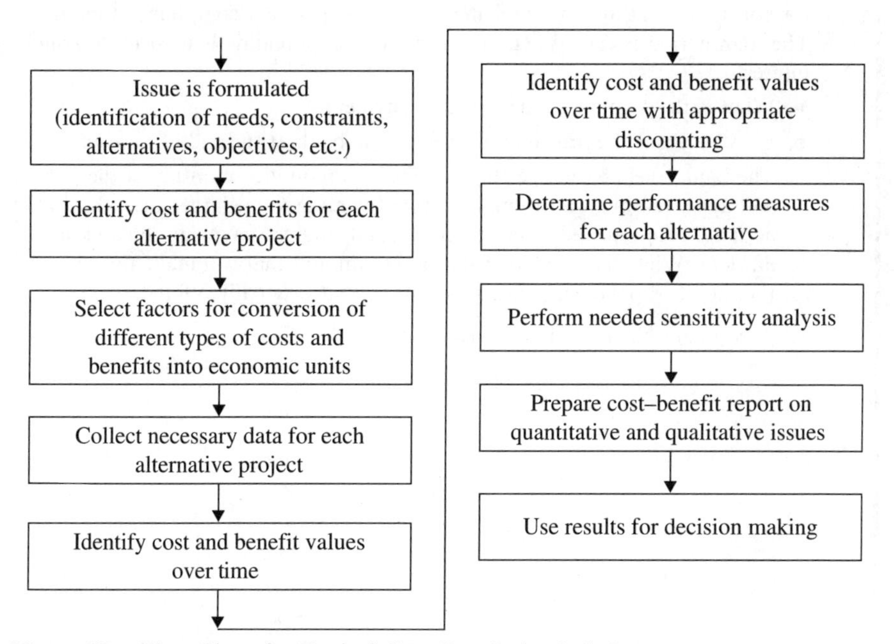

Figure 6.7. Flow Chart for Typical Cost–Benefit Analysis Process.

1. One or more proposed policies or projects have to be evaluated and compared with an existing situation.
2. Many of the costs and benefits associated with the alternative project can be quantified.
3. There is a legal mandate for a cost–benefit analysis and assessment, for example, by a government agency.
4. The distribution of costs and benefits is a concern, and it is desired to use cost–benefit analysis to determine the extent to which equity does or does not exist.

Cost–effectiveness analysis is, as we have seen, a very similar approach to analyze, assess, and compare the costs and benefits of various alternative projects. In principle, it is possible to adjust alternative plans of different generic types such that the same effectiveness index results from each project. Then the least costly project should be selected. In practice, this is difficult to do for many "discrete" projects in which alternatives have parameters that are continuously adjustable. Methods such as multiattribute utility theory can be used to evaluate effectiveness. In conjunction with cost analysis we are able to use the resulting effectiveness indices to assist in making trade-offs between the quantitative and qualitative attributes of the alternative projects.

PROBLEMS

1. A person borrows \$5000 at a bank to purchase a car. The loan agreement calls for constant monthly payments at 16% interest per year compounded monthly. The loan period is four years, and monthly compounding is used to determine interest.
   a. What is the amount of the monthly payment?
   b. How much of the principal is owed to the bank at the end of two years?
   c. The bank offers to charge 10% interest, compounded monthly, if the person will agree to assign a promissory note for an amount larger than the \$5000 actually received and make principal and interest payments on this larger amount. What is the value of this larger amount that will make this value of the modified payment plan equivalent to that originally offered?
2. An investment has the following cash flow:

Year	Amount
0	−1000
1	250
2	300
3	350
4	−500
5	400
6	450

The OCC, or discount rate, is 8%. What is the

a. present worth of benefits?

b. present worth of costs?

c. net present worth?

d. return on investment?

e. internal rate of return?

3. How much should your agreement to pay \$100,000 ten years from now be worth today if the interest rate is 12%?

4. A piece of equipment is purchased today for \$12,000. The life of the equipment is five years when it will be sold for \$2000. The discount rate is 12%. What is the depreciation each year using (a) straight line, (b) double declining balance, and (c) sum of the year's digits depreciation? What is the book value of the equipment each year for each of the three depreciation schedules? What is the net present worth of depreciation for the three depreciation methods?

5. This year your company had operating expenses of \$250,000, interest expenses of \$160,000, a depreciation of \$85,000, and a cost of merchandise of \$500,000. The income tax rates are 20% on the first \$25,000 of taxable income, 22% on all income over \$25,000, and 48% for all income over \$50,000. What are the tax due and the tax rate if sales volume was (a) \$1,000,000 or (b) \$1,280,000?

6. You have just borrowed \$100,000 to purchase a home. You are charged 3 points (3%) of the loan amount in the form of a discount such that you really receive \$97,000 but sign a note for \$100,000. The interest rate on this \$100,000 note is 14%. The mortgage is for 30 years, and you make equal monthly payments. Interest is deductible from income for income tax purposes, and your marginal tax rate is 40%. What is the effective rate of interest you are really paying?

7. What is the IRR for the projects listed in Problem 6? Rank the projects by IRR. What is the differential IRR (the IRR for the difference in cash flow between two projects) for projects A–B, B–C, and A–C? Are these consistent? Rank the projects by differential IRR. Contrast and compare the results of this example with those of Problem 2.

8. Suppose that an investment produces the following returns:

Year	Return
0	−20,000
1	10,000
2	15,000
3	−8,000
4	−7,000
5	20,000

What is the NPW as a function of interest rate? What is the IRR? Contrast and compare the IRR and NPW criteria when used as decision rules for this problem.

9. You have the opportunity to purchase a small manufacturing plant. Your planning horizon is 10 years. The price of your product will be \$100 for the next year and can be expected to increase by 10% a year. The plant costs \$2,500,000 and will be valueless at the end of 10 years. Fixed yearly costs are estimated at \$350,000 per year now and increase at 10% per year. The cost per item produced is \$45 for the next year and will increase at 10% per year. The effective tax rate is 48%. How many items need be sold each year to obtain an ROI of (a) 10% or (b) 20%? The OCC for your firm is 10%.

10. Many works that encourage use of the IRR criterion say that it is a simple criterion that does not need specification of an (externally determined) discount rate to enable prioritization of projects. Comment on this statement.

11. What is the rate of return for the investment of Problem 8 if the external discount rate is 10%? Contrast and compare this rate of return with the IRR.

12. You are having trouble selling your home and, consequently, offer a mortgage at zero interest for five years. You had hoped to issue a mortgage at 16% interest compounded monthly for 30 years. Equal monthly payments for each mortgage are required. By what factor would the initial mortgage have to be increased such that the net present value of the two schemes will be the same? What will be the monthly payments for each mortgage?

13. A frequently occurring problem in practice is the allocation of resources to several operating units from a central budgetary unit. Suppose that the net (discounted) present worth of an allocation of nonnegative resources a_j to the jth unit is $v_j(a_j)$. The units may be assumed to be independent in the sense that the value of a total allocation $\mathbf{a} = [a_1, a_2, \ldots, a_J]$ is additive:

$$V(\mathbf{a}) = \sum_{j=1}^{J} v_j(a_j)$$

There is a budget constraint, in that

$$\sum_{j=1}^{J} a_j \leq B$$

a. What are the necessary conditions for allocation of a budget B in the general case just posed?

b. Suppose that

$$v_j(a_j) = \alpha_j a_j - 0.5\beta_j a_j^2$$

What are the necessary conditions for optimality?

c. If $j = 2$, $\alpha_1 = 1$, $\alpha_2 = 1$, $\beta_1 = 0.5$, and $\beta_2 = 0.1$, what are the optimum allocations as a function of B?

14. Write a brief paper indicating modifications that need to and can be made to the NPW criterion to reflect a "borrowing" interest rate that is different from a "lending" interest rate.

15. Conduct a sensitivity analysis of Problem 1 for varying interest rates (stated at 16% in Problem 1). At what interest rate does the offer in part (c) of this problem cease to be attractive?

16. Conduct a sensitivity analysis of Problem 9 for changes in the inflation rate, indicated as 10% per year in Problem 9.

17. Consider a large privately owned resort. There are presently 5000 hotel rooms available whose prices are determined by competitive market conditions. Of the rooms occupied, 80% are occupied by residents of distant states. A price increase of 10% would reduce the number of rooms demanded by distant-state visitors by 15%, and would reduce it by 25% for near-state visitors. Examine the cost–benefit feasibility of a new 200-room hotel that should be expected to earn a revenue of $3000 per year. Describe any reasonable assumptions that you make.

18. A possible new firm employs 2000 workers for a 40-h work week. It is anticipated that opening of the firm will result in a wage increase from $5 per hour to $5.50 per hour, but will result in existing firms reducing their employment from 5000 to 4200 workers. No immigration of new workers is anticipated owing to the new firm. Suppose that the government takes a 10% income tax. What are the costs and benefits of the new firm to (a) employees, (b) owners of the new factory, (c) the aggregate industry in the region, and (d) the government?

19. A government agency is considering whether to allow the public to use a section of land for recreation. The costs of operating the facility will be $1,000,000 per year. The public will be charged $3 per person for use of the facility. It is (reliably) estimated that 400,000 people will visit the facility each year. What will be some of the concerns that enter into a cost–benefit analysis of the option to allow the public to use the facility?

20. Consider a typical linear supply–demand curve in a perfectly competitive economy. The government introduces a per unit tax collected from producers, and this shifts the production curve upward and results in a new equilibrium point. What are the social costs and benefits of the change to consumers, firms, and the government?

21. Repeat Problem 20 for the case where the per unit tax is replaced by a per unit subsidy.

22. Consider typical linear supply–demand curves in a perfectly competitive economy. Suppose that the government imposes a maximum price on the product of the competitive firm. What are the social cost and benefits of the changed conditions to consumers and firms?

23. Discuss the preparation and use of a benefit–cost analysis that will enable a corporation to determine whether to expand into a new product line.

24. Write a brief paper in which you contrast and compare effectiveness, as used in cost effectiveness, which is a measure of the degree to which objectives are achieved, and benefit as a measure of economic efficiency.

25. Write a critical review and discussion of one of the many formal cost–benefit analyses that you may find in the literature.

BIBLIOGRAPHY AND REFERENCES

Our first discussions in this chapter concerned the time value of money. A classic paper in this area, which first discussed some of the fundamental limitations with the IRR criterion and presented a Lagrange multiplier solution to the capital budgeting under constraint problem, is

Lorie J, Savage LJ. Three problems in capital rationing. J Business 1955;28(4):229–239.

The mathematical optimization approach to capital budgeting is discussed in

Peterson PP, Fabozzi FJ. Capital budgeting: theory and practice. Hoboken, NJ: Wiley; 2002.

Weingartner HM. Mathematical programming and the analysis of capital budgeting problems. Chicago: Markham Publishing; 1967.

The axiomatic development that justifies the net present worth concept for project evaluation may be found in

Williams AC, Nassar JJ. Financial measurement of capital investment. Manag Sci 1966;12(12):851–863.

There are many discussions of discount rates in the literature; especially recommended are

Baumol WJ. On the social rate of discount. Am Econ Rev 1968;58:788–802.

Marglin SA. The social rate of discount and the optimal rate of investment. Q J Econ 1963;77:95–111.

Contemporary works that present discussions of cost–benefit analysis and references to many cost–benefit studies that have been performed include

Brent RJ. Applied cost benefit analysis. 2nd ed. Northampton, MA: Edward Elgar; 2006.

Mishan EJ, Quah EH. Cost–benefit analysis. 5th ed. New York: Routledge; 2007.

Prest AR, Turvey R. Cost benefit analysis: a survey. Econ J 1965;75:683–735.

Sassone PG, Schaffer WA. Cost–benefit analysis—a handbook. New York: Academic; 1978.

Squire L, van der Tak HG. Economic analysis of projects. Baltimore: John Hopkins University; 1975.

Sugden R, Williams A. The principles of practical cost–benefit analysis. London: Oxford University; 1978.

Most of the aforementioned works concern cost–benefit analysis in the public sector. Works that discuss engineering project investments and decisions in the private sector and which are especially recommended include

Bussey LE. The economic analysis of industrial projects. Englewood Cliffs, NJ: Prentice-Hall; 1978.

Copeland TE, Weston IF. Financial theory and corporate policy. Reading, MA: Addison-Wesley; 2006.

Rose LM. Engineering investment decisions. Amsterdam: Elsevier; 1976.

The above-mentioned three texts discuss aspects of resource allocation under uncertainty considerations, an important topic that we have been unable to discuss here.

Two recent works that integrate public and private sector considerations are

Miller C, Sage AP. Application of a methodology for evaluation, prioritization and resource allocation to energy conservation program planning. Comput Electr Eng 1981;8(1):49–67.

Miller C, Sage AP. A methodology for the evaluation of research and development of projects and associated resource allocation. Comput Electr Eng 1981;8(2):123–152.

Especially valuable are two general works in systems science and engineering that provide much discussion concerning cost–benefit and cost–effectiveness analyses as well as related topics:

de Neufville Richard, Stafford JH. Systems analysis for engineers and managers. New York: McGraw-Hill; 1971.

Porter AL, Cunningham SW. Tech mining: exploiting new technologies for competitive advantage. Hoboken, NJ: Wiley; 2004.

Porter AL, Rossini FA, Carpenter SR, Roper AT. A guidebook for technology assessment and impact analysis. New York: North-Holland; 1980.

CHAPTER 7

COST ASSESSMENT

The accurate prediction of cost, including effort and schedule, required to accomplish various systems engineering activities in an organization is very important. Similarly, it is necessary to know the effectiveness that is likely to result from given expenditures of effort. When the costs of activities to be undertaken are underestimated, programs may encounter large cost overruns, which may lead to embarrassment for the organization. Associated with these cost overruns are such issues as delivery delays and user dissatisfaction. When costs are overestimated, there may be considerable reluctance on the part of the organization or its potential customers to undertake activities that could provide many beneficial results. Thus, cost is an important ingredient in risk management. In this chapter, we will examine costing in systems, software, and information technology development.

It is very important to note that a cost estimate is desired to predict or forecast the actual cost of developing a product or providing a service. There are many variables that will influence cost. The product or service scope, size, structure, and complexity will obviously affect the costs to develop it. The newness of the product or service to the development team will be another factor. The stability of the requirements for the product or service will be an issue as high requirements volatility over time may lead to continual changes in requirements and specifications that must be satisfied throughout the entire systems acquisition life cycle.

The integration and maintenance efforts that will be needed near the end of the acquisition life cycle surely influence costs. Something about each of these and other factors must necessarily be known to obtain a cost estimate. The more detailed and specific the *definition* of the product is, the more accurate we should expect the costs to be relative to the estimated costs for *development* and *deployment* of the product. One guideline we might espouse is to delay cost estimation until as late as possible in the product or service definition phase, and perhaps even postpone it until some development efforts have become known. If we wait until after deployment to obtain cost estimates, we should have an error-free estimate of the costs incurred to produce a product or system or service. This approach is hardly feasible or useful, as estimates need to be known early in the definition phase to determine whether it is realistic to undertake production.

Economic Systems Analysis and Assessment,
by Andrew P. Sage and William B. Rouse

Operating costs, once a solution is deployed, are important too. Operating costs can amount to 75% to 80% of the total costs of ownership. They can also be highly variable depending on the frequency of system use, the economy, inflation and wages, etc. Chapters 8 and 9 deal with operating costs to an extent, including uncertainties. Rouse (2010) addresses this issue in great depth as the economics of human systems integration are often dominated by these costs. Chapter 10 provides an overview of this material.

We also see that there is merit in decomposing an acquisition or production effort into a number of distinct components. A natural model for this is the life-cycle phases for production. We might, for example, consider the seven-phase software development life cycle of Boehm (1976a, b) or any of the systems engineering life-cycle models and develop a work breakdown structure (WBS) of the activities to be accomplished at each of these phases. The work elements for each of these phases can be obtained from a description of the effort at each phase, where we will concentrate on software effort:

Phase 1. Systems Requirements Definition. The specification of systems requirements is the first phase of effort. In implementing this phase, it is assumed that the system user and the systems analyst are sufficiently informed about what the new (or modified) system is intended to achieve so as to develop the system-level requirements to an acceptable extent such that they can be identified in sufficiently complete detail such that preliminary design can be initiated. All of this should be done before detailed design and coding may be initiated.

Phase 2. Software Requirements and Software Specifications. The development of the software requirements phase focuses on the outcomes of the system- or user-level requirements identification carried out in Phase 1 of this waterfall model of the software development life cycle. It is concerned with the nature and style of the software to be developed, the data and information that will be required, the associated structural framework, the required functionality, performance, and various interfaces. Requirements for both the system and the software are reviewed for consistency and then reviewed by the user to be certain that the software requirements faithfully interpret and produce the system requirements. A software requirements definition document is produced in this phase. It becomes a technology and a management guideline throughout all subsequent development phases, including validation and testing. These software requirements are then converted into a detailed software specifications, or requirements specifications, document.

Phase 3. Preliminary Design. The software specifications defined in Phase 2 are converted into a preliminary software product design in this phase, which is primarily aimed at further interpretation of the software specifications in terms of software system–level architecture. The product of this phase is an identification and microlevel definition of the data structure, software architecture, and procedural activities that must be carried out in the next phase. Data items and structures are described in abstract, or conceptual, terms as a guide to the detailed design phase. For this reason, this phase is often called preliminary conceptual design. Instructions that describe the input, output, and processing that are to be executed within a particular module are developed. Preliminary software design involves representing the functions of each software system in a way that these may readily be converted to a detailed design in the next phase.

Phase 4. Detailed Design. The preliminary design phase results in an insight into how the system is intended to work at a structural level and satisfy the technological system specifications. Detailed design phase activities involve definition of the program modules and interfaces that are necessary in preparation for the writing of code. Specific reference is made to data formats. Detailed descriptions of algorithms are provided. All of the inputs to and outputs from detailed design modules must be traceable back to the system and software requirements that were generated in Phases 1 and 2. In this phase, the system or software design is fine-tuned.

Phase 5. Code and Debug. In this phase, the detailed design is translated into machine-readable form. If the design has been accomplished in a sufficiently detailed manner, it may be possible to use automated code generators to perform all, or a major portion, of this task. After the software design requirements have been written as a set of program units in the appropriate high-level programming language, the resulting high-level code is compiled and executed. Generally, "bugs" are discovered, and debugging and recoding are done to verify the integrity of the overall coding operations of this phase.

Phase 6. Integration, Testing, and Preoperation. In this phase, the individual units or programs are integrated and tested as a complete system to ensure that the requirements specifications discussed in Phases 1 and 2 are met. Testing procedures center on the logical functions of the software. They assure that all statements have been tested and that all inputs and outputs operate properly. After system testing, the software is operated under controlled conditions to verify and validate the entire package in terms of satisfying the identified system requirements and software specifications.

Phase 7. Operations and Maintenance. This phase of the waterfall life cycle is often the longest from the perspective of the entire useful life of the software product. The phases involving detailed design, coding, and testing are usually the most effort intensive. In this phase, the system is installed at the user location and is tested, and then used, under actual operating conditions to the maximum extent possible. Maintenance commences immediately on detection of any errors that were not found during an earlier phase. The detection, diagnosis, and correction of errors are generally not the intended purpose of maintenance, however. Maintenance is primarily the proactive process of improving the system to accommodate new or different requirements as defined by the user after the initial product has been operationally deployed.

The phases enumerated above normally take place in the sequential manner described. However, considerable iteration and interaction occur between the several phases. Each software development organization tailors the process to meet the particular characteristics of the personnel of the organization and, potentially, the needs of users of the software product.

Figure 7.1 illustrates a hypothetical WBS for this development life cycle. We could base our estimates of cost on the detailed activities represented by this WBS, and a realistic WBS would be much more detailed than this, or we could develop a macrolevel model of development cost that is influenced by a number of primary and secondary drivers. One often-used approach is called activity-based costing (ABC). An unstated assumption in the development of Fig. 7.1 is that we are going to undertake a grand design type approach for system production. This figure and the associated discussion need some modification for any approach other than the grand design type approach.

We will examine model-based costing (MBC) and ABC approaches in this chapter. We will first discuss some generic issues relating to system and software productivity. This will include the nature of the variables that are important for these measures. Then, we will consider some early approaches. After this, we will examine some MBC approaches. Next, we will look at ABC approaches. Finally we will discuss some issues related to model and system evaluation. Much of our discussion concerns the engineering of information technology, systems, and services. In the end, we turn to a brief discussion of effectiveness.

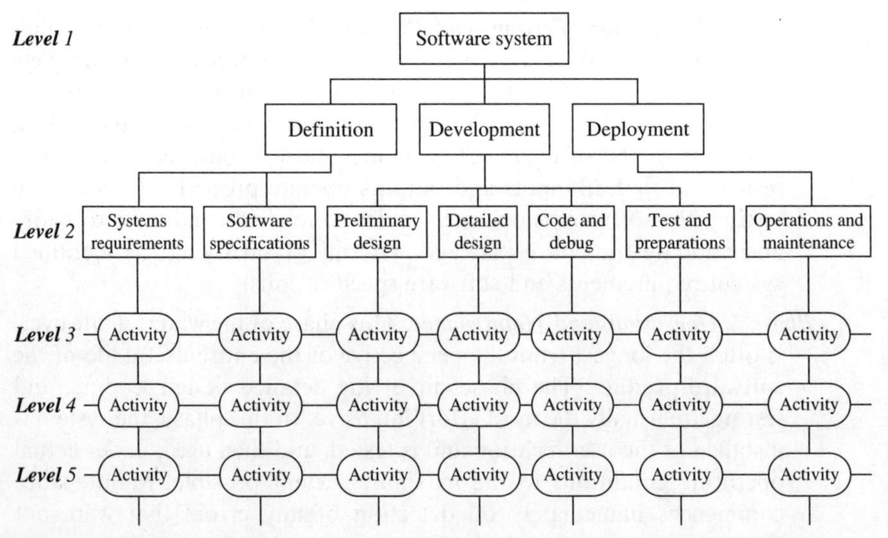

Figure 7.1. Work Breakdown Structure for the Grand Design Waterfall Software Development Life Cycle.

7.1 MODEL-BASED PARAMETERS FOR PRODUCTIVITY

There are a number of ways in which we could define software productivity. We might be tempted to say that software productivity is equivalent to the number of lines of source code that are delivered to the customer per person-month (PM) of effort. We immediately sense that a real problem with this definition is that complexity of the software does not enter into it. Also, there is an implied assumption that all that is important about software is programming. So we might be tempted to associate this definition more with programmer productivity than with software productivity.

Even with this more restricted interpretation, we still may not have a very complete definition of programmer productivity. Our definition involves "lines of source code." But what is a line of source code? A line of source code might represent a high-order language (HOL) line of source code. Even here, we must be concerned with whether we should include only retainable lines of code in the resulting compiled code or should also include such items as remarks statements that may be extraordinarily helpful in providing cues to the software structure. Also, we should perhaps include only those lines of source code that are delivered to the customer, and not all lines that were written and then perhaps discarded.

There are many concerns with an approach that suggests that the number of lines of code is an important and primary driver of software costs and schedule. Let us briefly examine some of these. An object line of code is a single machine language instruction either written as a machine language instruction or generated through some translation process from a different language. We immediately see a problem. Lines of code can be meaningful only at the end of a project after operational executable code is delivered and accepted by the client. Surely, delivered lines of code are not a meaningless measure of productivity. However, it is assumed that *one line of code is equal to and has precisely the same value as any other line of code*. Clearly, there is no reason whatsoever to accept this assertion.

Yet the number of source lines of code (SLOC) is often used as the primary driver of software cost estimation models. Often, this is modified to the term "delivered SLOC" (DSLOC) through the simple artifice of multiplying the estimate of SLOC by some fraction, which would generally depend on programmer experience and other variables. Boehm (1976a and b) delineates some of the difficulties with this metric, which are as follows:

1. Complex instructions, and complex combinations of instructions, will receive the same weight as a similarly long sequence of very simple statements.
2. This is not a uniform metric in the sense that similar length lines of machine-oriented language statements, higher order language statements, and very-high-order language statements will be given the same weight.

3. It may be unmotivated, truly productive work if programming teams learn that their productivity is measured in terms of the number of lines of code they write per unit time and then seek to artificially increase this productivity through writing many simple and near-useless lines of code.
4. It may encourage poor-quality code by encouraging rapid production of many sloppily structured lines of code.

We could continue this listing by noting that maintenance costs may well increase through the implied encouragement to produce many lines of code if these are produced without regard to maintainability. Transitioning to a new environment is especially difficult when many lines of carelessly structured code are involved.

Even if we could agree that SLOC or DSLOC is a worthwhile driver of costs, there would still remain a major problem in effectively estimating either quantity prior to the start of a software acquisition effort.

Productivity could also be defined in terms of an importance weighted sum of delivered functional units. This would allow us to consider the obvious reality that different lines of code and different software functional units will have different values. This is the basis for "function points" measurement method. It leads to an approach where software costs are measured in terms of such quantities as input transactions, outputs, inquiries, files, and output reports. These are software product quantities that can be estimated when the requirements definition document is developed. Such an approach will enable us to estimate the cost and schedule required to develop software, and perhaps even the size of the computer code that is a part of the software, in terms of:

1. the purpose to be accomplished, in terms of such operational software characteristics as the operating system used and application for which the software was developed, and development and test requirements for support software;
2. the function to be accomplished in terms of amount or size, complexity, clarity, and interrelation with other programs and computers;
3. use factors such as the number of users of code, sophistication of users, number of times code will execute, number of machines on which code will run, and hardware power;
4. development factors such as the operating system to be accommodated, development time available, development tools available, experiential familiarity of software development team, and the number of software modules.

Other taxonomies of factors for software development are possible. For example, we might identify seven cost and schedule drivers illustrated in Fig. 7.2. We could seek to develop a model for costing based on these drivers. In general, the important factors will include the people, process, environment,

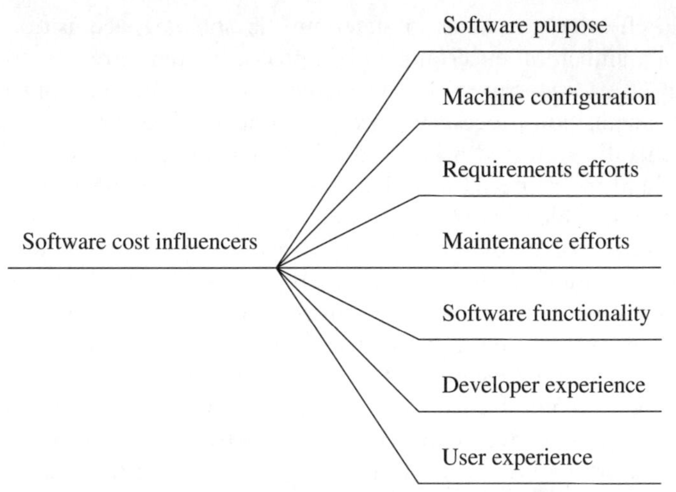

Figure 7.2. Some Cost Influences for Software Effort and Schedule Productivity.

product, and computer system to be used. The major requirement is that the factors be significant, general, measurable, observable, and independent. This poses major difficulties, both for software costing and for software acquisition, as we shall soon see. We now turn our attention to some early models for software costing. Then we look at some more contemporary efforts that are, to some extent, based on these early efforts.

7.2 EARLY MODELS FOR ESTIMATION OF COSTS, INCLUDING EFFORT AND SCHEDULE

Ideally, a software or system cost estimation model should describe the relationships among the characteristics of the overall acquisition effort so that we can determine the effort and schedule for acquisition from these relevant characteristics. The needed characteristics would usually exist at the level of the software or system specifications. It would be ideal if the estimation could be accomplished directly from user requirements, but this is generally not possible without many additional definitional efforts and assumptions.

Use of the software-level requirements specifications would also be satisfactory. If we had an automatic way of transitioning from system-level requirements to software requirements, and perhaps were able to solve the inverse problem as well, then knowledge of one would be equivalent to that of the other.

The features useful as input to cost and schedule estimation models should include descriptions of not only the software to be developed but also the characteristics of the developing organization, such as personnel experience and capability levels, microlevel and macrolevel software development tools, and program management approaches.

This effectively assumes a deterministic software acquisition process, whereas a number of uncertain and imprecise matters are associated with any large effort to acquire a new system. A deterministic approach to a stochastic acquisition process will generally not produce correct answers. The results generally obtained will be precise; in reality, the results should be stochastic and imprecise. Hopefully, however, the deterministic estimates will provide a reasonable approximation of the realized cost.

Software cost models usually involve at least two types of estimates. These are *effort* and *schedule*. These may be expressed according to the phases of the software life cycle. Many existing software cost models estimate the number of weeks or months needed to produce a given software system. A time limit on system development is often identified by the client as one of the system-level requirements. This is generally assumed as a constraint in the transition of system-level requirements to software-level requirements and associated-effort requirements. For this reason, most cost models focus on effort requirements over a prescribed schedule horizon. As we will discuss, generally some virtually irreducible minimum time is required to obtain a new system, and increasing the acquisition costs for the system may well not be associated with a reduction in the associated acquisition schedule.

We will use effort estimates as input variables that result in the calculation of costs of development in terms of the human resources required to build the target system. Typically, this estimate is expressed in terms of the number of people needed integrated over the time to be spent, so that the measure of effort is labor-months, PMs, or person-years. The actual cost in dollars can be generated from an effort estimate by multiplying the effort by the average cost of labor per unit time. More sophisticated models may be based on the fact that the cost per person-hour of effort depends on the type of labor involved. A development effort that involves much use of systems engineering talent for technical direction and rapid prototyping would thus be costed at a higher amount than an effort that involves little of this effort and a great deal of routine coding.

Many simple software cost models describe the size of the software development program only in terms of the size of the software product. This size is often measured in terms of *lines of code* delivered to the customer. Complexity figures, in terms of the number of programming operators and operands, and functionality measures are used in some models as we have noted. For example, Walston and Felix (1977) allow comment lines up to half of the delivered lines of code. Boehm (1976a) does not allow the inclusion of comment lines at all. Bailey and Basili (1981) include the total number of lines of new code, including remarks, but only one-fifth of the total number of lines of reused code, including comments. There is no problem with any of these approaches, although it does make routine use of the same information on all models, as well as comparison of the various approaches, more difficult. The meaning of the expression "lines of code" must simply be interpreted for each model before using the model.

An *experiential model* uses expert-based wholistic judgment as the primary input to the cost estimation or forecasting process. The accuracy of the prediction is a function of the experience and perception of the estimator. In its simplest form, an estimate based on expert judgment involves one or more experts who make educated guesses about the effort required for an entire software development program, or for a project thereof. The estimate may be derived from either a top-down or a bottom-up analysis of the proposed system. Often, experts are asked to make three predictions in a simple approach to forecasting: A, optimistic; B, pessimistic; and C, most likely. If A represents the optimistic cost estimate, B the pessimistic one, and C the most likely cost estimate, then the final estimate of cost or effort is presumed to be given by

$$\text{Cost} = \frac{A + 4C + B}{6} \tag{7.1}$$

This is simply a weighted average in which it is assumed that the weight of the most likely estimate has a worth of four times that of either the most optimistic or most pessimistic estimate. Alternately, the resulting estimate can be said to follow what is called a beta probability distribution. The Delphi technique (Porter *et al.*, 1991) can be used to generate A, B, and C as themselves averages of many individual responses. Of course, there is no scientific basis to support use of this cost relation. Many people would, in a judgmental situation such as this, simply provide optimistic and pessimistic estimates that are equally spaced about the most likely estimate. In this case, the most likely estimate is the average, which is half way between the optimistic and pessimistic estimates.

A number of other more formal models add more structure to the expert judgment approach. Wolverton (1974) developed a software cost matrix approach in which matrix elements represent the cost per line of code as calibrated from historical data. The multipliers in a vector product equation represent a phase–activity distribution as derived from the judgments of experts. The choice of element depends on expert judgment as to the type of software, novelty, and difficulty. This model contains as many as 8 phases and 25 activities per life-cycle phase. Use of the model is generally predicated on breaking the software acquisition effort into phases and estimating their costs individually. A software effort may represent a *new* or *old* development effort depending on the familiarity of the software developer.

Wolverton's software cost–effort model is based on subjective estimates of software complexity. These estimates are made by considering the different types of software to be developed and the development difficulty expected. A software cost matrix can also be used as an alternate to a graphic approach. For example, we have the data shown in Table 7.1 for the six types of software considered in this early work. Clearly, the cost for developing time critical software is large at $75 per line of code. Obviously, this includes all costs, not just the direct programming costs to write a line of code.

In this table, the costs are based on the type of software, represented by the row name, and the difficulty of developing such software, represented by the several columns. The difficulty is determined by two factors: whether the problem is old (O) or new (N), and whether it is easy (E), moderate (M), or hard (H). The matrix elements are the cost per line of code as calibrated from

TABLE 7.1. Wolverton Dollar Cost Estimates for Old and New Code Development

	Difficulty					
Type	OE	OM	OH	NE	NM	NH
Control	21	27	30	33	40	49
I/O	17	24	27	28	35	43
Pre/postprocessor	16	23	26	28	34	42
Algorithm	15	20	22	25	30	35
Data management	24	31	35	37	46	57
Time critical	75	75	75	75	75	75

historical data. To use the matrix, the software system under development may be partitioned into modules i, where $i = 1, 2, \ldots, n$. An estimate is made of the size S_i of each module, as measured in the number of lines of developed uncommented code, for each of the n modules. This estimate does not consider software process variables, except to the extent that they influence Table 7.1. Fundamentally, the software cost estimate is a function only of product characteristics unless one is to have a different graph or table for each change in process, and management, characteristic. It is not even clear, for example, how questions of acquisition management approach, software development organization experience, availability of CASE tools, and other process-related factors affect the resulting cost estimates.

The *cooperative programming model* (COPMO) (Thebaut and Shen, 1984) incorporates both software size and software development staff size by expressing effort in terms of team interactions. Effort is defined as

$$E = E_1(S) + E_2(W) \tag{7.2}$$

where E is overall effort, $E_1(S)$ is the effort required by one or more people working independently on a size S module development who require no interaction with other modules, $E_2(W)$ is the effort required to coordinate the development process with other software development team members, and W is the average number of staff assigned to software program development.

Additional needed relations are

$$E_1(S) = a + bS \tag{7.3}$$

$$E_2(W) = cW^d \tag{7.4}$$

The parameters a, b, c, and d are determined using past development efforts. These equations are not especially different from the effort estimation equations used in the other models. The value of d can be inferred to represent the amount of coordination, management, and communication needed on a project. Similarly, c is a measure of the weakness of the communication paths among the individuals working on the software effort.

A *resource estimation model* has been proposed by Fox (1985), who identifies a total of 27 contributors to software development costs. The eight major contributors to the cost of software development are identified in three major categories as follows:

1. ***Function***

   a. *scale* (1–8), the amount of function to be developed;

   b. *clarity* (1–10), the degree to which functions developed are understood;

   c. *logical complexity* (1–10), the number of conditional branches per 100 instructions;

   d. the need for *user interaction with the system* (1–5) and the intensity of this interaction.

2. ***Use Time Environments***

   e. *consequences of failure* (1–15), the effort required to meet reliability and recovery requirements;

   f. *real-time requirements* (1–5) in terms of how fast the various needed functions must be accomplished.

3. ***Development Time Factors***

   g. *stability of the software support tools* (1–10);

   h. *stability of the use phase computer hardware* (1–20).

Each of these factors is assumed to contribute to costs per line of delivered code. The code costs can vary up to $83.00 per line, as this is the sum of the largest numbers for each of these factors. Fox encourages use of a circle-like diagram on which is sketched the relative difficulty for each factor.

The Bailey and Basili (1981) model was developed using data from a database of 18 large projects from the NASA Goddard Space Center. They make no claims that the model, based on empirical evidence from specific projects, applies to projects other than those of the same general type. Most of the software in the NASA database used by Bailey and Basili was written in Fortran, and most of the applications they considered were scientific. Thus, the database is very homogeneous. The total number of lines of code is defined to be the total amount of new code written plus 20% of the old code that is reused. Comment lines of code are included in each.

The Walston–Felix (1977) model uses a size estimate adjusted by factors determined from the subjective answers to questions about 29 relevant software acquisition topics, as follows:

1. customer interface complexity;
2. user participation in the definition of requirements;
3. customer-originated program design changes;
4. customer experience with the application area of the project;
5. overall personnel experience and qualifications;

6. percentage of development programmers who participated in design of functional specifications;
7. previous experience with operational computers;
8. previous experience with programming languages;
9. previous experience with application of similar or greater size and complexity;
10. ratio of average staff size to project duration (people per month);
11. hardware under concurrent development;
12. access to development computer open under special request;
13. access to development computer closed;
14. classified security environment for computer and at least 25% of programs and data;
15. use of structured programming;
16. use of design and code inspections;
17. use of top-down development;
18. use of chief programmer team;
19. overall complexity of code developed;
20. complexity of application processing;
21. complexity of program flow;
22. overall constraints on program design;
23. design constraints on program's main storage;
24. design constraints on program's timing;
25. code for real-time or interactive operation or executing under severe time constraint;
26. percentage of code for delivery;
27. code classified as nonmathematical application and I/O formatting programs;
28. number of classes of items in the database per 1000 lines of code delivered;
29. number of pages of delivered documentation per 1000 lines of delivered code.

These process-related factors are quite specific. In the aggregate, they represent an attempt to measure the understanding of the development environment, personnel qualifications, proposed hardware, and customer interface. Walston and Felix used these factors to supplement their basic equation with a productivity index. Using a large database of empirical values gathered from 60 projects in IBM's Federal Systems Division, they produced the foregoing list of 29 factors that can affect software productivity. The projects were reviewed to determine parameters for the model that varied from

4000 to 467,000 lines of code, were written in 28 different high-level languages on 66 computers, and represented from 12 to 11,758 PMs of effort. For each mean productivity value x_i calculated from the database, a composite productivity factor P was computed as a multi-attribute utility theory (MAUT)-type calculation.

The Walston–Felix model of software production effort that results is fundamentally based on the equation $E = 5.25S^{0.91}$, where S is the number of SLOC, in thousands. A similar equation is obtained for schedule, in terms of time to project completion, as $T = 2.47E^{0.35} = 4.1S^{0.35}$, which is also obtained from a standard least-squares curve fit approach. The effort, or cost, S is estimated in PMs of effort and the schedule in months. We note that this implicitly specifies the project workforce. Many of the models that we consider assume that there is a reasonable minimum development time and that it is very unwise to compress this. Other equations of interest in this model are $L = 0.54E^{0.06} = 0.96S^{0.055}$ and $Z = 7.8E^{1.11} = 49S^{1.01}$, where L and Z represent the staffing requirements in people and the pages of documentation that will be needed, respectively.

The commonalities, and the differences, among these approaches are interesting. Table 7.2 summarizes the central and basic features of many of the estimators we have discussed in this section, and provides the basic COnstructive COst MOdel (COCOMO) results that we will discuss in the next section.

These show different economies of scale and multipliers. This is not unexpected since the data used to provide the estimators are different. The Boehm model equations are the most sophisticated of these, by far the most used, and allow for different environment assumptions. Each of these equations is necessarily associated with errors. According to DeMarco (1981), uncontrollable factors in the development process will almost always result in 10% to 20% error in the accuracy of this type of estimate. Sadly, estimates using models of this sort are rarely this good, unless very careful attention is paid to the parameters that go into the estimates. We will now examine a number of MBC, or effort and schedule, models to illustrate the similarities and differences among them. We will begin with the COCOMO models of Boehm.

TABLE 7.2. Basic Estimate Relations for Various Software Cost Models

$E = 5.5 + 0.73S^{1.16}$	Bailey–Basili
$E = 2.4S^{1.05}$	Boehm—basic organic COCOMO
$E = 3.2S^{1.05}$	Boehm—intermediate organic COCOMO
$E = 3.0S^{1.12}$	Boehm—basic semidetached COCOMO
$E = 3.0S^{1.12}$	Boehm—intermediate semidetached COCOMO
$E = 3.6S^{1.20}$	Boehm—basic embedded COCOMO
$E = 2.8S^{1.20}$	Boehm—intermediate embedded COCOMO
$E = 5.29S^{1.047}$	Doty
$E = 5.25S^{0.91}$	Walston–Felix

After discussing this and other approaches, we will examine some concerns relative to modeling errors.

7.3 THE CONSTRUCTIVE COST MODEL

Barry Boehm has been intimately involved in almost all of software engineering, and especially involved in developing and refining his COCOMO and associated databases. His book on this subject (Boehm, 1981) is generally regarded as a software engineering classic. There have been a number of extensions to this work for software, including COCOMO II, and more recently for systems engineering, COSYSMO and COSISIMO by Boehm and his doctoral students. There are a plethora of related papers as well. The COCOMO model, or in fact "models" as there are three fundamental models involved, was derived from a database of 63 projects that were active during the period 1964 to 1979. The database included programs written in Fortran, COBOL, PL/1, Jovial, and assembly language. They range in length from 2000 to 1,000,000 of code, exclusive of comments. The COCOMO database is more heterogeneous than most that have been used for software cost projection, incorporating business, science and engineering-based, and supervisory control software. In addition to estimating effort, COCOMO includes formulae for predicting development time and schedule and a breakdown of effort by phase and activity. This is a very extensive and thoroughly investigated model to which much serious thought has been applied. It is often used in practice. Also, there have been a number of extensions to the initial COCOMO model developments, and we will discuss some of these.

There are three COCOMO levels, depending on the detail included in, or desired in, the resulting cost estimate: basic, intermediate, and detailed. Each form of the COCOMO model uses a development cost estimate of the form

$$E = aS^{\delta} M(\mathbf{x}) \tag{7.5}$$

where $M(\mathbf{x})$ represents an adjustment multiplier. The units of E are PMs, or labor-months of effort. If we multiply the labor-months by the average cost of labor, we obtain the direct costs of the acquisition effort. Of interest also is the development schedule. The schedule time for development is given by

$$T = bE^{\Upsilon} \tag{7.6}$$

where E is the cost estimate just obtained. The parameters δ, Υ, a, and b are determined by the model and development mode, as we will soon discuss. In most representations of COCOMO, the units for S are thousands of lines of delivered instructions (KDSI), the units for E are PM or labor hours (LM) of effort, and the units for T, the time for development (TDEV), are months. A PM of effort is assumed to be comprised of 19 person days of 8 hours each, or 152 person-hours. This figure allows for time off for sick leave, holidays, and vacations.

For the basic COCOMO model, the cost driver is always 1.0. For the intermediate and detailed models, it is a composite function of 15 cost drivers x_1 through x_{15}. By setting these cost drivers to 1 in the intermediate model, we obtain an adjustment multiplier $M(\mathbf{x}) = 1$. Thus, the form of the intermediate COCOMO model reduces to that of the basic model when the cost drivers are at their nominal value of 1.0.

The values of a and δ *are not* obtained from least-squares regression, as is the case with many other cost estimation models. Boehm incorporates his own experience, the subjective opinion of other software managers, the results from use of other cost estimation models, and trial and error, to identify the structural relations that lead to the cost model as well as the parameters within the structure of this cost model. His initial parameters were fine-tuned using the TRW database. Thus, COCOMO is a composite model, basically combining experiential observations and wisdom together with statistical rule–based adjustments.

The parameters a and δ depend on what Boehm calls the *development mode* of a project. He labels a project as belonging to an organic, embedded, or semidetached mode, depending on the project's independence from other systems. In the initial effort, these first three modes were described. Later effort extended this to Ada programming language development efforts (Helm, 1992). These three modes may be described as follows:

1. The *organic mode* refers to the mode typically associated with relatively small programs that generally require relatively little innovation. They usually have rather relaxed software delivery time requirements, and the effort is to be undertaken in a stable in-house development environment. The external environment is generally also stable. An organic software development program can often almost run by itself, and will rarely require extraordinary systems management and technical direction efforts.
2. An *embedded mode* program is relatively large and has substantial operating constraints. Usually a high degree of complexity is involved in the hardware to be used as well as in the customer interface needs. These result in system and user requirements that need to be incorporated into the software development program. Often, the requirements for the to-be-delivered *embedded software* are exacting, and there is generally a need for innovative design in typically one-of-a-kind programs that have not been accomplished before and that will likely not be repeated.
3. A *semidetached mode* program has requirements and development characteristics that lie somewhere between organic and embedded.

The a and δ parameters for the basic and intermediate COCOMO model are presented in Table 7.3.

In the *basic COCOMO* model, the cost driver or adjustment multiplier $M(\mathbf{x}) = 1$ for all x_i. If we are considering semidetached mode development, the

TABLE 7.3. COCOMO Parameters for Basic and Intermediate Models

Mode	Basic model				Intermediate model			
	a	δ	b	Υ	a	δ	b	δ
Organic	2.4	1.05	2.5	0.38	3.2	1.05	2.5	0.38
Semidetached	3.0	1.12	2.5	0.38	3.0	1.12	2.5	0.38
Embedded	3.6	1.20	2.5	0.38	2.8	1.20	2.5	0.38

a and δ parameters are the same for both the intermediate and the basic COCOMO model. For organic and embedded development, the a parameters are different. This suggests that semidetached development is the nominal development mode for which the intermediate model COCOMO cost drivers are obtained. That the a parameter is different in the basic model than in the intermediate model, for other than semidetached developments, suggests that the meaning and interpretation of scaling of the cost drivers is different across these different development modes. In other words, the interpretation of the term "nominal" must be different for organic development efforts than it is for embedded development efforts. Using the basic model a parameters and the scaling indicated results in cost and effort values that are too low for organic mode development and too high for embedded mode development. This indicates that these cost-driver parameters are not independent of one another.

7.3.1 The Basic COCOMO Model

A single cost estimate, in terms of development effort, is obtained in the basic COCOMO model. This single effort measure is assumed to be distributed across the various phases of the software acquisition life cycle. In Boehm's COCOMO effort, a four-phase life cycle consisting of conceptual design, detailed design, programming and unit test, and integration and test is assumed to be used.

For the basic COCOMO model, no effort is assumed to be allocated to such deployment efforts as maintenance. The distribution of effort does vary across phases as a function of the size of the product, in terms of the number of lines of code that are developed—although the variation is not large. Figure 7.3 illustrates how this effort, in labor-months, is distributed across these four phases, for the three different development modes. As we would expect, the simpler organic mode of development results in a greater percentage of effort devoted to actual coding of a software product, and less effort associated with the conceptual architecture and detailed design phases. Of course, the simpler products also result in less effort being needed for integration and test since there is less need for a large number of development modules. Often the percentage effort for programming and unit test may be less with embedded software than it is for organic and semidetached software. However, the actual amount of development effort will generally be greater. Even if the number of

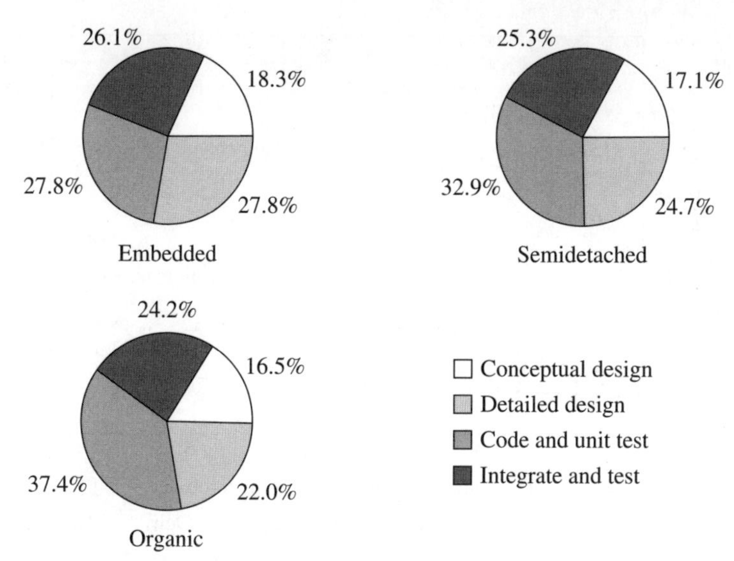

Figure 7.3. Effort Distribution Across the Four Life-Cycle Phases.

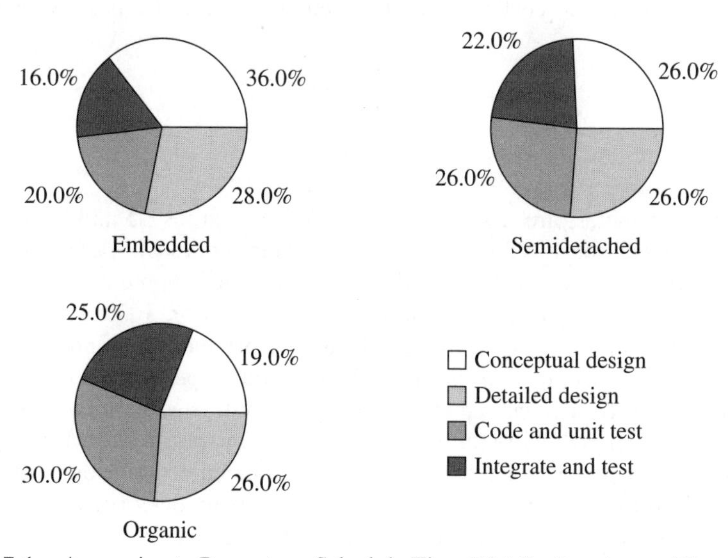

Figure 7.4. Approximate Percentage Schedule Time Distribution Across Phases for Basic COCOMO.

lines of code is the same, the fact that the *a* multiplier is larger for the embedded software will almost assure this. Figure 7.4 shows the companion results for the percentage distribution of schedule time across these four development phases. As we would expect, little effort needs to be devoted to conceptual design for organic software development. We can proceed almost directly to the detailed design and coding phases of development in this situation.

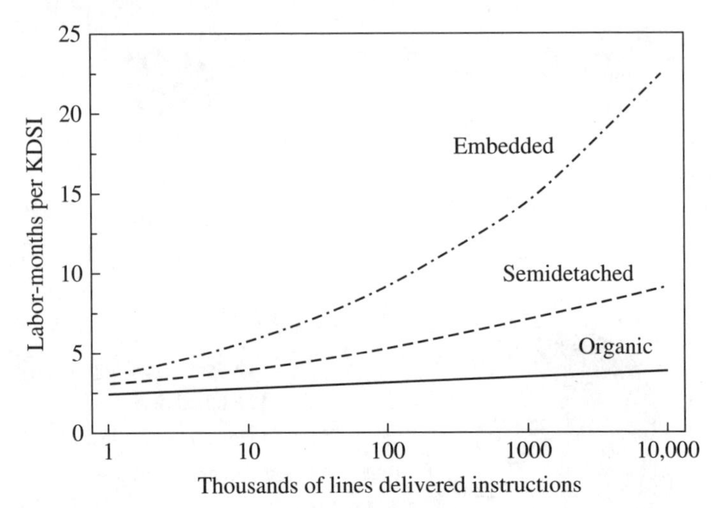

Figure 7.5. Labor-Months Required Per Thousand Lines of Delivered Instructions for Basic COCOMO.

The very simple cost and schedule equations for the basic COCOMO model are

$$E = aS^{\delta} \tag{7.7}$$

$$T = 2.5E^{\Upsilon} \tag{7.8}$$

where the parameters are as specified in Table 7.3. Figures 7.5 and 7.6 illustrate the development costs and schedule for the three modes of operation. Figure 7.5 illustrates how much more labor intensive development of embedded software is, as compared with organic software. For 1000 KSDI, which represents a very large software acquisition effort indeed, it takes approximately 4.23 times as much labor (14.33 labor-months per KSDI) to produce a delivered line of code in embedded software, as contrasted with the labor (3.39 labor-months per KSDI) required for organic software. As shown in Fig. 7.3, the total development time that *should* be allocated to development is not appreciably different across the three development modes. For a 1000-KDSI effort, the development time is approximately 54 months, or 4.5 years, regardless of the mode of development.[1] We can now use the effort and schedule distribution across phase results illustrated in Figs. 7.3 and 7.4 to determine phase-wise distributions of effort and schedule for the basic COCOMO model. This results in a relatively complete set of effort and

[1]It should be remarked that it would be unreasonable to expect organic software development of as large a system as would be represented by 1 million lines of delivered source instructions (1000 KDSI). In a similar manner, it is probably also unrealistic that there could be an embedded software development as small as 1000 lines of delivered source instructions (1 KDSI).

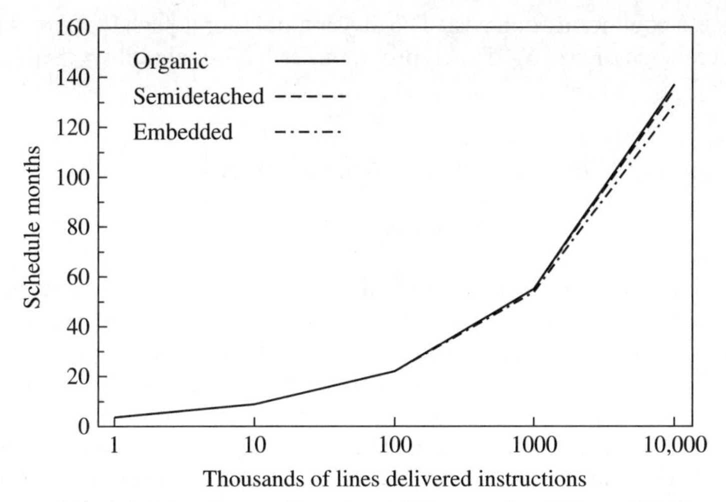

Figure 7.6. Schedule Months as a Function of Thousands of Lines of Delivered Instructions.

schedule plans for software development, as based on the assumptions inherent in the basic COCOMO model.

We might wonder why there is an amount calculated for effort, in labor-months, and an amount calculated in schedule time, months, through use of the basic COCOMO model. It might seem that it should be possible to obtain virtually any schedule time desired simply through adjusting the number of people working on an effort. Sadly, this turns out to be the case only within relatively narrow limits. A very classic work illustrates this specifically for software development. We digress briefly to discuss this important result here.

Fred Brooks, in a truly seminal work (Brooks, 1975), was perhaps the first to suggest a power-law relationship between lines of code, S, and software programming effort, E. The generic relationship for this is given by

$$E = dS^{\beta} \tag{7.9}$$

where the d and β parameters depend on the application being developed, programmer capability, and programming language used. In the work of Brooks, the parameter β was assumed to be 1.5, although few today would associate this large a diseconomy of scale with this relationship. This is the structure of the basic COCOMO model where β ranges from 1.05 to 1.20.

Brooks attempted to increase the power of this model by such efforts as adding terms to represent communication difficulties among the N programmers assumed to be working together on the programming project. This might result in a relation such as

$$C = dS^{\beta} + c\left[\frac{N(N-1)}{2}\right] \tag{7.10}$$

where c is a communication cost. The division of labor is such that programmer i produces code of size S_i. If each programmer has equal ability, then we have

$$E_i = dS_i^{\beta} \tag{7.11}$$

Since a total of S lines of code are produced, we have

$$S = \sum S_i = S_1 + S_2 + \cdots + S_N \tag{7.12}$$

where the summation extends over all N programmers. It is interesting to obtain the optimum allocation of effort among the programmers. To do this we use elementary variational calculus principles. We adjoin the constraint equation of Equation 7.12 to the cost function of Equation 7.10 by means of a Lagrange multiplier. We assume that each programmer has the same cost of communications. We also assume that each programmer produces the same amount of code such that $S = NS_i$. Then, we minimize the effort relation

$$E = NdS_i^{\beta} + c\left[\frac{N(N-1)}{2}\right] \tag{7.13}$$

where we have automatically satisfied the equality constraint of Equation 7.12. In this particular case, it is very interesting to sketch the relationship

$$\frac{J(N)}{d} = N\left(\frac{S}{N}\right)^{\beta} + E\left[\frac{N(N-1)}{2d}\right] \tag{7.14}$$

for several values of the economy of scale factor β and the communications effort ratio E/d.

The general results for negative economies of scale look like that shown in Fig. 7.7. The very interesting observation that the amount of effort required does not continually decrease with an increase in the number of programmers

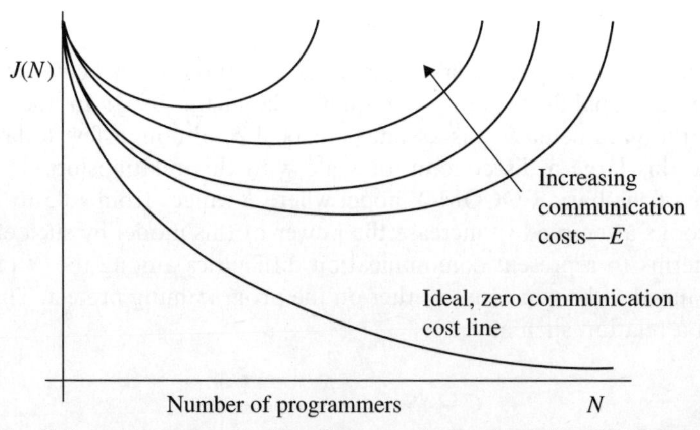

Figure 7.7. Effects of Communication Costs on the Productivity of N Programmers.

involved in the effort is evident. Only for very small values of the communication effort ratio would this be the case. We would like to have the ideal result noted in the figure. But we will get this only if these are perfect communications such that $E = 0$. In general, there will be an optimum number of programmers for maximum code production. Increasing the number of programmers results in a decrease in the amount of code that they can produce in a given amount of time.

All of this is potentially rather distressing. Not only do we have diseconomies of scale, but also the addition of programmers to aid an already late software development project may well result in further delays! While there is nothing generic to software in this, software is perhaps more likely to suffer from diseconomies of scale than other production technology efforts.

What this sort of simple analysis suggests is that we simply must bring about automated software production technologies, that is to say, those technologies that take advantages of the macroenhancement efforts of rapid prototyping, reusable software, and automatic code generation approaches to code production if we are to cope with truly large systems efforts. It seems highly doubtful that there are many major advances left to be obtained simply through only increases in programmer productivity. Of course, this is, in no way, an argument against programmer productivity, which is very necessary—but not at all sufficient. We note that we need both a cost equation and a schedule equation to deal with these issues.

7.3.2 Intermediate COCOMO

The basic COCOMO model is very simple and assumes a large number of nominal development parameters. The intermediate COCOMO model allows for judgment relative to 15 important cost drivers. These cost drivers are sometimes called environmental factors.

For the *intermediate COCOMO* model, the adjustment multiplier is calculated as the product of individual cost drivers, as given by

$$M(x) = \bigcap_{j=1}^{15} m(x_j) \tag{7.15}$$

The 15 cost drivers or environmental factors are grouped into 4 categories as illustrated in Fig. 7.8. Each of the 15 categories illustrated in this figure is associated with an appropriate multiplier factor $m(x_i)$. In the basic COCOMO model, each cost factor $m(x_i)$ is assumed to be equal to 1, and so each of the adjustment multipliers is equal to 1. In the intermediate and detailed COCOMO models, the cost factors $m(x_i)$ vary as a function of the quantitative values of the various cost drivers. For each category, a scale of five or six values is used to represent the possible choices for each category. For many of the cost drivers, values were obtained for the COCOMO database using the classical Delphi technique to elicit expert opinion. Reliability cost, for example, can be rated from very low to very high, in accordance with the equivalence between

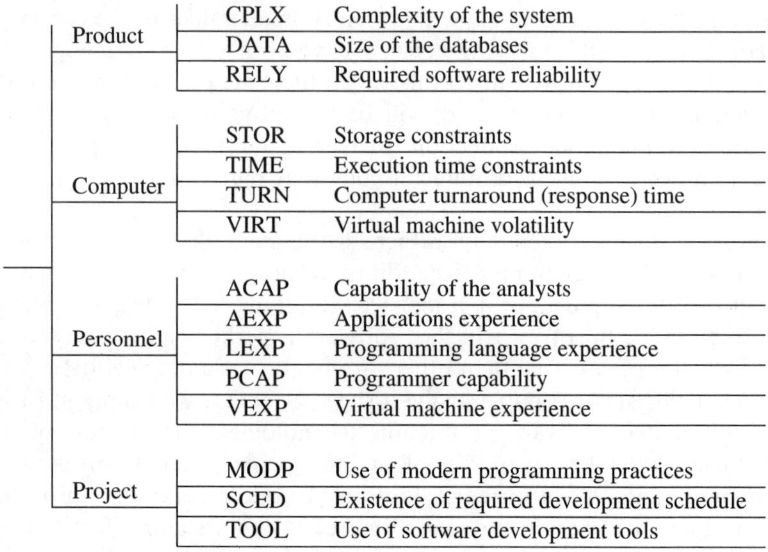

Category	Driver	Description
Product	CPLX	Complexity of the system
	DATA	Size of the databases
	RELY	Required software reliability
Computer	STOR	Storage constraints
	TIME	Execution time constraints
	TURN	Computer turnaround (response) time
	VIRT	Virtual machine volatility
Personnel	ACAP	Capability of the analysts
	AEXP	Applications experience
	LEXP	Programming language experience
	PCAP	Programmer capability
	VEXP	Virtual machine experience
Project	MODP	Use of modern programming practices
	SCED	Existence of required development schedule
	TOOL	Use of software development tools

Figure 7.8. Intermediate-Level COCOMO Cost Drivers or Multipliers.

TABLE 7.4. Reliability Categories and Multipliers (RELY)

Rating	Effect of lack of factor	Multiplier
Very low	Slight inconvenience if there is a problem	0.75
Low	Losses easily recovered due to problem	0.88
Nominal	Moderate difficulty in recovery	1.00
High	Major financial loss if problems develop	1.15
Very high	Risk to human life	1.40

verbal or semantic descriptors and a corresponding numerical driver shown in Table 7.4.

There is, in effect, a table like Table 7.4 for each of the 15 cost drivers in the database. It is assumed that one of the specific values called for in these tables will be used in that the model is validated only for these values. This should not preclude careful use of parameter values within the given ranges.

Appropriate descriptions for each of the 15 cost drivers in the intermediate COCOMO model are as follows:

CPLX—Software Product Complexity. CPLX rates the complexity of the software to be developed on a scale varying from very low (0.70) to very high (1.65). It is directly related to the type of processing required and the distance of the software from the hardware, or the compatibility between the software and the hardware. It turns out that no metrics have proved universally acceptable in this area, so the rating is often done by analogy. A recommended input value for CPLX of HI would mean that the software development needs involve hardware input/output interfaces, advanced data routines, and the use of such tools such as third-generation compilers. A CPLX rating of HI translates into a value of 1.15 as an environmental factor or multiplier.

DATA—Database Size. DATA rates the extra effort required to create software that involves manipulation and maintenance of a sizable database. This factor is evaluated by considering the ratio of the database size (in bytes) to the number of deliverable source instructions (DSI). DATA varies from low (0.94) to very high (1.16).

RELY—Required Software Reliability RELY rates the criticality of the software to be acquired in performing its intended function in an accurate and timely manner. It may vary from very low (0.75) to very high (1.40).

STOR—Storage Constraints. STOR measures the additional work for software development that results from constraints on memory availability in the target computer. Memory is defined as that random access storage or read-only memory (ROM) that is directly accessible to the central processing unit (CPU) or I/O processor. STOR may vary from nominal (1.00) to very extra high (1.56).

TIME—Execution Time Constraints. TIME relates to measures of the approximate percentage of the available CPU execution time that will be used by the software. It may vary from nominal (1.00) to extra high (1.6). A nominal recommended input value for TIME, NM, is associated with a multiplier of 1.00 and reflects that there are no constraints on the CPU execution time.

TURN—Computer Turnaround. TURN rates the average time spent waiting for the host or developmental computer to complete an action, such as compile a section of code or print a listing. It may vary from low (0.87) to very high (1.15).

VIRT—Virtual Machine Volatility. VIRT rates the degree of change that the host and target machines are projected to undergo during the development phases of the software acquisition effort. Machine here refers to the hardware and software that the project calls upon to complete its task. VIRT may vary from low (0.87) to very high (1.30). A low (LO) value of VIRT would result when no changes are expected across the development phases.

ACAP—Analyst Team Capability. ACAP measures the capability of the analyst team doing the following: defining and validating the system requirements and specifications, preparing preliminary design specifications, inspecting the software and testing plans, consulting efforts during detail design and code and unit test phases, and participation in the integration and test phase. ACAP may vary from very low (1.46) to very high (0.71). Thus, this metric decreases as performance improves. A recommended input value for a very high (VH) ACAP might be based on a rating of high for previous program performance, a rating of extra high for analyst team educational background, and a rating of good for analyst team communication.

AEXP—Project Application Experience. AEXP rates the project development team with respect to their level of experience in working on other development projects of similar scope and difficulty. It may vary, in the COCOMO model, from very low (VL = 1.29) to very high (VH = 0.82). A recommended input value for high (HI) AEXP might be based on the fact that the development team has much experience on projects of similar complexity that varies from 5 to 12 years for team members.

LEXP—Language Experience. LEXP measures the design analyst and programmer teams' experience with the programming language that will be used to design the software. It may vary from very low (VL = 1.14) to high (HI = 0.95). Thus, the value of this metric also decreases as performance improves.

PCAP—Programming Team Capacity PCAP measures the capability of the programmers performing the detailed software module design during the critical design phase of development, and the writing and testing of the physical code during the coding and integration testing phases. It may vary from very low (VL = 1.42) to very high (VH = 0.70). A recommended input value for nominal (NM = 1.0) PCAP might be based on a rating of nominal for previous program performance, of average for programming team educational background, and of average for programming team communication.

VEXP—Virtual Machine Experience. VEXP measures design analyst and programmer experience with the hardware and software of the host and target computers. Software refers to both the operating system and the application software. VEXP may vary from very low (VL = 1.21) to high (HI = 0.90). A recommended input value for HI VEXP might be based on the fact that the acquisition contractor is not using an HOL in a machine-independent fashion, and is experienced with the host and target machines.

MODP—Modern Programming Practice. MODP quantifies a contractor's use of and commitment to modern programming practices. It includes top-down requirements analysis and design, structured design notation, design and code walkthroughs, and structured code. It may vary from very low (VL = 1.24) to very high (VH = 0.82). A recommended value for VH MODP might be based on the fact that modern programming practices are prescribed by company policy and an education and training program is in effect.

SCED—Existence of Required Development Schedule. SCED reflects the extent to which there exists a required acquisition schedule that is realistic. It measures schedule compression and extension. Attempting to develop software when the time required for development is too short will result in a high rating for this multiplier. While it might seem that a very relaxed time schedule would result in a low rating, the value is actually increased above the nominal value of 1.0 because of inefficiencies that may be introduced when the amount of slack in development is very high. The value of SCED varies from very low (VL = 1.23), which relates to too short a development time, to nominal (NO = 1.0), to very high (VL = 1.10), which relates to too slack a development schedule.

TOOL—Use of Software Development Tools. TOOL relates to the contractor's use of automated software tools and the extent to which these tools are integrated throughout the software development process. It may vary from very low (VL = 1.24) to very high (VH = 0.83). A value of VH might be associated with a contractor's use of automated software tools in a moderately integrated development environment consisting of UNIX or a Minimum Ada Programming Support Environment (MAPSE), configuration management (CM), extended design tools, automated verification systems, cross-compilers, display formatters, and data entry control tools.

Table 7.5 illustrates the values for the intermediate COCOMO model effort multipliers as initially presented by Boehm. The overall or aggregate effects of these cost-driver multipliers may be large if they are set close to the extreme positions. If each driver is set at the lowest possible value given in this table, the product of the cost drivers is 0.0886. The largest possible product of the cost drivers is 72.379. It is, of course, unreasonable to expect this sort of range in practice. The lowest possible value would be associated with a truly excellent development team attacking an extraordinarily simple acquisition effort. The largest value would be associated with a virtually incompetent team attacking an extraordinarily difficult effort.

At these extremes, there would be little reason to suspect that the COCOMO model would have great validity. It is more reasonable that there would be correlation between product, project and computer complexity, and development team capability. If this is the case, then the multiplier products vary from 2.601 for a very-low-capability team working on a simple problem to 2.706 for a high-capacity team working on a most difficult acquisition effort.

Intermediate-level COCOMO uses the same approach as basic COCOMO to allocate costs, and schedules, across a four-phase development life cycle.

Let us describe the use of the intermediate COCOMO model in some detail. The steps given here can be shortened such that they are applicable to the basic COCOMO model. The intermediate COCOMO model estimates the effort and cost of a proposed software development in the following manner:

1. A *nominal development effort* is estimated as a function of software product size S in thousands of delivered source lines of instructions.

TABLE 7.5. Cost Drivers for COCOMO Function Multipliers

Cost driver	Multiplier values $M(x_i)$					
	Very low	Low	Nominal	High	Very high	Extra high
Product						
CPLX	0.70	0.85	1.00	1.15	1.30	1.65
DATA		0.94	1.00	1.08	1.16	
RELY	0.75	0.88	1.00	1.15	1.40	
Computer						
STOR			1.00	1.06	1.21	1.56
TIME			1.00	1.11	1.30	1.66
TURN		0.87	1.00	1.07	1.15	
VIRT		0.87	1.00	1.15	1.30	
Personnel						
ACAP	1.46	1.19	1.00	0.86	0.71	
AEXP	1.29	1.13	1.00	0.91	0.82	
LEXP	1.14	1.07	1.00	0.95		
PCAP	1.42	1.17	1.00	0.86	0.70	
VEXP	1.21	1.10	1.00	0.90		
Project						
MODP	1.24	1.10	1.00	0.91	0.82	
SCED	1.23	1.08	1.00	1.04	1.10	
TOOL	1.24	1.10	1.00	0.91	0.83	

TABLE 7.6. Effort Multipliers for Requirements Volatility (RVOL)

Rating	Effect	Multiplier
Low	Essentially no requirements volatility	0.91
Nominal	Infrequent, noncritical redirection efforts needed	1.00
High	Occasional, moderate redirection efforts are needed	1.19
Very high	Frequent, moderate redirection efforts are needed	1.38
Extra high	Frequent, major redirection efforts are needed	1.62

To make this calculation, we use the basic COCOMO effort determination equation $E = aS^{\delta}$, where a and δ are as given in Table 7.3.

2. A set of effort multipliers, or environmental factors or cost drivers, is determined from the rating of the software product on the set of 15 cost-driver attributes that we describe in Table 7.6. Each of the 15 cost drivers has a rating scale and a set of effort multipliers that indicates the amount by which the nominal estimate needs to be multiplied to accommodate the additional or reduced demands associated with these cost drivers.

3. The actual estimated development effort is determined by multiplying the nominal development effort estimate, using the a and Υ parameters for intermediate COCOMO, by all of the software product effort multipliers. For a specific software development project, we compute the product of all of the effort multipliers for the 15 cost-driver attributes. The resulting estimated effort for the entire project is the product of all of these terms times the nominal effort determined in step 1.
4. Additional factors are then used to obtain more disaggregate elements of interest from the development effort estimate, such as dollar costs, development schedule, phase and activity distribution, and annual maintenance costs. In addition to calculating the development effort E, in terms of PMs of effort, it is often desired to obtain the development schedule in months in terms of either the number of lines of code or the total PMs of effort. This turns out to be given by $T = bE^{\Upsilon}$, where b and Υ are also given numerical values, for the three software modes, as indicated in Table 7.3.

Figure 7.9 illustrates the general set of steps to be followed in obtaining a cost and schedule estimate using COCOMO. It suggests, as does this discussion, the initial use of basic COCOMO and then a switch to intermediate COCOMO to accommodate environmental factors that differ from the nominal.

The amount of variability of the 15 drivers about the nominal value of 1 varies considerably, as we see in Fig. 7.10. In this situation, the order of the

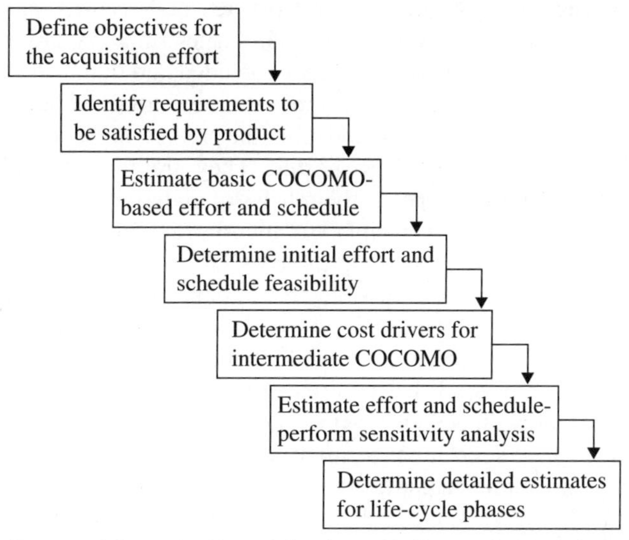

Figure 7.9. Suggested Steps for Determination of Effort and Schedule Using Intermediate COCOMO.

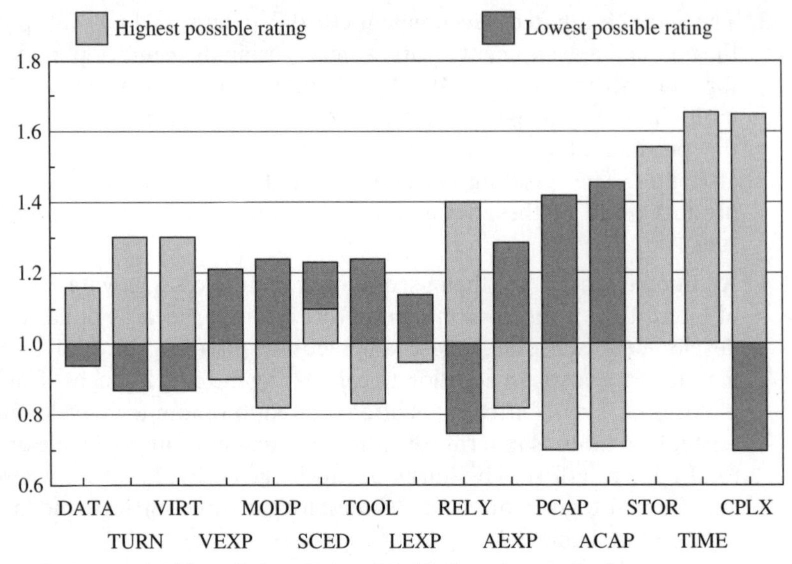

Figure 7.10. Variability of Cost-Driver Multipliers Across Ratings.

drivers is arranged such that those associated with greater variability are to the right of the curve. These tend to be the drivers associated with human capabilities and problem complexity, not the ones associated with the availability of automated tools.

There are a number of possible cost drivers that are not included in the intermediate COCOMO model. Among these are such factors as requirements volatility, management quality, customer (interface) quality, amount of documentation, hardware configuration, and privacy and security facets. Boehm discusses these factors, and several others. He suggests that parsimony is among the major reasons why other factors are not included and that inclusion of many of these factors would unnecessarily complicate the analysis while not providing for greatly enhanced fidelity of cost and schedule predictions. This does not mean that these factors are necessarily neglected. They are assumed to be at some nominal values and that changes in them will not materially affect results.

It is not difficult, in principle, to incorporate other parameters. For example, we might wish to modify intermediate COCOMO to a two-level version such that it becomes possible to consider *requirements volatility*, or requirements creep, which is interpreted to mean the tendency for requirements to change after they have been initially identified and where software development has proceeded beyond the requirements phase.

The first intermediate COCOMO 1 model or level is restricted to the cost drivers listed in Fig. 7.8. Our new intermediate COCOMO 2 model or level would include an additional driver for requirements volatility, with values ranging from a low of 0.91 to a high of 1.62. This assumes that these multipliers are experientially determined to be correct for given amounts of "volatility." Table 7.6 presents the effort multipliers for requirements volatility. This is then assumed to be a 16th cost driver in the COCOMO model equations obtained earlier. We can describe this 16th cost driver as follows:

RVOL—Requirements Volatility. RVOL measures the amount of redesign and redevelopment that is the result of changes in customer-specified requirements. The nominal value for this parameter is 1.0, which indicates no volatility or creep in requirements. This assumes that there will be only small and generally noncritical redirections.

An interesting question arises concerning whether or not the other drivers in the intermediate COCOMO model need to be changed to accommodate this new driver. If we can assume that the drivers are independent, the answer is no. While there is no strong reason to assume independence, we can assume near independence for small enough changes in the driver parameters.

There are a number of similar extensions of the COCOMO model. One very useful one has led to what is known as a Revised Intermediate COCOMO (*REVIC*) model. REVIC predicts the development of life-cycle costs for software acquisition across the three basic phases of definition, development, and deployment. It encompasses costs and schedule from requirements analysis to completion of the software acceptance testing and later maintenance. As with COCOMO, the number of DSLOC is the basic cost driver.

There are two major extensions to the development life cycle assumed in intermediate COCOMO that enables REVIC to include many aspects of definition and deployment. Six issues are of importance, which are discussed next:

1. The life cycle for COCOMO does not include the software requirements engineering phase. The amount of effort expended in this phase can vary dramatically. REVIC predicts the effort and schedule in the software requirements engineering phase by taking a percentage of the development phases. It utilizes a default value, 12% for effort and 30% for schedule, for the percentage to be used for nominal software acquisition programs, and allows the user to change this percentage, if desired. Of course, the very initial portions of the definition phase, perhaps accomplished during proposal preparation or specified as part of the request for proposals (RFP), must contain some user requirements. REVIC estimates the cost of transitioning from a set of user requirements and system specifications to software specifications.
2. The COCOMO life cycle really ends at the end of the software development phase, when the integration and test phase has been successfully completed. This phase is characterized by successful integration of computer software components (CSCs) into computer software configuration items (CSCIs) and associated testing of the CSCIs against the test criteria developed during the software acquisition program. This does not include the system-level integration of CSCIs and associated system-level testing to ensure that the system-level requirements have been met. This is generally accomplished as part of operational test and evaluation (OT&E) phase, as this is

sometimes called. REVIC predicts effort and schedule for this phase similar to the requirements specification phase, by taking a percentage of the development phase estimates. It provides a default percentage, 22% for effort and 26% for schedule, for this phase as nominal amounts and allows a user to change this percentage if desired.

3. REVIC provides an estimate for the maintenance of software over a 15-year period by implementing the following equation:

$$M = aS^{\delta} A\eta \tag{7.16}$$

where M is the maintenance effort in labor-months, A is an annual percentage code change factor (expressed as a fraction) or annual change traffic, and η represents a set of maintenance environmental factors. REVIC provides a default percentage for the annual change traffic and allows this to be changed, if desired. It assumes the presence of a transition period after the initial deployment of the software, during which time residual errors are found. After this, there exists a near-steady-state condition that allows a positive, and declining over time, fraction to be used for the annual change traffic during the first three years of deployment. Beginning with the fourth year, REVIC assumes that the maintenance activity consists of both error corrections and new software enhancements.

4. The essential difference between REVIC and intermediate COCOMO is the set of environmental factors used as cost drivers in the equations for C and T. An enhanced set of drivers is used, as we will soon describe. REVIC was calibrated using more recent data than available for COCOMO, and this results in different quantities for some of the environmental factors. On average, the parameter values used by the basic effort and schedule equations are slightly higher in REVIC than they are in COCOMO.

5. One of the additional cost drivers relates to the use of reusable code. When reusable code is present, an adjustment factor needs to be calculated and used to determine the size of a number of modules that are developed. A separate estimate for cost and schedule is determined through use of REVIC, or COCOMO, for each of these modules. The overall effort or cost is the sum of these. We describe this adjustment factor in more detail in our treatment of the detailed COCOMO model as it is the same factor that is used there.

6. Other differences occur in obtaining distribution of effort and schedule for the various phases of the development life cycle and in an automatic calculation feature for the standard deviation associated with risk assessment efforts. COCOMO provides a table for distributing effort and schedule over the development phases, based on the size of the code being developed and the mode of development, as illustrated in Figs. 7.3 and 7.4. REVIC provides a single weighted

"average" distribution for the overall effort and schedule, and has the ability to allow a user to change the distribution of effort allocated to the specifications and OT&E phases. Thus, the value of the SCED parameter is set equal to nominal or 1.0. The software user can set the actual development schedule through forcing it below that which results from the nominal calculation. This will always increase both labor-months of effort and associated costs.

There are 19 environmental factors in REVIC. Those four not included in the intermediate COCOMO model are as follows:

- RVOL—requirements volatility;
- RUSE—required reusability;
- SECU—DoD security classification; and
- RISK—risk associated with platform.

We have already described requirements volatility. Appropriate descriptors of the other three are as follows:

RUSE—Required Reusability. RUSE measures the extra effort that is needed to generalize software modules when they must be developed specifically for reuse in other software packages. The nominal value for reuse, NM, is 1.0, which means that there is no contractually required reuse of the software.

SECU—Classified Security Application. SECU measures the extra effort required to develop software in a classified area or for a classified security application. We note that this factor does not relate to the certification of security processing levels by the National Security Agency.

RISK—Management Reserve for Risk. RISK is the REVIC parameter that provides for adding a percentage factor to account for varying levels of program risk. If it is determined that the software development has a very low program risk, we use very low for the risk factor, and this translates into a numerical value of 1.00 for the cost-driver or environmental factor.

Thus, REVIC is essentially intermediate-level COCOMO with the addition of four additional environmental factors, an updated database, and considerable ability for "what if" type analysis, assessment, and sensitivity studies.

A standard set of two-letter descriptors associated with each of the seven cost multipliers, one more (extra–extra high) than generally described in intermediate COCOMO, is used to describe the extent to which each environmental factor differs from the nominal amount:

- VL—very low;
- LO—low;

- NM—nominal;
- HI—high;
- VH—very high;
- XH—extra high;
- XX—extra–extra high.

Some anchoring of the scoring for the factors is used to calibrate the use of the REVIC algorithms. This anchoring is particularly needed as many of the environmental factor terms need specific definition across the people responsible for providing estimates, if these estimates are to be compatible and consistent. The environmental factors and calibration presented in Table 7.7 were used in version 9 of the REVIC software. Many of the parameters are precisely the same as those used in the intermediate COCOMO model.

TABLE 7.7. REVIC Environmental Factor Calibration and Scoring

Product attributes		
Software product complexity—CPLX		
VL	Simple	0.70
LO	Data processing	0.85
NM	Math routines	1.00
HI	Advanced data structures	1.15
VH	Real-time and advanced math	1.30
XH	Complex scientific math	1.65
Database size—DATA		
LO	DB bytes/program SLOC, $D/P < 10$	0.94
NM	$10 \leq D/P < 100$	1.00
HI	$100 \leq D/P < 1000$	1.08
VH	$D/P \geq 1000$	1.16
Required software reliability—RELY		
VL	Slight	0.75
LO	Easy	0.88
NM	Moderate	1.00
HI	MIL-STD/high finance	1.15
VH	Loss of life	1.40
Computer attributes		
Main storage constraints—STOR		
VL	None	1.00
LO	None	1.00
NM	None	1.00
HI	70% utilization	1.06
VH	85% utilization	1.21
XH	Utilization > 95%	1.56

TABLE 7.7. (Continued)

Execution time constraints—TIME		
VL	None	1.00
LO	None	1.00
NM	60% utilization	1.00
HI	70% utilization	1.11
VH	85% utilization	1.30
XH	Utilization > 95%	1.66
Computer turnaround (response) time—TURN		
VL	<6 min	0.79
LO	<30 min	0.87
NM	<4 h	1.00
HI	>4 h	1.07
VH	>12 h	1.15
Virtual machine volatility—VIRT		
VL	No changes	0.87
LO	1 change in 6 months	0.87
NM	1 change in 3 months	1.00
HI	1 change every month	1.15
VH	Several changes every month	1.30
XH	Constant changes	1.49
Personnel attributes		
Analyst capability—ACAP		
VL	15th percentile	1.46
LO	35th percentile	1.19
NM	55th percentile	1.00
HI	75th percentile	0.86
VH	90th percentile	0.71
Project application experience—AEXP		
VL	Less than 4 months	1.29
LO	1 year	1.13
NM	3 years	1.00
HI	6 years	0.91
VH	Greater than 12 years	0.82
Language experience—LEXP		
VL	None	1.14
LO	Less than 1 year	1.07
NM	From 1 to 2 years	1.00
HI	2 years	0.95
VH	Greater than 2 years	0.95
Programming team capability—PCAP		
VL	15th percentile	1.42
LO	35th percentile	1.17

TABLE 7.7. (Continued)

NM	55th percentile	1.00
HI	75th percentile	0.86
VH	90th percentile	0.70
Virtual machine experience—VEXP		
VL	None	1.21
LO	Less than 6 months	1.10
NM	Less than 1 year	1.00
HI	From 1 to 2 years	0.90
VH	Greater than 2 years	0.90
Project attributes		
Modern programming practices—MODP		
VL	None	1.24
LO	Beginners	1.10
NM	Some	1.00
HI	In general use	0.91
VH	Routine	0.82
Existence of required development schedule—SCED		
NM		1.00
Use of software development tools—TOOL		
VL	Very few primitive	1.24
LO	Basic microcomputer tools	1.10
NM	Basic minicomputer tools	1.00
HI	Basic mainframe tools	0.91
VH	Extensive, but little tool integration	0.83
XH	Moderate, in an integrated environment	0.73
XX	Full use, in an integrated environment	0.62
Additional drivers in REVIC		
Requirements volatility—RQTV		
VL	None	0.91
LO	Essentially none	0.91
NM	Small, noncritical redirection	1.00
HI	Occasional, moderate redirection	1.19
VH	Frequent moderate, occasional major	1.38
XH	Frequent major redirection	1.62
Required reusability—RUSE		
NM	No reuse	1.00
HI	Reuse of single mission products	1.10
VH	Reuse across single mission products	1.30
XH	Reuse for any and all applications	1.50
Classified security application—SECU		
NM	Unclassified	1.00
HI	Classified (secret, or top secret)	1.10

TABLE 7.7. (Continued)

Management reserve for risk—RISK		
VL	Ground systems	1.00
LO	MIL-SPEC ground systems	1.20
NM	Unmanned airborne	1.40
HI	Manned airborne	1.60
VH	Unmanned space	1.80
XH	Manned space	2.00
XX	Manned space	2.50

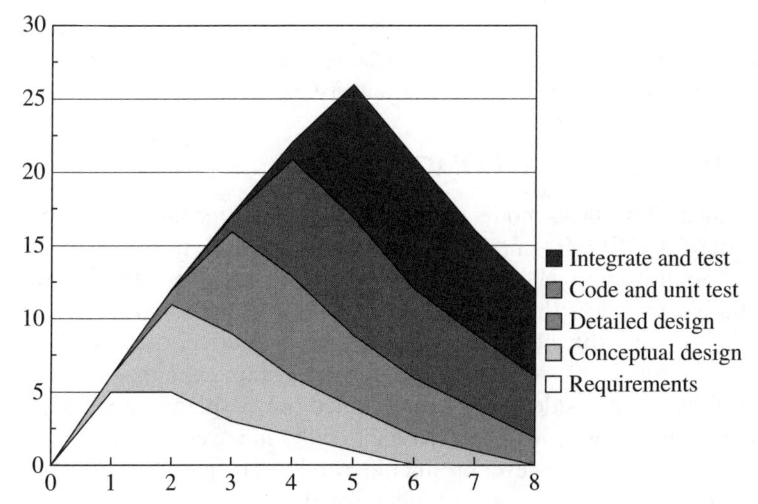

Figure 7.11. Generic Distribution of Labor-Months Over Time.

What results from this is a somewhat enhanced version of intermediate COCOMO and a software package especially suited for systems management prediction of the effort and schedule needed for software development. The possibility of performing various sensitivity analysis studies for changing parameters is especially attractive. We can also reuse the model to develop the several estimates needed for incremental and evolutionary development of software.

Of course, it is easily possible to implement intermediate COCOMO on any of the currently available microcomputer spreadsheet programs. One of the objectives in any of these implementations is to predict costs and schedules for the various life-cycle phases, such as to obtain the sort of results identified in Figs. 7.11 and 7.12.

There have been a considerable number of efforts to evaluate the accuracy of COCOMO models. In one of the more detailed studies (Helm, 1992), based

Time after project start

Life-cycle activity	*Costs*	*LM*	0 1 2 3 4 5 6 7 8 9
Requirements	\$60K	6	
Conceptual design	\$90K	9	
Detailed design	\$180K	18	
Code and unit test	\$240K	24	
Integrate and test	\$310K	31	

Figure 7.12. Gantt Chart of Schedule and Phase-Based Estimates of Costs and Effort.

on a study of seven new software development efforts, it was found that COCOMO was able to predict costs and schedule for all seven projects with less than 20% relative error. To make predictions of this quality requires detailed familiarity with past organizational practices and careful calibration of the COCOMO model to reflect these characteristics.

7.3.3 The Detailed COCOMO Model

The detailed COCOMO model adds refinements not included in the basic and intermediate COCOMO models. Detailed COCOMO introduces two more components to the model. Phase-sensitive effort multipliers are included, as a set of tables, to reflect the fact that some phases of the software development life cycle are more affected than others by the cost-driver factors. This phase-wise distribution of effort augmentation is the primary change over intermediate COCOMO. It provides a basis for detailed project planning to complete the software development program across all of the life-cycle phases.

In addition, a three-level product hierarchy is employed, so that generally different cost-driver ratings are supplied for the assumed four phases of effort for 4 cost drivers that primarily affect software modules (CPLX, LEXP, PCAP, and VEXP) and for the 11 drivers that primarily affect subsystems (ACAP, AEXP, DATA, MODP, RELY, SCED, STOR, TIME, TOOL, TURN, and VIRT). These drivers are presumably used at the level at which each attribute is most susceptible to variation. As in basic and intermediate COCOMO, the value of all nominal cost drivers is unity.

Multipliers have been developed for detailed COCOMO that can be applied to the total project effort and total project schedule completion time to allocate effort and schedule components to each phase in the life cycle of a software development program. Four distinct life-cycle phases are assumed to exist. The effort and schedule for each phase are assumed to be given in terms of the overall effort and schedule by the same general equations as for intermediate COCOMO, but with the use of phase-sensitive effort multipliers. Figures 7.13 and 7.14 illustrate the way in which two of these multipliers vary across the four assumed life-cycle phases. In general, there is usually not much difference in the multipliers for the detailed design phase and those for the code and unit test phase between detailed and intermediate COCOMO.

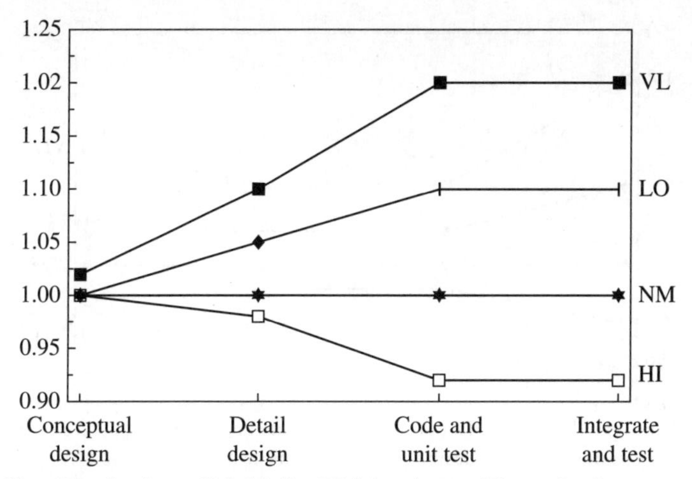

Figure 7.13. Distribution of Multiplier Weights Across Phases for Language Experience (LEXP) Factor in Detailed COCOMO.

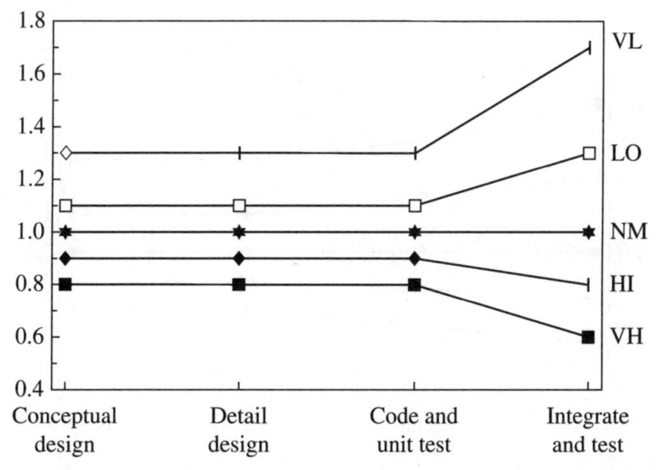

Figure 7.14. Distribution of Multiplier Weights Across Phases for Reliability (RELY) Factor in Detailed COCOMO.

For some of the cost drivers, such as the two illustrated in Figs. 7.13 and 7.14, there is not much difference in the conceptual design phase (denoted as the requirements and product design phase by Boehm) between intermediate and detailed COCOMO. There is a major difference, however, in the integration and test phase. We would expect this with respect to such factors as programmer capability, modern programming practice, use of software tools, and needed system reliability. For others, such as analyst team capability and project application experience, we would expect a considerable difference in the conceptual design phase as this is the phase in which these experiences are most critical. Figures 7.15 and 7.16 illustrate the variations in these multipliers over

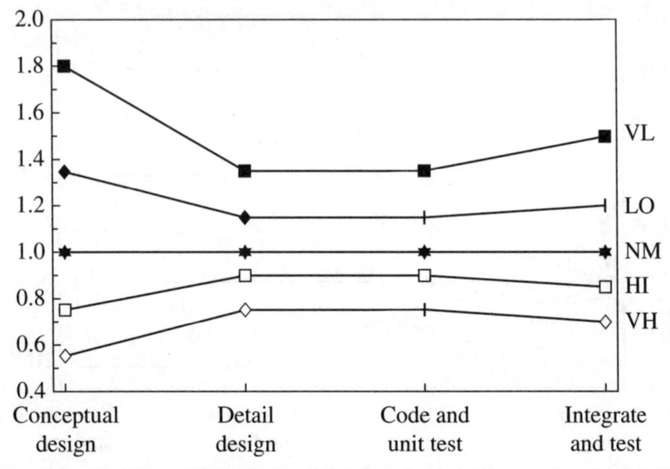

Figure 7.15. Distribution of Multiplier Weights Across Phases for Capability of Analysts (ACAP) Factor in Detailed COCOMO.

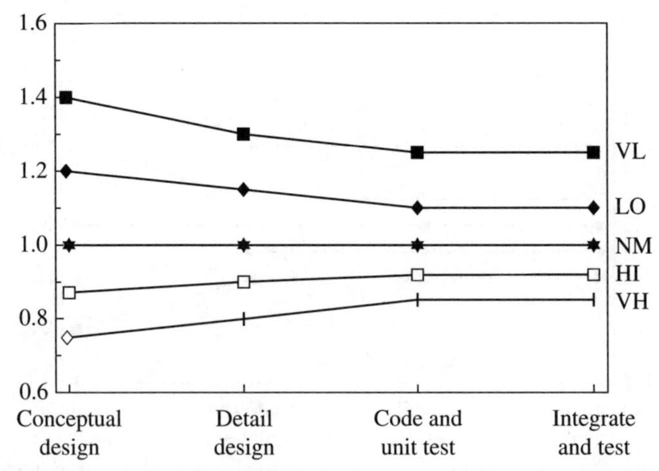

Figure 7.16. Distribution of Multiplier Weights Across Phases for Application Experience (AEXP) in Detailed COCOMO.

the four development phases. In each case, the multiplier associated with the intermediate COCOMO is a composite of these multipliers across the four phases shown. Much more detail concerning implementation of the detailed COCOMO model is available in Boehm (1981).

7.3.4 Other COCOMO-Based Models

The COCOMO model also has provisions for adjusting estimates of software costing when the software is developed, in part, from existing code. This involves a reuse COCOMO model. The adjustment is made at the module level, and each module is evaluated to determine how much code will be used without modification and how much will be modified.

Another variation of COCOMO is called Ada COCOMO (Boehm and Royce, 1987). This model is very similar to the original COCOMO model with only a few differences to allow for explicit incorporation of rapid prototyping, the spiral life cycle, risk management, and requirements volatility.

In general, the COCOMO model appears most appropriate for the grand design waterfall life cycle in which software is built right and completely in a single pass through the life cycle. It is easily modified to allow for other nongrand design life-cycle models and modes of development. In this approach, we calculate an effort for each of the successive, iterative, or evolutionary builds. This has been suggested by Balda and Gustafson (1990) and by a number of other researchers.

The previous model developments were macrolevel models based on the assumption of a given structure for the model and then determining parameters within this structure to match observed data. An alternate approach is possible in which we postulate a theoretically based model for resource expenditures in developing software.

It is not unreasonable to postulate that development time will influence the overall project effort and cost. Norden (1963) was perhaps the first to suggest a time sequencing of typical project effort. In investigating projects at IBM, he observed that the staffing of these research and development projects resembled a Rayleigh distribution. This sort of distribution is what might be empirically observed in practice, or obtained as a result of the sort of modeling effort that might have led to Fig. 7.11.

Putnam and Myers (1992) used the observations of Norden (1963) for project effort expenditures to develop a model of software costing. Putnam initially studied 50 projects from the Army Computer Systems Command and 150 other projects, thus developing one of the largest databases to support cost estimation. This resource allocation model, which has been developed under the name *S*oftware *Li*fecycle *M*anagement (SLIM), includes several effort factors: the volume of work, a difficulty gradient that measures complexity, a project technology factor that measures staff experience, and delivery time constraints. An investigation of the accuracy of SLIM (Conte *et al.*, 1986) indicates that the model exaggerates the effect of schedule compression on development. Another potential problem evolves from rather heavy reliance on the project lines of code and development time characteristics. Kemerer (1987) has also performed an evaluation of the SLIM model, as well as COCOMO and several other software packages. Jensen (1984) has proposed a useful variation to Putnam's SLIM model.

There are a number of alternative approaches to costing. ABC, which is not really a COCOMO variant and thus could be discussed in a new section if we presented a detailed treatment of this, is another approach to costing (Johnson and Kaplan, 1991; Kaplan, 1990; Kaplan and Cooper, 1990; Cooper and Kaplan, 1988; O'Guin, 1991). The ABC approach is based on activities required to produce a product that consume resources and that have value (or lack thereof) for customers. One of the major objectives of an ABC system is that of associating costs with the activities that lead to products. This is usually accomplished by first assigning all resources to either products or customers. The resources of an organization include direct labor, materials, supplies, depreciation, and fringe benefits. The organizational support units are separated into major functional units. Each of these functional units should have a significant cost associated with it, and each should be driven by different activities. The support unit, or department, costs are separated into a number of functional cost pools. These department costs should be fully burdened (one goal with ABC is to dramatically decrease undifferentiated burden), and all costs should be associated with the set of departments. The costs for each department are assigned to the various functional units such that each function has its own cost pool. The functional unit cost pools are next assigned, using first-stage cost drivers, to each of the activity centers. Measures of activity, such as customer orders, are very important first-stage cost drivers in that they are also measures of productivity. The organizational support costs are assigned to products or customers through these activity centers. An activity center is a functional economic grouping that contains similar or homogeneous sets of processes. The product-driven activity centers are those to which costs associated with the product lines or processes are assigned. The customer-driven activity centers are those driven by customers, such as sales and order department costs, and those that represent the costs of supporting customers and markets. The substance of an ABC system is associated with the set of second-stage cost drivers. The total cost of every activity center is separated into a number of cost-driver pools. The costs associated with each cost-driver pool are assigned to products using a second-stage cost driver. These second-stage cost drivers lead to resource consumption.

COCOMO is now available as COCOMO II and represents an extension of the initial COCOMO model described here. It is described in detail in Boehm et al. (2000). The URL http://sunset.usc.edu/csse/research/COCOMOII/cocomo_main.html is relatively extensive with much information of value, especially in terms of cost estimations efforts by Barry Boehm and/or his colleagues at the University of Southern California.

There have been several efforts to extend cost modeling COCOMO-like efforts to systems engineering. The papers by Lane and Valerdi (2007) and Lane and Boehm (2008) are noteworthy. The *Handbook of Systems Engineering and Management* (Sage and Rouse, 2009) describes a number of these as well, especially its Chapter 6 on cost management by Blanchard and Chapter 29 on project planning by Buys. Many contemporary systems engineering subjects that are described in detail may not formally involve costing in their initial description but must necessarily be incorporated to bring about realization of implementable systems concepts of cost analysis and assessment. One of the many such instances might involve the role of cost analysis and assessment in incorporating service-oriented architectural concepts for model-based systems engineering efforts (Andary and Sage, 2010).

7.4 SUMMARY

In this chapter, we have examined a number of issues surrounding the costing of software. While most of our efforts concerned costing, we did provide some comments concerning valuation, effectiveness, and other important issues. Modeling and estimation of software effort and schedule is a relatively new endeavor. Modeling and estimation of systems engineering, including systems engineering effort and schedule, is an even newer endeavor. Most such models are validated from databases of cost factors that reflect a particular user and developer organization characteristics. Such models are useful when developer organizations are stable and continue to produce systems, or services, similar to those developed in the past for users who have similarly stable environments and needs. However, models have little utility if the mixture of personnel experience or expertise changes or if the development organization attempts to develop a new type of system or a user organization specifies a new system type.

It would be desirable to build models that are independent of either user or developer organization, so that the models can be transported to other development teams and user situations. This is clearly foolish, however, as it assumes context-free solutions to problems, and this is very unrealistic. *The major conclusion from this is that it is absolutely necessary to consider developer and user organization characteristics in the cost estimation models that we build.*

A second problem is the use of cost-influencing factors and the relationship expressed among them. It is not clear that the appropriate factors have been identified; nor is it clear that they have been related correctly mathematically. Many of the cost models use adjustment multipliers in which the cost factors are summed or multiplied together.

A third issue involves changes in technology, tools, and methods. The contribution made to cost by these three aspects of project development is not as well understood as it should be for high-quality valuation purposes. Consequently, the appropriate change required in the cost equation is not known when technology, tools, and methods differ from those in place when the supporting database for a model was generated. Many of the databases used for contemporary cost and effort models use decades-old information, which is surely not reflective of current practices. Very few of the cost-estimating models are able to incorporate the potential benefits (or disbenefits) brought out by the use (misuse) of CASE tools. Nor do they consider the macroenhancement approaches of reusability, prototyping, and knowledge-based systems for development.

The development process itself presents a fourth set of needs. Some models examine the product or service rather than the *software development process* to obtain information about cost- and effort-influencing factors. An examination of only product- or service-related variables will generally never provide sufficient information to allow us to estimate effort and cost. More work needs to be done to determine what aspects of the development process contribute most to cost. The major improvements in productivity will come about through better development processes, and not at all through only implementing methods that ensure higher worker programmer productivity.

Most of the available methods select a single important deliverable SLOC for software, and use this variable as the core variable for estimation purposes. Various multipliers of this variable are proposed that presume to represent realities of conditions extant as contrasted with the nominal conditions under which the effort relationship to SLOC was determined. While this is not necessarily an appropriate way to go, it might be more meaningful to develop a program development plan in terms of client needs and then cost each of the phase-distributed efforts of the plan.

It is vitally important to develop cost estimation, including effort and schedule estimation, models that can be used very early in the development process. The purpose of these models is to predict development life-cycle costs and efforts as soon as possible and to provide information about the costs and effectiveness of various approaches to the acquisition process and its management. This will allow us to develop accurate information about development programs that clients might wish to fund and to allocate the appropriate resources to these developments. It will also result in feedback of important development information that will predict how changes in user requirements (software, system, or service), specifications, design, maintenance, and other factors will affect cost, effort, performance, and reliability.

A study of cost estimation (Lederer and Prasad, 1992) provides the following nine guidelines for cost estimation:

1. Assign the initial cost estimation task to the final system developers.
2. Delay finalizing the initial cost estimates until the end of a thorough study of the conceptual system design.
3. Anticipate and control user changes to the system functionality and purpose.
4. Carefully monitor the progress of the project under development.
5. Evaluate progress on the project under development through use of independent auditors.
6. Use cost estimates to evaluate project personnel on their performance.
7. Management should carefully study and appraise cost estimates.
8. Rely on documented facts, standards, and simple arithmetic formulas rather than guessing, intuition, personal memory, and complex formulas.
9. Do not rely on cost estimation software for an accurate estimate.

While these guidelines were established specifically for information system software, there is every reason to believe that they have more general applicability. This study identifies the use of cost estimates as selecting projects for implementation, staffing projects, controlling and monitoring project implementations, scheduling projects, auditing project progress and success, and evaluating project developers and estimators. Associated with this, we suggest, is evaluating the costs of quality, or the benefits of quality, and the costs of poor quality.

PROBLEMS

1. Use the basic COCOMO model under all three operating modes (organic, embedded, semidetached) and determine performance relations for the ratio of delivered source lines per person-month of effort. Examine this relation and its reasonableness for several types of software projects.

2. How would you go about including software engineering using rapid prototyping in the COCOMO models discussed here?

3. The decision to make or buy software is often a very important one. For many software engineering programs, there are opportunities to purchase an already available package that will function, perhaps with some amount of reengineering, as desired. Consider the following situations:

   a. You are the manager of a software engineering organization. It costs your organization an average of $35 per line of code to completely engineer software through all phases of the life cycle. Your group estimates that the desired functionality will require 10,000 lines of delivered source code. One of your software engineers notes that it is possible to buy a software package with the desired functional capability for $20,000. Should the software product be purchased or should it be engineered by your organization, and why?

   b. Further study indicates that 2000 lines of source code will have to be written and added to a possibly purchased software product to make it functionally useful. How does this alter the buy or make decision?

   c. How do your responses to (a) and (b) change if you are the Chief Information Officer (CIO) for a third firm that has the opportunity to purchase already developed software or pay to have it developed especially for your firm?

4. How should the COCOMO models be used for cost estimation in an engineering effort that uses reusable software?

5. How would COCOMO be used for engineering a system using reusable software?

6. How would COCOMO be used to estimate costs using structural, functional, and purposeful prototyping?

7. Please prepare an analogous discussion to that in this chapter for COCOMO II. You should find useful the URL given earlier for the Center for Systems and Software Engineering at USC.

8. Please prepare a discussion analogous to that presented in this chapter for costing of systems engineering products and services. You should find useful the URL given earlier for the Center for Systems and Software Engineering at USC.

9. The cost of ownership of a system invariably involves much more than just the acquisition cost. A good example is the acquisition of a personal car. The price one pays, including the "deal" struck, destination charges, sales tax, etc., begins a time series of cash flows as long as one owns and operates the vehicle. These costs include fuel, maintenance, insurance, property taxes (or equivalent), tolls, parking, and so on.

a. Estimate the yearly total costs of owning and operating a private vehicle, both as an overall total per year and in terms of costs per hour of usage.

b. Projecting the time series of costs over the life of the vehicle, calculate the net present cost of the vehicle using the methods discussed in Chapter 6.Your answers here include your direct out-of-pocket total costs of ownership. There are also indirect costs. Closest to home are the costs of garaging the vehicle, which includes the costs (or mortgage payments) of the garage, insurance against fire or other damage, and maintenance of the garage. Another cost relates to the construction and maintenance of roads and highways on which you drive. A portion of your tax dollars supports the availability of roadways for your use. Not as close to home are the costs of defending energy sources, for example, in the oil-rich nations in the Middle East. A portion of your income tax dollars supports this defense. Even less direct are the costs of dealing with the pollution created by your car. Your contribution to environmental damage and global warming will eventually create costs that taxpayers will have to support.

c. Estimate your total indirect costs of vehicle ownership and operation, including any taxes you pay to enable your abilities to drive your vehicle.

d. Project your total annual and hourly costs, and discuss the insights you have gained in performing this cost analysis.

10. Assume that you have been commissioned to develop a software application, perhaps for an iPhone, that will calculate the net present total cost of ownership, both direct and indirect, of a new vehicle. The software language that you must use is highly simplified. It only allows unformatted inputs and outputs, one variable at a time. The only computations possible are addition, subtraction, multiplication, division, and exponentiation. The user of this software will need to input costs in each of the categories discussed in the previous problem, for each of the 10 years. Alternatively, the user might enter the costs for the first year with an escalation factor for subsequent years. The software will need to sum the costs and then compute the discounted present cost using a discount rate chosen by the user. Using COCOMO, estimate the expected effort needed to program this software application. *Hint*: This will require a very much reduced version of COCOMO. Make your assumptions very explicit as to how you have simplified the model.

BIBLIOGRAPHY AND REFERENCES

Andary JF, Sage AP. The role of service oriented architectures in systems engineering. Inf Knowl Syst Manag 2010;9(1):47–74.

Bailey JW, Basili VR. A meta-model for software development resource expenditures. Proceedings of the Fifth International Conference on Software Engineering; 1981. p 107–116.

Balda DM, Gustafson DA. Cost estimation models for the reuse and prototype software development life-cycles. ACM SIGSOFT Software Eng Notes 1990;15(3):1–18.

Boehm BW. Improving software productivity. IEEE Comput 1976a;20(9):43–57.

Boehm BW. Software engineering. IEEE Trans Comput 1976b;25(12):1226–1241.

Boehm BW. Software engineering economics. Englewood Cliffs, NJ: Prentice-Hall; 1981.

Boehm BW, Royce W. Ada COCOMO: TRW IOC version. Proceedings of the Third COCOMO Users Group Meeting; November 1987. Carnegie Mellon University, Software Engineering Institute; Pittsburgh PA.

Boehm B, *et al.* Software cost estimation with COCOMO II. Englewood Cliffs, NJ: Prentice-Hall; 2000.

Brooks FP. The mythical man-month. Reading, MA: Addison-Wesley; 1975.

Conte SD, Dunsmore HE, Shen VY. Software engineering metrics and models. New York: Benjamin Cummings; 1986.

Cooper R, Kaplan RS. Measure costs right: making the right decisions. Harv Business Rev 1988;66 (5):96–103.

DeMarco T. Controlling software projects. New York: Yourdon Press; 1981.

Fox JM. Software and its development. Englewood Cliffs, NJ: Prentice Hall; 1985.

Helm, JE. The viability of using COCOMO in the special application software bidding and estimating process. IEEE transactions on engineering management 1992; 39 (1):42–58.

Jensen RW. A comparison of the Jensen and COCOMO schedule and cost estimation models. Proceedings of the International Society of Parametric Analysis; 1984. p 96–106.

Johnson HT, Kaplan RS. Relevance lost: the rise and fall of management accounting. Boston, MA: Harvard Business School Press; 1991.

Kaplan RS. Measures for manufacturing excellence. Boston, MA: Harvard Business School Press; 1990.

Kaplan RS, Cooper R. Cost management systems. Englewood Cliffs, NJ: Prentice Hall; 1990.

Kemerer CF. An empirical validation of software cost estimation models. Commun ACM 1987;30 (5):416–429.

Lane JA, Boehm B. System of systems lead system integrators: where do they spend their time and what makes them more or less efficient. Syst Eng 2008;11(1):81–91.

Lane JA, Valerdi R. Synthesizing SoS concepts for use in cost modeling. Syst Eng 2007;10 (1):297–308.

Lederer AL, Prasad J. Nine management guidelines for better cost estimating. Commun Assoc Comput Machinery 1992;35(2):51–59.

Norden PV. Useful tools for project management. In: Operations research in research and development. New York: Wiley; 1963.

O'Guin MO. The complete guide to activity based costing. Englewood Cliffs, NJ: Prentice Hall; 1991.

Porter AL, Roper AT, Mason TW, Rossini FA, Banks J. Forecasting and management of technology. New York: Wiley; 1991.

Putnam LH, Myers W. Measures for excellence: reliable software on time, within budget. Englewood Cliffs, NJ: Pergamon Press; 1992.

Rouse WB, editor. The economics of human systems integration: valuation of investments in people's training and education, safety and health, and work productivity. New York: Wiley; 2010.

Sage AP. Systems engineering. New York: Wiley; 1992.

Sage AP. Systems management for information technology and software engineering. New York, Wiley; 1995.

Sage AP, Palmer JD. Software systems engineering. New York: Wiley; 1990.

Sage AP, Rouse WB, editors. Handbook of systems engineering and management. 2nd ed. New York: Wiley; 2009.

Thebaut SM, Shen VY. An analytic resource model for large-scale software development. Inf Processing Manag 1984;20(1–2):293–315.

Walston C, Felix C. A method of programming measurement and estimation. IBM Syst J 1977;10 (1):54–73.

Wolverton RW. The cost of developing large scale software. IEEE Trans Comput 1974;23 (6):615–636.

CHAPTER 8

APPROACHES TO INVESTMENT VALUATION

Enterprises typically have to address many opportunities for investments. Often they create these opportunities via R&D or market research. Sometimes these opportunities represent the convergence of the enterprise's aspirations and an unexpected way to fulfill these aspirations—in other words, serendipity. Regardless of the source, a central question concerns the worth or value of the opportunity relative to the cost of pursuing it. In part due to uncertainty, enterprises usually want the assessed value to substantially exceed the projected costs.

This chapter does not address the noneconomic attributes of investments, which are often very important. For example, when a firm invests in upgrading its business processes, or in R&D to create new, proprietary technologies, it usually has other concerns that extend beyond only the economic value of the monetary investment. Typical additional concerns include strategic fit, leveraging of core competencies, and the extent to which the investment is likely to create a sustainable competitive advantage (Rouse, 2001). An investment not compatible with the firm's business strategy, or one that requires new competencies or that creates only temporary advantage, is less likely to garner investment despite its potential investment value greatly exceeding its costs.

This chapter addresses several alternative methods for valuation of investments including discounted cash flow (DCF), the capital asset pricing model (CAPM), Holt's cash flow ROI (CFROI) valuation (discussed in Madden, 1999), Stern–Stewart economic value added (EVA) (Stern, Shiely, Ross, 2001), and various competing methodologies. Two case studies are employed. One addresses investing in the efficiency of the processes for delivering current market offerings. The other one concerns investing in R&D to create and deploy new market offerings.

This chapter also addresses the limitations of these valuation methods. They all were developed with a particular perspective, depend on several critical assumptions, and often require parameter estimates that can be difficult to make. Several of the assumptions underlying the methods in this chapter are relaxed in Chapter 9. However, the methods in Chapter 9 involve a variety of new assumptions relative to the phenomena associated with relaxing the assumptions in this chapter.

Economic Systems Analysis and Assessment,
by Andrew P. Sage and William B. Rouse

8.1 MICROECONOMICS

It is important to place these valuation methods in the context of earlier expositions in this book, particularly in Chapters 2 to 4. Thus, this chapter begins by revisiting the theory of the firm and the theory of the consumer. The formulations from earlier chapters are used to pursue a notional example focused on a firm's pricing decisions. These decisions are then linked to alternative investment decisions. This leads to a discussion of fundamental investment issues.

8.1.1 Theory of the Firm

Recall from Chapter 2 that an enterprise's production function, f, is a specific mapping from or between the M input variables $\mathbf{x}$ to the production process and the output quantity produced, denoted by q:

$$q = f(\mathbf{x}) = f(x_1, x_2, \ldots, x_M) \tag{8.1}$$

If the firm prices these products or services at p per unit, then revenue is given by pq. If each input x_i has "wage" w_i, then profit $\prod$ is given by

$$\Pi(\mathbf{x}) = pf(\mathbf{x}) - \mathbf{w}^{\mathrm{T}}\mathbf{x} - \text{fixed costs} \tag{8.2}$$

8.1.2 Theory of the Consumer

In Chapter 3, it was noted that consumers consider multiple attributes, which we can denote by $y_1, y_2, \ldots, y_M$. Multi-attribute utility theory considers the relationship between the set of attributes of an alternative $\mathbf{y}$ and the consumer's relative preferences for this alternative:

$$U(\mathbf{y}) = U[u(y_1), u(y_2), \ldots, u(y_L)] \tag{8.3}$$

There may be K stakeholders whose preferences are of interest. This leads to the multi-attribute, multistakeholder model:

$$U = U[U_1(\mathbf{y}), U_2(\mathbf{y}), \ldots, U_K(\mathbf{y})] \tag{8.4}$$

8.1.3 Example: Optimal Pricing

Consider how these simple models can provide insights into investment issues. First, assume that the consumer's desires Q_D can be characterized by

$$Q_D(t) = Q_0[1 - \alpha P(t)] = \text{Price sensitivity} \tag{8.5}$$

Note that this assumes that the consumer values only one attribute—price. It further assumes that the consumer's utility function is a linear function of price. In reality, one would expect a nonlinear utility function, but over small price ranges this is a reasonable approximation.

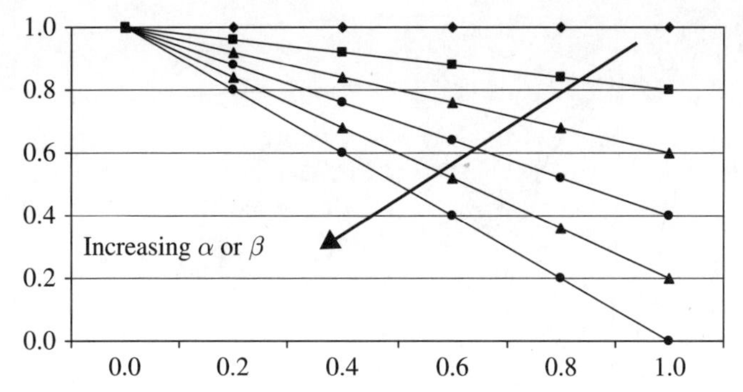

Figure 8.1. Price Sensitivity and Economy of Scale Models.

The profitability of the enterprise that provides these offerings, in quantities denoted by Q_P, can be characterized by

$$\pi(t) = [P(t)Q_D(t) - \mathrm{VC}(t)Q_P(t)] - \mathrm{FC}(t) = \text{Profit} \tag{8.6}$$

where the variable and fixed costs can be modeled by

$$\mathrm{VC}(t) = V_0[1 - \beta Q_P(t)] = \text{Economy of scale} \tag{8.7}$$

$$\mathrm{FC}(t) = [1 - \mathrm{GM}(t)]P(t)Q_P(t) = \text{Fixed costs} \tag{8.8}$$

$$\mathrm{GM}(t) : \text{gross margin after G\&A, R\&D, etc.} \tag{8.9}$$

The linear models for price sensitivity and economy of scales are shown in Fig. 8.1. These linear approximations assume that variations in prices and quantities do not take on extreme values where nonlinearities would certainly emerge.

To find the optimal pricing, one takes the first partial derivative of profit with respect to price ($\partial\pi/\partial P$), sets this result equal to zero, and solves for the price. To determine whether this price is a maximum or minimum, one takes the second partial derivative with respect to price ($\partial^2\pi/\partial P^2$), substitutes in the optimal price, and examines the sign of the result. If the second derivative is negative, the price is a maximum; if it is positive, the price is a minimum.

Performing these operations, along with quite a bit of algebra, yields the price that optimizes profit (Rouse, 2010):

$$P_{\mathrm{OPT}} = \frac{2\alpha\beta V_0 Q_0 - \mathrm{GM} - \alpha V_0}{2\alpha^2\beta V_0 Q_0 - 2\alpha\mathrm{GM}} \tag{8.10}$$

This represents the maximum profit when $\alpha\beta V_0 Q_0 \leq \mathrm{GM}$, and the minimum loss when $\alpha\beta V_0 Q_0 > \mathrm{GM}$. The latter condition holds when the parameters of the model are such that it is not possible to make a profit.

Figure 8.2 shows the sensitivity of the optimal price to market price sensitivities and economies of scale. Market price sensitivity, reflected in α, has a

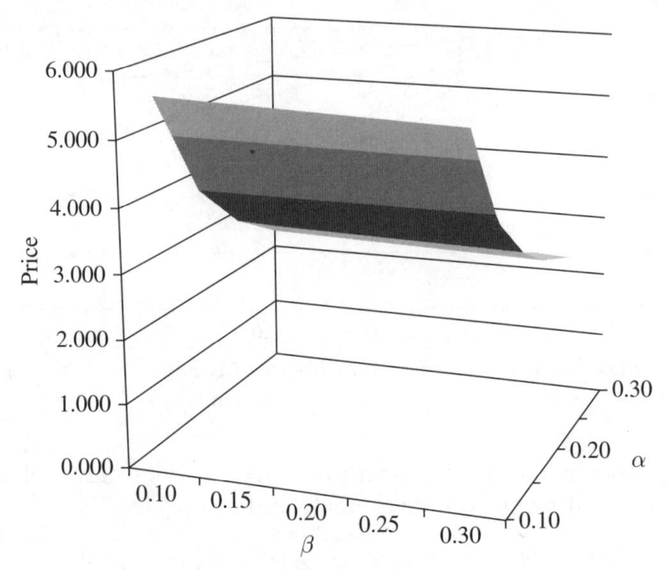

Figure 8.2. Optimal Price as a Function of α and β.

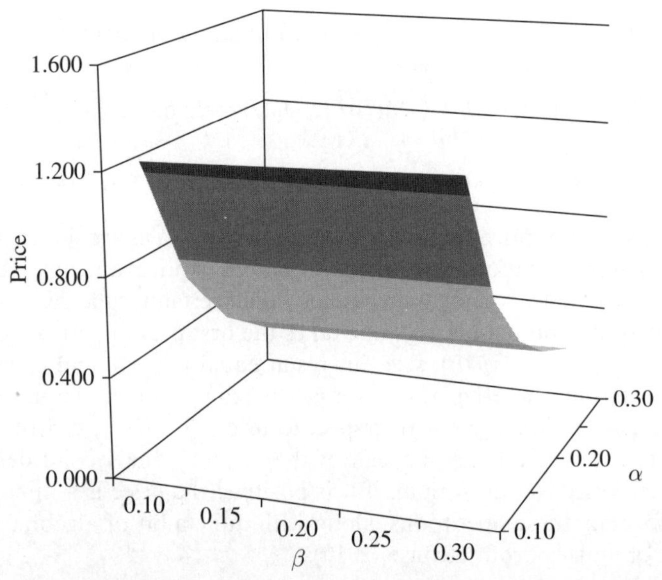

Figure 8.3. Maximum Profit as a Function of α and β.

very strong effect, while economy of scale, captured by β, has a much more modest effect. Figure 8.3 shows the sensitivity of profit, where the disparate effects of α and β are again evident. Thus, one can see that price and profits are highly affected by the price sensitivity of the market. In contrast, investments in improvements of economies of scale do not yield the same magnitudes of returns. Thus, a firm's discretionary monies may be better spent on convincing consumers, for example, via advertising that they want the firm's product or service rather than invest these monies in efficiency.

Consider the firm's investment choices in terms of the model parameters. The firm could increase prices and profits by investing in decreasing α. They might do this by increasing advertising and/or addressing broader markets. This would tend to increase general and administrative (G&A) expenses and, hence, decrease GM that would, in turn, decrease profits. Thus, this may or may not be a good idea.

Another way to decrease α would be to invest in creating new proprietary products and services. The required R&D investments would have to be balanced against potential profits to determine whether these investments made sense. Such investments are the focus of a case study later in this chapter.

The firm could invest in improving process efficiency and, consequently, increase β. As noted above, such investments do not tend to yield the magnitude of returns possible from decreasing α. However, in some markets, for example, military aircraft, advertising does not provide much leverage, the range of customers is constrained, and new proprietary offerings are rarely initiated without customer investments.

Another approach to increasing prices and improving profits is to increase GM. This can be accomplished by shrinking G&A, disinvesting in R&D, and minimizing other costs not directly related to current offerings. This can have an immediate and significant impact. However, it also tends to mortgage the future in the sense that the investments that are curtailed may be central to success in the future.

The application of this model to price controls in healthcare provided a variety of insights (Rouse, 2010). Identifying the criteria for increased advertising, minimizing discretionary costs, or even market exit—as opposed to investing in increased efficiency—enabled predictions of these behaviors that were subsequently validated via compilations of studies where healthcare providers exhibited these exact behaviors.

Thus, both in this model and in reality, it is clear that there are important trade-offs among the possible ways to improve the firm's profitability. The ability to address and resolve these trade-offs is central to a firm's competitiveness and success.

8.1.4 Fundamental Investment Issues

Achieving an enterprise's objectives requires increasing the efficiency of existing processes and/or launching new solutions that require new processes and/or technologies. In a broad sense, success depends on running the current enterprise well while also investing in and creating the future enterprise. Many senior executives find this a difficult balancing act.

It is reasonable to argue that the economic value of potential future cash flow streams should inform decisions regarding the appropriate levels of investments in these processes and technologies. The series of initiatives needed to yield these future cash flow streams can determine the processes and technologies needed and the costs of providing them. The remainder of this chapter addresses the economic valuation of such initiatives.

As noted earlier, it is important to keep in mind that noneconomic attributes can also be very important. Strategic fit, leveraging of core competencies, and sustainable advantage are all great examples of such attributes. It is very important that economic analyses and assessments be appropriate and accurate. However, this is only a necessary condition for good decision making, not a sufficient condition.

8.2 THE INVESTMENT PROBLEM

One can think about an enterprise's investment problems in terms of three time series, with typical time increments of one year. There is the investment time series including those in R&D and capacity creation. Once capacity is created, there is the time series of operating costs for creating and delivering products and services. The third time series is of profits, which can be characterized as free cash flow. For simplicity, we will not consider issues like taxes and depreciation.

There are many uncertainties associated with these time series. There are technical uncertainties related to the success of the R&D and the creation of capacity. There are market uncertainties related to demands for products and services, both related to the nature of these offerings and the competition, as well as the economic climate. Investment valuation needs to be concerned with these types of uncertainties.

One way to mitigate these uncertainties is to stage investments (Rouse *et al.*, 2000). For example, the enterprise might delay committing to capacity creation until the R&D has proven to be successful. It might also delay this commitment until there are clearer market signals that the demand for the offerings will be sufficient to justify investments. In Chapter 9, the economic value of being able to appropriately stage investments is discussed.

The general investment valuation problem can be stated as follows. Given the three time series described above, are the required investments worth the projected returns? Alternatively, what are the required investments worth? Clearly, one would like the worth of investments to far exceed the costs of investments.

Why do people tend to seek investments whose worth far exceeds the costs? This is due, in part, to substantial uncertainties being associated with projected returns. A significant difference between worth and cost provides a margin for error. Another reason is the typical constraints on the investment budget. If only a few investments can be made, then those alternatives whose worth most exceeds costs are a good place to start. Of course, the noneconomic attributes have to also be considered in the decision-making process.

8.2.1 Two Case Studies

Two case studies will be employed to illustrate the investment valuation methods discussed in this chapter as well as in Chapter 9. In this section, these case studies are outlined in terms of returns sought, the costs incurred, and associated uncertainties and risks.

8.2.1.1 Efficiency Case Study The first case study focuses on investing in the efficiency of the processes for delivery of current market offerings. The returns sought include decreased variable costs as well as decreased fixed costs. There may be a trade-off between these two types of costs in that decreasing variable costs may require increasing fixed costs. The costs incurred in pursuit of these efficiencies include increased engineering costs and possibly increased capital costs for new or upgraded capacity.

Typical examples of such investments include process automation, supply chain management, sales automation, customer support, and information technology efforts more broadly. These investments often include equipment, facilities, and engineering services for installation, tailoring, and testing. There may also be training costs for the personnel whose workflow is affected by these investments.

The uncertainties and risks associated with this case study include abilities to sustain prices for current offerings, including the risk of commoditization whereby all competitors are providing effectively identical offerings, resulting in steadily decreasing prices and profit margins. There is also uncertainty associated with sustained demand for current offerings. At some point, new types of offerings will displace current offerings. Finally, there are uncertainties and risks associated with the strategies of competitors.

8.2.1.2 R&D Case Study The second case study concerns investing in R&D to create and deploy new market offerings. The returns sought include new offerings over time, hopefully yielding price and profit premiums. The costs incurred include increased R&D costs and, assuming the R&D is successful, increased engineering costs and increased capital costs to create the needed capacity to deliver the new offerings. The estimation of such costs was addressed in Chapter 7.

Examples of R&D investments include discovery, development, and translation of new technologies into new market offerings in terms of products and services. If these investments result in proprietary capabilities (e.g., a new chip design or a new approach to process automation), the capabilities may enable price and profit premiums. In some situations, patents can prevent other firms from developing similar product, service, or process innovations.

The uncertainties and risks associated with this case study include R&D costs and the probability of technical success. There is usually considerable uncertainty in the likely demand for new offerings. Finally, there are often substantial uncertainties associated with competitors' strategies with regard to the types of offerings envisioned and technologies pursued.

Table 8.1 shows a "pro forma" financial summary for these two case studies. The shaded areas indicate where the nonzero entries typically appear. The notes at the bottom of the table refer to whether or not costs are scalable. This refers to whether or not the demand for offerings affects these costs.

TABLE 8.1. Pro Forma Financials

Revenues/costs	Year 1	Year 2	Year 3	...	Year *N*
R&D					
Labor costs					
Capital costs					
Deployment					
Labor costs					
Capacity costs					
Operations					
Revenues					
Operating costs					
Free cash flows					

Notes:
1. R&D capital costs include equipment, prototypes, etc.— seldom scalable.
2. Deployment capacity costs include manufacturing, assembly, test, etc.— may be scalable.
3. Operating costs vary with volume of products or services delivered—usually scalable.

Operating costs usually scale with quantity produced. Capacity costs may or may not be scalable depending on whether capacity is "fungible" relative to other uses. R&D costs are rarely scalable in that these costs are usually incurred whether or not any of the subsequent offerings are successful in the marketplace.

8.3 INVESTMENT VALUATION

This section describes and compares the following four different, but related, methods for valuation of investments:

- discounted cash flow (DCF);
- capital asset pricing model (CAPM);
- cash flow ROI valuation (CFROI); and
- economic value added (EVA).

Following a discussion of these methods, the methods are compared by applying them to one of the two case studies. The limitations of these approaches are also discussed, thereby setting the stage for Chapter 9.

The basic nomenclature that is employed is as follows. The time series of costs is denoted by c_i, $i = 0, 1, \ldots N$. The time series of returns is denoted by r_i, $i = 0, 1, \ldots N$. N denotes the number of time periods of interest. Often point estimates of c_i and r_i are employed. Of course, the use of probability distributions for these variables would usually be more appropriate, especially for returns. This possibility is pursued in more depth in Chapter 9.

8.3.1 Discounted Cash Flow

DCF is concerned with the time value of money, which was discussed in detail in Chapter 6. A given amount of money in the future is worth less than the same amount of money now. Why is this the case? The basic idea is that a given amount of money now should be worth more in the future. For example, $100 now should equal $105 dollars one year from now, assuming a 5% annual interest rate. Thus, $100 one year from now will be worth $95.24 now (i.e., $100/1.05), again assuming a 5% annual simple interest rate.

Thus, future profits or cash flows should be discounted by an interest rate that reflects the earnings forgone by delaying receipt of these cash flows. In general, the current value of money is the sum of the discounted values of future cash flows assuming a given interest rate over the period during which these cash flows will occur. This interest rate is termed the discount rate (DR) because it is the rate at which future cash flows are discounted.

The choice of DR involves significant subtleties, as elaborated in Chapter 6. One needs to decide what returns the money would have earned had the money been received now rather than in the future. Would this money have been invested in government bonds? Then the DR might be 5% or, in light of recent events, much lower. However, would the firm really have invested in government bonds?

More likely, enterprises would also have invested in other lines of business. Thus, the DR might be set as the rate of return experienced with the firm's mature lines of business. Of course, this assumes that these funds could have been productively invested in this matter. It also assumes that the future of these lines of business will mimic the past.

Another approach to choosing the DR is to use the interest rate the firm pays to borrow funds. The idea is that delaying receipts of cash flows will, in effect at least, require the firm to borrow the amounts involved now. The firm will then have to pay interest on the borrowed monies until the future cash flows are received and the loans are repaid.

Yet another common practice is to adjust the DR to reflect perceived uncertainties associated with future cash flows. In effect, investors may feel that the projected returns, usually expressed in terms of point estimates, are actually means of distributions with very significant variances. These point estimates may also be perceived as inflated estimates of these means. Thus, it is not unusual for venture capitalists to employ DRs of 50% or more.

Clearly, investors tend to convolve many phenomena in their use of this single parameter. This is an understandable, but inappropriate, practice. The time value of money reflects the fact that future cash flows are worth less than current cash flows, for the reasons outlined above. The uncertainties and risks associated with these cash flows are related but different phenomena. This issue is revisited later in this chapter and in Chapter 9.

8.3.1.1 Net Present Value (NPV) NPV and internal rate of return are the classic metrics associated with DCF (Brigham and Gapenski, 1988). NPV is given by

$$\text{NPV} = \sum_{i=1}^{N} \frac{r_i - c_i}{(1 + \text{DR})^i} \tag{8.11}$$

NPV reflects the amount one should be willing to pay now for benefits received in the future. These future benefits are discounted by the interest paid now to offset the delays in receiving these later benefits.

Note that NPV assumes that costs continue to be incurred over the set of time periods of interest regardless of the results in previous periods. Thus, multistage investments are assumed to succeed at each stage and justify continued investments. This assumption is relaxed in Chapter 9.

8.3.1.2 *Internal Rate of Return*

$$\text{IRR} = \text{DR} \qquad \text{such that} \sum_{i=1}^{N} \frac{r_i - c_i}{(1 + \text{DR})^i} = 0 \tag{8.12}$$

This metric enables comparing alternative investments by forcing the NPV of each investment to zero. Note that this assumes a fixed interest rate and reinvestment of intermediate returns at the internal rate of return. This can be problematic when there are no investment opportunities available that provide this rate of return. This difficulty is explained in detail in Chapter 6.

NPV and IRR are used pervasively across types of investments and industry domains. As discussed above, these metrics are rather assumption laden. Unfortunately, users of these metrics often forget the assumptions they have, in effect, made in adopting these metrics.

8.3.2 Example—Even Returns Versus Lump Sum Returns

Consider a time series of free cash flow of \$1000 per year for 10 years versus \$0 for 9 years and *X* in year 10: what value of *X* is needed for these two series to be equivalent for DRs of 5% and 10%?

Using Equation 8.11, the NPV of the even returns (i.e., \$1000 per year) is \$7722 and \$6145 for DRs of 5% and 10%, respectively. Thus, the higher the DR, the lower the NPV. For equal NPV for the lump sum returns (i.e., \$*X* at year 10), the value of *X* must be \$12,578 and \$15,938 for DRs of 5% and 10%, respectively.

Thus, one can see that the DR has a very significant impact. To see this, recalculate the NPV and *X* for a DR of 20%. The value of *X* needed to provide an equivalent NPV for both types of returns is \$25,955. Thus, one needs 260% greater return if one has to wait 10 years to get this return.

8.3.3 Capital Asset Pricing Model

The CAPM provides a classic approach to estimating the likely returns on an asset (Luenberger, 1997). It relates the return on an asset (RA) to the risk-free

TABLE 8.2. Examples of β and Volatility

Company	β	Volatility (σ_{iM}), %
3M	1.00	17
Coca-Cola	1.19	18
General Electric	1.26	15
Eastman Kodak	1.43	34
Texas Instruments	1.46	23
McDonalds	1.56	21
Hewlett-Packard	1.65	21
Disney Productions	2.23	22
Holiday Inns	2.56	39
Lockheed Martin	3.02	43

rate of return (RF) and the correlation (β) of asset returns to market returns (RM):

$$\mathrm{RA}_i = \mathrm{RF} + \beta_i(\mathrm{RM} - \mathrm{RF}) \tag{8.13}$$

$$\beta_i = \frac{\sigma_{iM}}{\sigma_M^2} \tag{8.14}$$

Once RA is determined, one can project future cash flows by multiplying RA times the asset value. One can then calculate the NPV of these cash flows.

Aggressive investments or highly leveraged investments will have high β, while conservative investments whose performance is unrelated to general market behavior will have low β. Table 8.2 shows examples of β and volatility, excerpted from Luenberger (1997).

Note that CAPM is concerned with how the market is likely to view an asset, for example, an equity share in a company. This reflects, of course, past company performance. It also is highly affected by the investors' perceptions of the company and its prospects.

8.3.4 Holt's Cash Flow ROI Valuation

A deeper assessment focuses not just on share price but also on cash flows over time. Holt's Value Associates has developed such a finer-grain metric, which they term CFROI (Madden, 1999). They estimate the *warranted* value of a firm using

$$\begin{aligned}\text{Firm value} &= \sum_{i=1}^{N} \frac{(\text{Forecasted net cash receipts from existing assets})_i}{(1+\mathrm{DR})^i} \\ &\quad + \sum_{i=1}^{N} \frac{(\text{Forecasted net cash receipts from future assets})_i}{(1+\mathrm{DR})^i}\end{aligned} \tag{8.15}$$

Here, the first term in this equation reflects the returns on existing assets. The second term reflects returns on future investments.

This distinction is important. Classic economic theory asserts that the share price of a firm should be the NPV of future earnings, multiplied by the price–earnings ratio of the firm. For high-growth firms, this metric significantly underestimates share value because such firms have "options" to enter future markets with new offerings that will provide returns much greater than those likely from current market offerings.

If one looks at high-growth companies such as Apple, one is likely to find that their share price cannot be justified even if everyone on the world buys an iPod, iPhone, and iPad. However, the reason the company's stock is valued so highly is that people believe the company will provide continued new innovations in future years—in other words, Apple has options on future offerings that will provide strong revenue and profit growth. We return to the distinction between returns on current and future assets in Chapter 9 and discuss metrics that go beyond NPV.

The term "warranted" is italicized in the above discussion. CFROI is concerned with what the value of a firm should be rather than what it is. If the warranted value of a firm is greater than the market value of the firm, then investing in the shares of the company is probably attractive. On the other hand, if the warranted value is less than the market value, then one should disinvest in the company.

CFROI is essentially the IRR calculated over streams of gross cash flows and current values of gross assets. Cash in equals gross cash flows per year (income, depreciation, interest, rental, etc.), while cash out equals the current value of gross assets of capital (monetary, inventories, plant, land, etc.). As with all IRR calculations, this assumes that each year all profits are reinvested with a rate of return equal to the IRR, which is often difficult to justify. This will inflate the estimated returns when one cannot get the same return from free cash flows.

We note that the warranted value includes the realizable value of nonoperating assets. Thus, it includes the liquidation value of the firm as well as its abilities to generate free cash flows. This is especially important when a firm has a significant portion of its assets that are not associated with generating its operating income.

8.3.5 Stern–Stewart Economic Value Added

CAPM is concerned with projecting the market value of an asset, while CFROI is concerned with assessing the warranted value of an asset. Another issue concerns the profit generation abilities of a firm. Stern *et al.* (2001) and Stewart (1991) have addressed this issue and developed a metric termed EVA as given by

$$\mathrm{EVA}_i = \mathrm{NOPAT}_i - \mathrm{WACC}_i \times K_i \tag{8.16}$$

where NOPAT is the net operating profit after tax, WACC is the weighted average cost of capital, and K is the capital employed. The market value added over time is given by

$$\text{MVA} = \sum_{i=1}^{\infty} \frac{\text{EVA}_i}{(1+\text{WACC})^i} \tag{8.17}$$

More specifically, NOPAT is the free cash at year-end, adjusted for R&D, inventory, depreciation, amortization of goodwill, etc. WACC is a complex function of capital structure, that is, proportion of debt and equity on the balance sheet; stock volatility, as measured by its β; and market risk premium.

Use of this metric involves addressing two challenges. First, one has to estimate the capital employed to earn the NOPAT. If this just involves financial assets, this is more straightforward than if it involves physical assets, for example, plant and equipment, or overhead assets that have to be attributed to lines of business. Second, one needs to estimate WACC that, as indicated above, can also be complicated.

Considering the nature of Equation 8.16, it is quite possible for EVA to be negative. This means that the earnings on an investment are less than the interest being paid on the capital needed to make the investment. In this case, the capital would have been better deployed elsewhere. Of course, this requires that the capital be fungible (i.e., freely usable for other purposes), which is possible with money but not necessarily with facilities and equipment. Thus, it is possible that investors may accept a negative EVA if that is the only way to stay in business and employ existing assets. Another interpretation is that the value of the assets employed is not as high as management estimates. Given that they are not fungible, perhaps they are worth less than estimated.

8.3.6 Comparison of Methods

To better understand the differences among these four metrics (i.e., DCF, CAPM, CFROI, and EVA), consider the example R&D investment summarized in Table 8.3. In this example, the firm invests $100 million in years 1 and 2 to perform the R&D to enable new market offerings. Assuming the R&D is successful, the firm then projects it will need to invest $500 million in years 3 and 4 to develop and refine the capacity to bring the new offerings to market. In year 5, the firm intends to launch the new offerings and expects annual revenues will grow from $1 billion to $3 billion over six years, while free annual cash flows grow from $200 million to $600 million.

The input data are shown in the top rows of Table 8.3. In the rows below this input data, one can find the results of using the four metrics discussed earlier. Using DCF, we find that NPV equals $435 million and IRR equals 20%. This appears to be a very attractive investment. CAPM has the same NPV as the cash flow time series has not changed. RA equals 25%, based on an assumed risk-free rate of 5%. Thus, we would project an annual return, or appreciation, of $109 million (i.e., 25% times the asset value of $435 million).

TABLE 8.3. Example R&D Investment

	Year 1	Year 2	Year 3	Year 4	Year 5	Year 6	Year 7	Year 8	Year 9	Year 10	TV
R&D											
Labor costs	90	90									
Capital costs	10	10									$17
Deployment											
Labor costs			100	100							
Capacity costs	0	0	400	400							$331
Operations											
Revenues					1000	2000	2500	2500	3000	3000	
Operating costs					800	1600	2000	2000	2400	2400	
Free cash flows					200	400	500	500	600	600	0
Total outflows	100	100	500	500							
Operations free cash flow				0	200	400	500	500	600	600	
Net free cash flow	−100	−100	−500	−500	200	400	500	500	600	600	0
Rate	10%										
β	3.0										
DCF											
NPV	$435										
IRR	20%			DCF emphasizes the value of a stream of costs and returns							
CAPM	−100	−100	−500	−500	200	400	500	500	600	600	0
NPV	$435										
RA	25%			CAPM emphasizes the market valuation of a firm							
CFROI (adds asset values)	−90	−90	−100	−100	200	400	500	500	600	600	0
NPV	$2420										
IRR	47%			CFROI emphasizes the warranted market value of a firm							
EVA (deducts use of assets)	−101	−102	−542	−582	118	318	418	418	518	518	0
NPV	$101										
IRR	12%			EVA emphasizes the profit generation capacity of a firm							

TABLE 8.4. Comparison of Methods

	Financial metrics	Comments
DCF	\$435, IRR = 20%	DCF emphasizes the value of a stream of costs and returns
CAPM	\$435, RA = 25%	CAPM emphasizes the market valuation of a firm, that is, stock price × number of shares outstanding
CFROI	\$2420, IRR = 47%	CFROI emphasizes the warranted value of a firm, including both cash flows and asset value
EVA	\$101, IRR = 12%	EVA emphasizes the profit generation capacity of a firm, adjusted for the cost of capital of the firm

CFROI adds asset values into the calculation, including the capital investments on the balance sheet for years 1 to 4. These assets might, of course, depreciate but perhaps not significantly over so few years. Including these assets increases the NPV to \$2420 million and the IRR to 47%. This makes the investment very attractive. The assumed fungibility of the capital investments multiplies the NPV by a factor of over five.

EVA results in an NPV of \$101 million and an IRR of 12%. This is due to free cash flow being decreased by the charge for use of assets. Contrary to CFROI, investing in and leveraging assets decreases EVA. The basic premise is that these resources were, in effect at least, borrowed to enable the project, and the carrying cost of this loan needs to be paid first. Only earnings above this amount should be counted in valuation of this investment. The value added by this investment is the monies earned beyond the cost of borrowing the funds to make the investment.

Table 8.4 provides an overall comparison of results of employing these four different methods for investment valuation. Clearly, EVA is the most conservative and CFROI is the most liberal. DCF and CAPM fall in the middle. Which method is right? One could argue that the answer depends on whether one is buying or selling the asset in question. However, the real answer is that the choice depends on one's intent for valuation of the asset.

If one is concerned with the likely market valuation of an asset, then CAPM is likely to be a good choice, assuming estimates of β and volatility are available. On the other hand, if the concern is with the warranted value of an asset, CFROI provides an estimate of the value of projected earnings plus the value of assets, assuming that one can make reasonable estimates of the value of liquidating assets. Finally, if one wants to own the asset to gain the cash flows it can yield beyond the interest payable on loans secured to provide the monies to invest in creating these earnings, then EVA is a reasonable metric, assuming that one can attribute use of assets appropriately and determine the weighted cost of capital.

Yet, NPV and IRR are very straightforward and much less laced with assumptions, at least none that are not already underlying the other three metrics. Thus, it is reasonable to argue that one should use NPV and/or IRR unless the special purposes of CAPM, CFROI, and/or EVA are important to the investment opportunity at hand.

One concern with all four approaches is what one should assume for year 11 and beyond. As this example stands, we are assuming that the asset is of no value after the 10 years shown in Table 8.3. In fact, one may have an asset that can be monetized, but how should it be valued?

The value at the end of the period of interest is referred to as the terminal value (TV). One way to represent TV is as an annuity that provides a growing, diminishing, or constant return in perpetuity, that is, forever. The best estimate of the return for year 11 is the return for year 10, that is, $600 million. If we assume that this holds constant in perpetuity, the NPV in year 11 is $600 million divided by the DR, in this case 10%. Thus, the TV equals $6000 million. We can either hold this annuity or sell it for its NPV at that time.

Adding this TV to year 11 and recomputing the overall NPV and IRR for years 1 to 11, we obtain an NPV of $2538 million and IRR of 36%. Thus, the TV dominates the overall value of this investment, regardless of which assessment method we employ. For this reason, most investors are very wary of TV projections. They think it is most unwise to have a valuation that is dominated by an assumed value so far in the future when markets and technologies are very likely to have substantially changed. Many investors have a practice of assuming that TV equals zero. Any value that eventually emerges in year 11 and beyond is viewed as a bonus.

8.3.7 Limitations of Approaches

As indicated above, the approaches discussed and illustrated in this chapter differ in the intent for which they were developed. They have in common, however, two limitations. First, sources of uncertainty are not explicitly addressed. The DR is often used to compensate for this, but this requires one parameter to handle too many phenomena.

Ideally, one would like the three times series of interest—investments, operating costs, and profits—to be represented as probability distributions at each point in time with means and variances based on some agreed-upon logic or rationale. These inputs would then be used to compute the probability distribution of NPV or IRR, or one of the other metrics. Variations in the input assumptions would be used to assess the sensitivity of the outputs to these variations. In this way, one could project both returns and risks—a topic considered in more detail in Chapter 9.

Beyond needing a more robust approach to uncertainty and risk, the approaches in this chapter are limited by the fact that the multistage nature of decisions is not considered. Staging of decisions provides enormous value for hedging downside risks. Such staging is of economic value because it limits, perhaps truncates, the downside of the valuation probability distributions. If after a particular stage, one finds that the investment no longer makes sense, the plug is pulled and potential losses, or mediocre returns, are stopped. The approaches presented in this chapter assume that, once started, an investment is always continued. This assumption is removed in Chapter 9.

It is important to note that many situations do not warrant multiple stages of decisions. When the lion's share of the investment must occur upfront, then the assumptions underlying NPV and IRR are not as troublesome. In contrast, when the initial investment is relatively small and later investments are relatively large, the possibility for multiple stages can make an enormous difference to the investment valuation.

8.4 SUMMARY

This chapter first reviewed some basic principles of microeconomics from earlier chapters to frame the discussion of investment valuation. The investment problem was then outlined and two case studies described—these case studies are revisited in Chapter 9. Four standard approaches to investment valuation were then discussed—DCF, CAPM, CFROI, and EVA. A comparison of these methods, using one of the case studies, indicated substantial differences in assessment results. This can be attributed to the different intents for which each method was developed. Two important limitations of all these methods were discussed: the representation of uncertainty and the possibilities for multistage decisions. In Chapter 9, the approaches we discuss are not burdened by these limitations.

It is important to also emphasize the fact that noneconomic attributes are often also very important in investment analysis, perhaps of even greater importance than the economic attributes. The second or third best alternative from an economic point of view might be the best alternative once one considers strategic fit, leveraging of core competencies, and possibilities of sustainable competitive advantage. Nevertheless, while the numbers are not all that counts, one needs to count the numbers right.

BIBLIOGRAPHY AND REFERENCES

Brigham EF, Gapenski LC. Financial management: theory and practice. Chicago, IL: Dryden; 1988.

Luenberger DG. Investment science. Oxford, UK: Oxford University Press; 1997.

Madden BJ. CFROI valuation: a total system approach to valuing the firm. Woburn, MA: Butterworth-Heinemann; 1999.

Rouse WB. Essential challenges of strategic management. New York: Wiley; 2001.

Rouse WB. Impacts of healthcare price controls: potential unintended consequences of firms' responses to price policies. IEEE Syst J 2010;4(1):34–38.

Rouse WB, Howard CW, Carns WE, Prendergast EJ. Technology investment advisor: an options-based approach to technology strategy. Inf Knowl Syst Manag 2000;2(1):63–81.

Stern JM, Shiely JS, Ross I. The EVA challenge: implementing value added change in an organization. New York: Wiley; 2001.

Stewart GB. The quest for value. New York: Collins; 1991.

CHAPTER 9

REAL OPTIONS FOR INVESTMENT VALUATION

This chapter considers investments that can be staged such that decisions can be made at each stage whether or not to continue investment for the next stage. This multistage approach provides decision makers with considerably flexibility. This flexibility enables hedging against the downside possibility that an investment no longer makes sense because either an earlier stage has failed or the opportunity for which the investment was targeted no longer makes sense.

Framing investment decisions in this way results in each decision to proceed with the next stage as equivalent to purchasing an option on the subsequent stages. The option is the right, but not the requirement, to invest in the next stage at a later time. If the results of the current stage are positive, and the opportunity embodied in the next stage still looks attractive, then one can exercise this option and proceed.

The decision to proceed with investing in the next stage may, in effect, result in the purchasing of an option on the subsequent stages. In this way, the series of investment decisions may be represented as a stream of option purchase decisions, followed by option exercise decisions. Only at the last stage is the asset deployed in the marketplace. However, as later discussion shows, this deployment can also be represented as creating another option for future market offerings.

This chapter proceeds as follows. First, the nature of options is elaborated. This includes the logic of options, option pricing models, and computational methods, with emphasis on Black–Scholes formulations. Strategic metrics for characterizing the worth of investments are then discussed, building systematically on the discussion in Chapter 8. Investment decision making then becomes the focus, including the notions of investment portfolios and discussion of a wealth of investment case studies. Finally, the construct of value-centered investing is elaborated where the core notion is option-based strategic thinking about the contingent needs of an enterprise.

Economic Systems Analysis and Assessment,
by Andrew P. Sage and William B. Rouse

9.1 NATURE OF OPTIONS

An option provides the right to do something. Purchasing an option provides a "chit" that can later be used, or exercised, if one decides that doing what the option enables still makes sense. For example, one might buy an option that enables buying a fixed number of shares of an enterprise for a given price at any time within the next five years. If the market value of these shares exceeds the option price, it is said that the option is "in the money" and worth exercising. Depending on the type of option (see below), one might use the chit and buy the shares for an immediate gain.

Options do not always pertain to shares of stock. One might buy an option to buy a fixed amount of commodities such as corn, pork bellies, or fuel at a particular price at a specific point of time. In this way, one could hedge the downside risks of these commodities becoming much more expensive and undermining the competitiveness of one's products and services. If the price of one's products and services cannot be raised, perhaps due to intense competition, then such an option greatly decreases the risk of having to sell products and services for less than the costs of producing and/or delivering them.

The models and methods discussed in this chapter originated in the financial industry. This industry deals with a variety of types of options or, in general, derivatives that attach value to the right to something rather than the thing itself. A "call" option is the right to buy something on or before a particular point in time for a particular price. In contrast, a "put" option is the right to sell something on or before a particular point in time for a particular price.

There are European and American versions of calls and puts. European options can only be exercised at the end point of the time period during which the option is valid. An American option can be exercised at any time up to and including that point in time. European call options are the best representation of the types of investments discussed in this chapter, as well as in Chapter 8.

An overarching question for any of these types of options is their value. This determines how much one should be willing to pay for an option. Black and Scholes (1973) and Merton (1973) developed an analytic solution for European call options. There was widespread adoption of these models throughout the financial industry within one year, a rather amazing rate of diffusion of an innovation.[1]

In the 1990s, the idea of "real" options received considerable attention—see Dixit and Pindyck (1994), Trigeorgis (1996), Luenberger (1997), Amram and Kulatilaka (1999), and Boer (1999). The idea was to use the models, methods, and tools applied for financial options to investment problems associated with tangible assets such as factories, products, and technologies.

[1]Robert Merton and Myron Scholes won the Nobel Prize in Economic Sciences for this work in 1997. Fischer Black, who died in 1995, was not eligible to be included but was noted in the Nobel citation.

9.1.1 Multistage Options

Figure 9.1 illustrates two multistage real options. The top example includes two stages. The first stage involves, for example, an investment in R&D to create a new technology to enable some function of importance to the investor. If this R&D is successful, and the need for the function continues, the investor may choose to exercise the option gained by funding the first stage and, therefore, invest in the second stage.

The bottom example in Fig. 9.1 includes three stages. The first stage involves investing in research to create a potential new function. If this research is successful, and the function now possible remains attractive, the investor may decide to invest in the second stage to develop the research outcomes into a full capability. If this capability is successfully developed, and the functionality enabled still remains desirable, the investor may invest in the third stage to deploy the capability.

The options associated with the first stages of both of the examples in Fig. 9.1 have economic value regardless of whether or not these options are exercised in later stages. This may seem counterintuitive. Why would one buy something that one does not subsequently use? This possibility may better align with intuition if one thinks about options such as insurance. Owning an option ensures that one has the right to an attractive opportunity *if* the opportunity emerges in the future. Life insurance is attractive if one dies, but few people would regret not exercising their life insurance last year! Nevertheless, people are usually glad to have an option—an insurance policy—that meets the contingent need if one were to die.

9.1.2 Fundamental Logic

There is a fundamental logic for addressing the types of investments embodied in the two case studies introduced in Chapter 8 (Rouse *et al.*, 2000). The idea is that technology and process investments create contingent opportunities for later solution investments. In other words, they create options that can later be

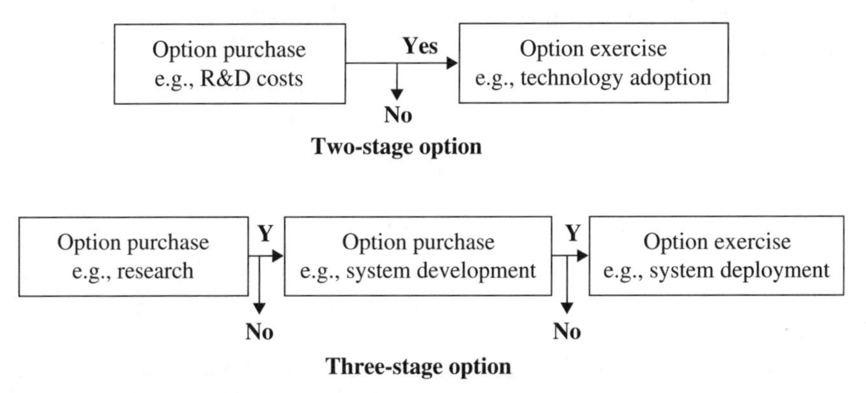

Figure 9.1. Example Multistage Options.

exercised and result in deployment of upgraded or new market offerings. Thus, these technology and process investments not only yield new processes and technologies but also yield options for later investments—all for the costs of the initial investments.

There are usually significant uncertainties regarding likely cash flows from later investments to exercise options and deploy upgraded or new offerings. If there were not, one would likely commit to deployment up front. However, the greater the uncertainty or volatility, the more attractive it is to own options rather than being committed to the uncertain future now. The options give one the right but not the requirement to make the deployment decision later.

Thus, delaying investment decisions—rather than deciding now—can be of substantial value. The farther into the future that a decision is to be made, the more valuable the associated option. Intuitively, one would rather not make decisions now that could be made (much) later, especially when there is significant uncertainty. Therefore, the value of an option (OV or Option Value) typically increases with delay time for exercising it.

In summary, the OV increases with projected cash flows should one exercise the option, uncertainty (volatility) associated with these cash flows, and time until the decision to exercise it need be made. It may seem unusual that uncertainty and time increase value, but the key point is that they increase the OV on a future that may not materialize. That is exactly why one prefers to own an option.

9.2 OPTION PRICING THEORY

The question we address here is how to estimate or assess the economic OV on uncertain future cash flows that may or not be realized. Black and Scholes (1973) addressed this by envisioning a "replicating portfolio" that consists of some number of owned shares of stock and borrowed capital with interest paid at the risk-free rate. This replicating portfolio will have the same payoff as the call option at expiration and therefore, by the fundamental theorem of finance, the portfolio value must equal the call option value. They constructed this portfolio to be entirely self-financing and thus deterministic.

This conceptual insight led them (Black and Scholes, 1973), with contributions from Merton (1973), to derive the now-famous Black–Scholes equation starting with

$$dS = \mu S dt + \sigma S dz \tag{9.1}$$

$$dB = rB dt \tag{9.2}$$

where S is the price of the underlying security, z is a standard Brownian motion or Wiener process over $[0, T]$, and B is the value of a risk-free bond carrying an interest rate of r over $[0, T]$.

The stochastic process employed to represent the time variation of stock prices is often referred to as geometric Brownian motion or exponential

Brownian motion where the logarithm of the random variable follows a Brownian motion process. It is commonly used as an approximation of stock price dynamics. μ is termed the percentage drift and σ the percentage volatility, and both values are assumed constant.

An important implication of assuming geometric Brownian motion is the lognormality of the random value, which means that the probability distribution of the variable is skewed and cannot take on negative values. This is appropriate because negative asset values will result in an option not being exercised as the value of the option is zero. Thus, and again essentially by definition, options are only exercised when they are "in the money." Otherwise, they are discarded.

A security that is derivative to S is one where its price is a function of S and t, given by $f(S, t)$, and can be obtained by solving the Black–Scholes equation (Black and Scholes, 1973)

$$\frac{\partial f}{\partial t} + \left(\frac{\partial f}{\partial S}\right) rS + \left(\frac{1}{2}\right)\left(\frac{\partial^2 f}{\partial S^2}\right)\sigma^2 S^2 = rf \tag{9.3}$$

The solution of this equation for European call options—which is the designation used for options that can only be exercised at time T—is given by the Black–Scholes call option formula as follows:

$$f(S,t) = C(S,t) = SN(d_1) - Ke^{-r(T-t)}N(d_2) \tag{9.4}$$

$$d_1 = \frac{\ln(S/K) + \left(r + (\sigma^2/2)\right)(T-t)}{\sigma}\sqrt{T-t} \tag{9.5}$$

$$d_2 = d_1 - \sigma\sqrt{T-t} \tag{9.6}$$

where S is the net present value (NPV) of the asset of interest, K is the NPV of the option exercise price (OEP), and $N(\cdot)$ is the cumulative distribution function of the standard normal distribution. The NPVs noted require a discount rate that reflects the investor's cost of capital.

To provide another perspective on valuation of options, consider the following formulation by Smithson (1998). The OV equals the discounted expected value of the asset (EVA) at maturity, conditional on this value at maturity exceeding the OEP, minus the discounted OEP, all times the probability that, at maturity, the asset value is greater than the OEP. Net option value (NOV) equals the option value calculated in this manner minus the discounted option purchase price (OPP). In equation form, OV is thus given by

$$\text{OV} = [(\text{EVA at maturity} \mid \text{value} > \text{OEP}) - \text{OEP}]\text{Prob.}(\text{Value} > \text{OEP}) \tag{9.7}$$

As noted, the Black–Scholes solution is for European call options. What if the investment problem at hand does not fit this formulation? Different formulations may require computational rather than analytic solutions.

Common computational methods include Monte Carlo simulation, finite difference methods, and binomial and trinomial lattices. See Dixit and Pindyck (1994), Trigeorgis (1996), and Luenberger (1997) for treatments of these methods.

9.2.1 Example—Transforming Shipbuilding

Assume that the U.S. Navy would like to transform the way it acquires ships and, therefore, proposes several changes that will streamline the development and design process and reduce rework. Thus, the Navy has the option to transform its ship acquisition enterprise. To determine whether or not the Navy should initiate transformation, an option model was developed (Pennock *et al.*, 2007).

To mitigate technical risks of unsuccessful transformation, it was assumed that there would be a three-stage process:

- **Stage 1:** Concept development and feasibility analysis. This stage is relatively short and inexpensive. If the transformation idea proves to be infeasible in this stage, the Navy can terminate the project at no additional cost.
- **Stage 2:** Pilot testing the changes on the acquisition of a single ship. If the project fails in this stage, rework costs will be required to rectify the situation and complete the acquisition of the ship.
- **Stage 3:** Implementing the transformation across the whole shipbuilding enterprise. If the transformation fails in this stage, a substantial cost in rework is incurred.

Table 9.1 summarizes the staging parameter values for this example.

Note that as a three-stage model, the solution can only be approximated using the Black–Scholes approach. Thus, Pennock employed the binomial lattice method. A binomial lattice is a large decision tree that represents the decision problem as a series of stages in time. At each stage, the asset price can either increase or decrease by a fixed amount, somewhat larger on the upside than the downside due to upward drift of the assumed geometric Brownian motion. The full lattice extends from $t = 0$ to T, the expiration time of the option. Once the full tree is formulated, one works backwards from $t = T$ to 0, using backward induction to determine the best decision at each stage.

Using the binary lattice model developed, Pennock found the NOV of this transformation option to be approximately $0.61 billion. If one were to calculate

TABLE 9.1. Stage Parameter Values

Stage	Stage cost, $ billion	*P*(Success)	Rework cost, $ billion	Duration, years
1	0.001	0.4	0	0.5
2	0.01	0.6	1	3
3	0.1	0.8	10	N/A

the traditional NPV when considering this technical risk, they would find that the value of the transformation project is approximately −$6.43 billion. This means that one would expect to incur a substantial loss by initiating this project. Here one can see the discrepancy between the NOV and the NPV. The NPV is too conservative because it fails to account for the risk mitigation inherent in staging. So, in this example, a decision maker using NPV as the decision criterion would reject a potentially beneficial program.

The example can be expanded by introducing increased market risk, that is, allowing for uncertainty in cash flows. Option values will inherently increase because options will be exercised only if the upside occurs. If the downside occurs, options will simply not be exercised. The resulting NOV is $5.94 billion, a value that is almost 10 times greater than that without the market risk. Hence, risk can be quite valuable if one can take advantage of the upside while also avoiding the downside.

9.3 OPTION CALCULATOR

Table 9.2 illustrates a Black–Scholes option calculator programmed in Microsoft Excel. The leftmost column includes labels for the rows. The next column (to the right) includes the model parameters and all the Black–Scholes calculations. All of the rows to the right of this column provide the inputs to the NPV and Black–Scholes calculations. Note that Equation 9.4 is employed using S equals the asset value NPV and K equals the option exercise NPV. The parameters for Equations 9.4 to 9.6 are shown above the NPV results.

The example in Table 9.2 employs the data from Table 8.3 for the R&D investment case study. Whereas the NPV in Table 8.3 was $435 million, the NOV in Table 9.2 is $694 million. The difference of $259 million represents the value of being able to hedge the downside risk that either the R&D is not successful or the projected market demand for the new offerings does not emerge as projected.

As shown in Fig. 9.2, NOV is quite sensitive to volatility and discount rate. Increasing volatility leads to increasing NOV. It may seem counter-intuitive for value to increase with uncertainty. However, increased volatility results in greater upside potential for an option. Of most importance here, the downside does not matter because, if it occurs, one need not exercise the option. In other words, the option hedges the downside risk while preserving the possibility of upside gain.

Increasing the discount rate, used to calculate S and K in Equations 9.4 and 9.5, leads to decreasing NOV. This should not be surprising as the discount rate has the same effect on NPV. As shown in Fig. 9.2, this effect is quite strong. Thus, it is important that one employs a discount that appropriately reflects the cost of capital. At the same time, one should not increase the discount rate to reflect the uncertainty represented by the volatility. In this way, option models provide us with a better way to address uncertainty than is possible with traditional discounted cash flow models.

TABLE 9.2. Black–Scholes Calculator

Black–Scholes call option calculator											
Expiration (T)	2										
Volatility (σ)	0.9										
Risk-free rate (r)	5%										
Discount rate (DR)	10%										
	NPV	Year 1	Year 2	Year 3	Year 4	Year 5	Year 6	Year 7	Year 8	Year 9	Year 10
Option purchase	\$174	\$100	\$100								
Option exercise	\$717	\$0	\$0	\$500	\$500						
Asset value	\$1326	\$0	\$0	\$0	\$0	\$200	\$400	\$500	\$500	\$600	\$600
Option value	\$867										
Net option value	\$694										
d_1	1.197610296										
d_2	−0.07518191										
$N(d_1)$	0.884465617										
$N(d_2)$	0.470034989										
$C(S, 0)$	\$867										

Notes:

1. Option purchase: NPV of investment to purchase option.
2. Option exercise: NPV of investment to exercise option.
3. Asset value: NPV of projected free cash flow from owning asset (revenues − costs).
4. Option value: BS calculation assuming no dividends and constant interest compounded at rate r.

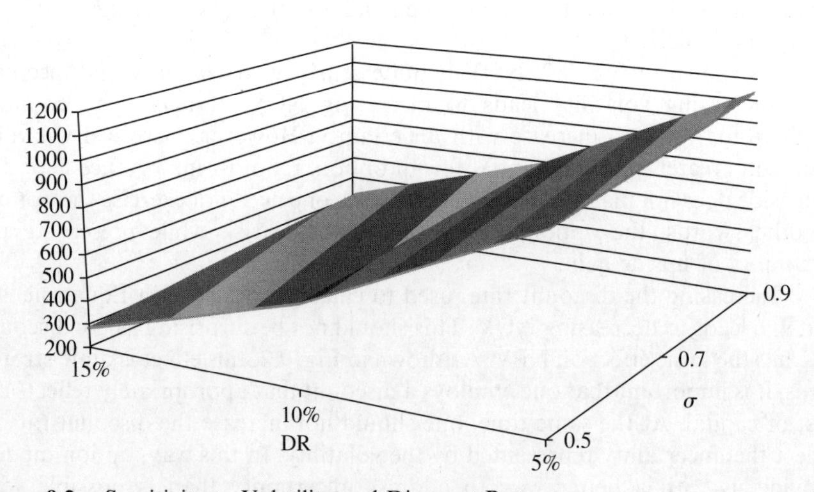

Figure 9.2. Sensitivity to Volatility and Discount Rate.

9.3.1 Monte Carlo Analysis

It is quite common for people to feel uneasy about their projections of option purchase costs, option exercise costs, and free cash flows in terms of the difference between revenues and operating costs. We need to better understand the impact of these uncertainties.

This understanding can be gained by representing input data as probability distributions rather than point estimates. These distributions can then be sampled, and each sample can be used to compute NOV. The results of each computation can be compiled into a resulting probability distribution for NOV. Typically, 1000 samples are used in this compilation.

This approach is termed Monte Carlo simulation or analysis because of its use of random number generators to create the samples of the input data for each calculation. Table 9.3 shows a spreadsheet-based option calculator with Monte Carlo analysis capabilities. The 1000 samples of the Monte Carlo simulation occur in a second worksheet (not shown) using standard functions within Microsoft Excel.

As inputs to the Monte Carlo simulation or analysis, one chooses the mean and standard deviation of the probability distributions for option purchase, option exercise, and asset value. These are expressed as percentages. The mean and standard deviation percentages define a normal distribution from which random samples are drawn. These samples are then multiplied by the appropriate NPVs (i.e., option purchase, option exercise, or asset value) to define the inputs to the Black–Scholes calculation, along with the other parameters from the upper left of the spreadsheet.

Note that use of the normal distribution for this type of Monte Carlo is reasonable for small to moderate variations of parameters. For large variations, for example standard deviation of 30% to 50% or more, the normal distribution presents problems in that negative percentages can result, which, of course, are meaningless. If such large variations are of interest, one should employ something like an exponential or lognormal sampling distribution.

The approach to Monte Carlo analysis just described involves varying the three NPVs that are input to the Black–Scholes calculation. One could instead vary each of the point estimates toward the top of the spreadsheet, or perhaps vary the finer-grained point estimates that served as inputs to these estimates. Monte Carlo analysis can be pursued at any level of detail one chooses. This can require, of course, much more detailed modeling of uncertainties in input parameters.

The bottom of Table 9.3 shows the probability distribution for NOV on the left, with the cumulative distribution on the right. These histograms reflect means of 100% and standard deviations of 10% for all three NPVs. Choosing means of 100% results in the mean values being the same as the above point estimates. Thus, the average NOV calculated from the Monte Carlo results is almost exactly the same as that calculated without random variations, that is, $692 versus $694. However, the 10% standard deviations yield an NOV standard deviation of $122. Thus, small component variations can add up to very substantial overall variations.

TABLE 9.3. Option Calculator with Monte Carlo

Black–Scholes call option calculator with Monte Carlo simulation

Expiration (T)	2										
Volatility (σ)	0.9										
Risk-free rate (r)	5%										
Discount rate (DR)	10%										
	NPV	Year 1	Year 2	Year 3	Year 4	Year 5	Year 6	Year 7	Year 8	Year 9	Year 10
Option purchase	$174	$100	$100								
Option exercise	$717	$0	$0	$500	$500						
Asset value	$1326	$0	$0	$0	$0	$200	$400	$500	$500	$600	$600
Option value	$867										
Net option value	$694										
d_1	1.197610296										
d_2	−0.07518191										
$N(d_1)$	1										
$N(d_2)$	0										
$C(S, 0)$	$867										

Notes:

1. Option purchase: NPV of investment to purchase option.
2. Option exercise: NPV of investment to exercise option.
3. Asset value: NPV of projected free cash flow from owning asset (revenues − costs).
4. Option value: BS calculation assuming no dividends and constant interest compounded at rate r.
5. Net option value: option value minus option purchase.

Monte Carlo simulation

	Mean, %	Standard deviation, %		
Option purchase	100	10	Average	$692
Option exercise	100	10	Standard deviation	$122
Asset value	100	10	Sample size	1000

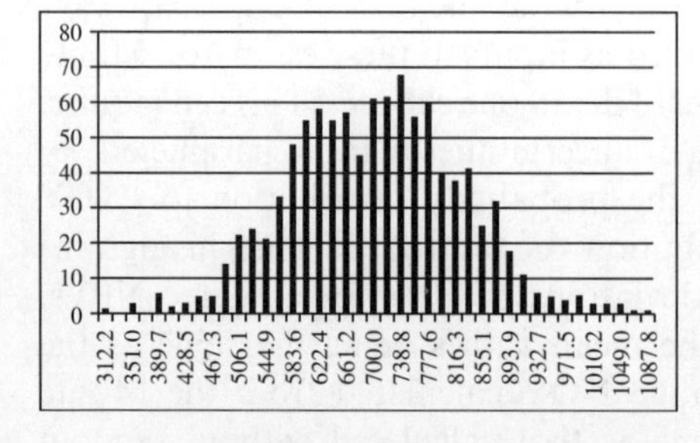

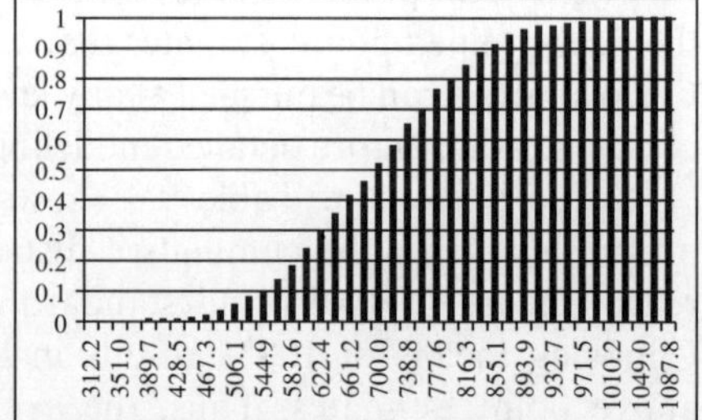

Figure 9.3 portrays the sensitivity of NOV to percentage means that are greater than 100%. Increasing the option purchase NPV decreases the NOV, as one would expect; whereas increasing the option exercise NPV has much less effect. Increasing the asset value NPV has a dramatic effect, again, as one

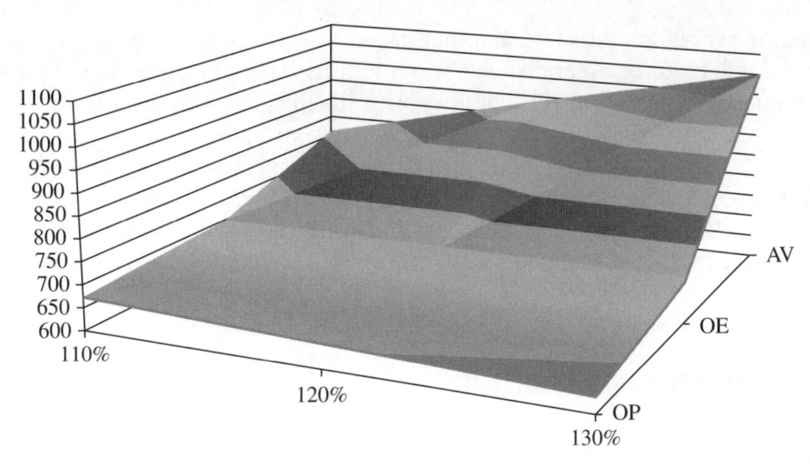

Figure 9.3. NOV with Increasing NPV.

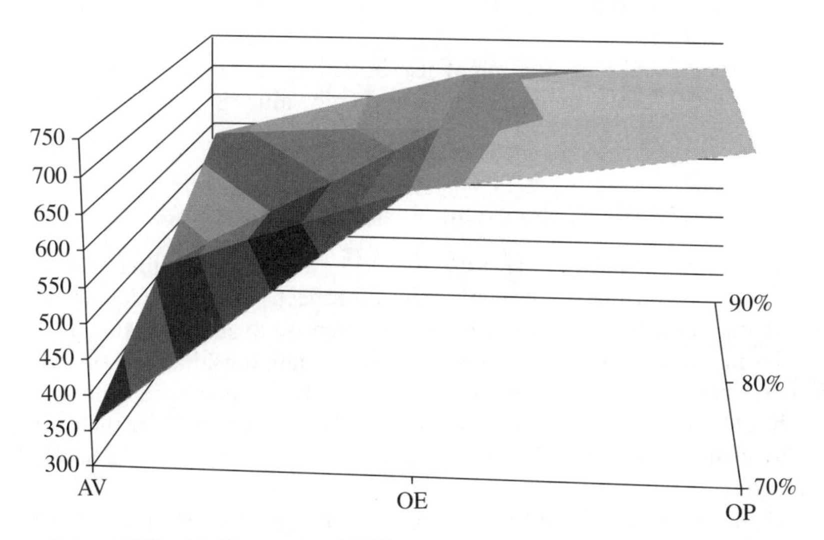

Figure 9.4. NOV with Decreasing NPV.

would expect. This suggests that the investment analyst should pay careful attention to projections of the asset value.

Figure 9.4 portrays the sensitivity of NOV to percentage means less than 100%. Decreasing the option purchase NPV increases NOV, as one would expect; decreasing the option exercise NPV has much less effect. Decreasing the asset value NPV has a dramatic effect, again, as one would expect. This adds emphasis to the suggestion that the investment analysis pay careful attention to the projections of the asset value.

It is very important to note that the effects of varying the NPVs in this way are heavily dependent on the particular parameters of the overall model.

Thus, one should not assume that the specific sensitivities portrayed in Figs. 9.3 and 9.4 will hold for other problem formulations. Therefore, one should pursue such analyses for each investment problem independent of previous analyses.

This discussion thus far illustrates how straightforward it can be to formulate option pricing models and create associated option calculators. What is less straightforward is the appropriate framing of options. The case studies presented later in this chapter illustrate the range of types of investments that can be formulated as option investments. Later discussion presents some guidelines for framing of options.

9.4 STRATEGIC METRICS

In Chapter 8, NPV and related metrics were discussed. This chapter has discussed NOV—option value minus the cost of purchasing the option. Which metric should we use? In situations where the lion's share of the overall investment occurs up front, NOV will approach NPV. In this case, we should use the simpler metric.

In other situations, one might use both metrics. Boer (1998, 1999) has argued for a composite metric, such as strategic value (SV), given by

$$\text{SV} = \text{NPV} + \text{NOV} \tag{9.8}$$

$$\text{NOV} = \text{Option value} - \text{Option purchase price} \tag{9.9}$$

To avoid double counting, the NPV should be based on the financials of the existing lines of business while NOV should reflect new lines of business that may, or may not, be entered in the future based on available options.

To illustrate, a large corporation had a small subsidiary in the wireless LAN business in the commercial sector. They were concerned with whether they should sell this business because the NPV of the financials for their existing lines of business equaled \$3 million—pretty small potatoes for a \$20-billion corporation.

Using the methods and tools discussed in this chapter, it was estimated that the NOV associated with the possibility of entering the consumer market for wireless LAN equaled \$300 million. Thus the SV for this subsidiary was \$303 million. This insight resulted in the corporation retaining this small subsidiary due to the attractive SV.

Consider the efficiency case study investment summarized in Table 9.4. In this example, the firm invests \$300 million in years 1 and 2 to perform the engineering and acquire the equipment and facilities to enable substantial efficiency increases. These capabilities are deployed in year 3 and lead to increasing profitability in the subsequent years. The bottom of Table 9.4 shows the metrics from Chapter 8 as follows:

- NPV is a modest \$55 million, with an IRR of 12%.
- CAPM is the same as for the R&D investment, 25%.

- CFROI is $1000 million due to the assumed ability to monetize assets.
- EVA is −$172 million due to the asset-intensive nature of this investment.

There is no option for this investment as the $600-million commitment in years 1 and 2 represents all of the investment needed. There is no downstream investment needed to deploy the capabilities.

In contrast, revisiting Table 9.2, this R&D investment represents a classic real option using the data from Table 8.3. As indicated earlier, the staging of this investment results in an additional $259 million in economic value. Staging

TABLE 9.4. Example Efficiency Investment

	Year 1	Year 2	Year 3	Year 4	Year 5	Year 6	Year 7	Year 8	Year 9	Year 10	TV
Engineering											
Labor costs	100	100									
Capital costs	200	200									$347
Operations											
Revenues			1000	1000	1000	1000	1000	1000	1000	1000	
Operating costs			950	925	900	875	850	800	750	700	
Free cash flows			50	75	100	125	150	200	250	300	0
Total outflows	300	300	0	0							
Operations free cash flow				75	100	125	150	200	250	300	
Net free cash flow	−300	−300	0	75	100	125	150	200	250	300	0
Rate	10%										
β	3.0										
DCF											
NPV	$55										
IRR	12%		DCF emphasizes the value of a stream of costs and returns								
CAPM	−300	−300	0	75	100	125	150	200	250	300	0
NPV	$55										
RA	25%		CAPM emphasizes the market valuation of a firm								
CFROI (adds asset values)	−100	−100	0	75	100	125	150	200	250	300	0
NPV	$1000										
IRR	36%		CFROI emphasizes the warranted market value of a firm								
EVA (deducts use of assets)	−320	−340	−40	35	60	85	110	160	210	260	0
NPV	($172)										
IRR	4%		EVA emphasizes the profit generation capacity of a firm								

TABLE 9.5. Comparison of Methods

	Efficiency case study	R&D case study	Comments
NPV	$55	$435	R&D investment $259 undervalued
NOV	$0	$694	Efficiency investment not staged
SV = NPV + NOV	$749 = $55 + $694		

of the efficiency investment might have increased its valuation, although such staging may not have made sense technically. For example, if the technologies and methodologies involved are quite mature, then there is little chance that the engineering stage will fail. Further, if the goal is to make a greater margin on offerings to existing customers, due to greater efficiencies rather than increased prices, then the market risks may be quite small.

Table 9.5 compares the results for the two case studies using the three strategic metrics—NPV, NOV, and SV. The efficiency case study is best addressed with NPV. In contrast, addressing the R&D case study with NPV results in undervaluation by $259 million. Thus, NOV is the metric of choice for this case study. The SV of this portfolio of two investments is $749 million, equaling the NPV for the efficiency investment plus the NOV for the R&D investment.

This analysis shows that both investments are economically justified. However, it is quite possible that the firm may not have the economic resources, or human resources, to pursue both investments. The choice, then, is between investing in getting better at what they are already doing and investing in doing new things. As elaborated in Chapter 8, each of these choices has associated uncertainties and risks. The specific nature of these uncertainties and risks is likely to dictate the final investment decision.

However, another perspective can help to resolve this. The firm could decide to proceed with the efficiency initiative while also proceeding to buy the option on the new offerings embodied in the R&D initiative. In other words, the firm need not at this point decide to proceed with the new market offerings. It only needs to assure that it has the capability to pursue these new offerings if they still make sense two years from now. Of course, this is the essence of option-based strategy and shows why the real options approach is so valuable.

9.5 INVESTMENT DECISION MAKING

This section focuses on two issues. First, the need to consider potential investments in the context of the portfolio of investments is discussed. The goal is to balance returns and risks across the portfolio of candidates. Second, this section focuses on 14 case studies of actual investment decisions where these decisions were made using the models and methods elaborated in this chapter.

9.5.1 Investment Portfolios

Typically, investment decisions are not made in the isolation of a single investment opportunity. Such decisions are usually made in the context of several alternatives. The decisions of importance concern which alternatives receive investments and which do not. This involves considering the portfolio of alternatives.

The portfolio of technology options can be portrayed as shown in Fig. 9.5. Return is expressed by NPV or NOV. The former is used for those investments where the lion's share of the commitment occurs upstream and subsequent downstream "exercise" decisions involve small amounts compared to the upstream investments. NPV calculations are close enough in those cases.

Risk (or confidence) is expressed as the probability that returns are below (risk) or above (confidence) some desired level—zero being the common choice. Assessment of these metrics requires estimation of the probability distribution of returns, not just expected values. In some situations, this distribution can be derived analytically, but more often Monte Carlo analysis or equivalent is used to generate the needed measures.

The line connecting several of the potential investments (P_A, P_B, P_H, and P_Z) in Fig. 9.5 is termed the "efficient frontier." Each potential investment on the efficient frontier is such that no other potential investment dominates it in terms of *both* return and confidence. In contrast, candidates interior (below and/or left) to the efficient frontier are all dominated by other candidates in terms of both metrics. Ideally, from an economic perspective at least, the candidates in which one chooses to invest—purchase options—should lie on the efficient frontier. Choices from the interior are usually justified by other, typically noneconomic attributes.

A primary purpose of a portfolio is risk diversification. Some investments will likely yield returns below their expected values, but it is very unlikely that

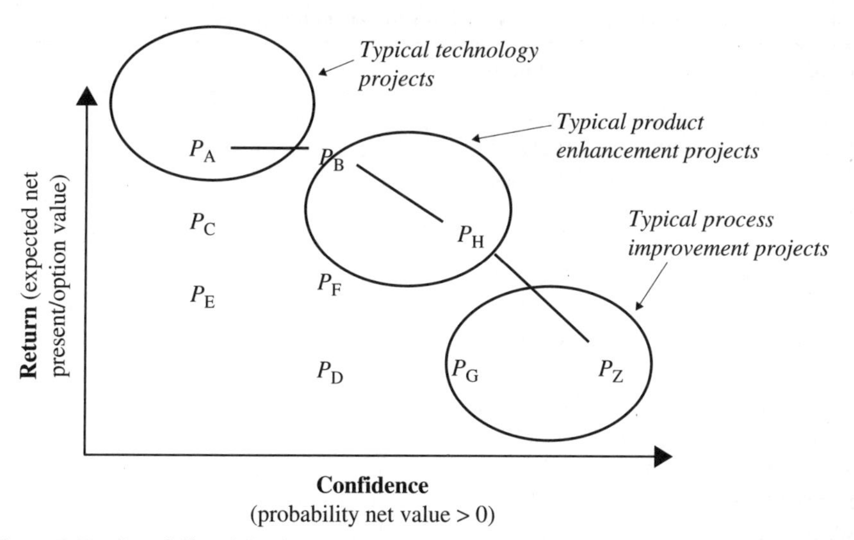

Figure 9.5. Portfolio of Options.

all of them will—unless, of course, the underlying risks are correlated. For example, if the success of all the potential R&D investments depends on a common scientific breakthrough, then despite a large number of investments, risk has not been diversified. Thus, one usually designs investment portfolios to avoid correlated risks.

While this makes sense, it is not always feasible—or desirable—for R&D investments. Often multiple investments are made because of potential synergies among these investments in terms of technologies, markets, people, etc. Such synergies can be quite beneficial, but must be balanced against the likely correlated risks.

Further, it is essential to recognize that options are not like certificates that are issued on purchase. Considerable work is needed once "purchase" decisions are made. Options often emerge piecemeal and with varying grain size. Significant integration of the pieces may be needed before the value on which the investment decisions were based is actually available and viable.

9.5.2 Investment Case Studies

Table 9.6 summarizes the key metrics for 14 case studies of real investment decision making from a wide range of domains. These examples represent a total of $4.2 billion of NOV. Put differently, these enterprises—in both private and public sectors—made investments that "bought" options worth $4.2 billion more than they had to invest to gain these options.

The ways in which these investors bought options fall roughly in four categories. The nature of these categories provides insights into how options

TABLE 9.6. Example Options-Based Valuations of Technology Investments

Technology	Option purchase	Option exercise	Net option value
Aircraft (manufacturing)	R&D	Deploy improvements	8
Aircraft (unmanned)	R&D	Deploy system	137
Auto radar	Run business	Expand offerings	133
Batteries (lithium ion)	R&D	License technology	215
Batteries (lithium polymer)	R&D	Acquire capacity	552
Fuel cell components	R&D	Initiate offering	471
Microsatellites	R&D	Deploy system	43
Optical multiplexers	R&D	Expand capacity	488
Optical switches	Run business	Expand offerings	619
Security software	Run business	Add market channels	267
Semiconductors (amplifiers)	Invest in capacity	Expand offerings	431
Semiconductors (graphics)	R&D	Initiate offering	99
Semiconductors (memory)	R&D	Initiate offering	546
Wireless LAN	Run business	R&D	191

can be "purchased" as by-products of investments primarily intended for more near-term purposes.

9.5.2.1 Investing in R&D In many cases, options are bought by investing in R&D to create "technology options" that may, or may not, be exercised in the future by business units. Ideally, the nature of the technology options of interest will be driven by the future aspirations of the business units. The characteristics and competitiveness of these aspirations are typically laced with uncertainties. These uncertainties are the basis for needing options, usually more options than will later be exercised as the business needs to hedge against alternative future market situations.

9.5.2.2 Running the Business These examples are situations where the option was bought by continuing to operate an existing line of business, perhaps with modest current profitability but substantial future opportunities. Running the business, which may have resulted from exercising an earlier option, can lead, in effect, to purchasing an option for what the business can become in the future. Indeed, as the wireless LAN example illustrates, the NOV of the potential future business may be much larger than the NPV of the current business. Thus, SV may be dominated by NOV and may justify sustaining a small NPV business.

9.5.2.3 Acquiring Capacity These examples involve the possibility of acquiring capacity to support market growth that is uncertain in terms of timing, magnitude, and potential profitability. Building or acquiring capacity that is eventually not needed could be an enormous waste of scarce resources. Purchasing an option on capacity may be a much better investment than actually investing in acquiring capacity, despite the fact that the option may not be exercised. The option may take the form of a contingent contract with an owner of capacity for the use of the capacity if needed. Another form could be a contingent contract to purchase capacity, perhaps at a price that includes a discount for the amount paid for the original option.

9.5.2.4 Acquiring Competitor These examples involve the possibility of acquiring competitors to support growth in current and adjacent markets that is uncertain in terms of timing, magnitude, and potential profitability. Such acquisitions can provide market offerings, capacity, and customers. Interestingly, acquisitions can also provide new technology options. Thus, it is quite possible that purchasing this type of option may provide one or more instances of the other three types of options.

9.5.3 Interpretation of Results

Consider the example of semiconductor memory in the second row from the bottom of Table 9.6. For $109 million of R&D, this company "purchased" an option to deploy this technology in its markets four years later for an expected investment of approximately $1.7 billion. The expected profit was roughly $3.5 billion. The NOV of over $0.5 billion reflects the fact that they bought this option for much less than it was worth.

The option value of over $600 million (i.e., the R&D investment plus the NOV) represents roughly one-third of the net present difference between the expected profit from exercising the option and the investment required to exercise it (i.e., 600/(3500 − 1700)). This is due to considerable uncertainties in the >10-year time period when most of the profits would accrue.

In the second row from the top of Table 9.6, a government agency invested $420 million in R&D to "purchase" an option on unmanned air vehicle technology that, when deployed 10 years later for $72 million, would yield roughly $750 million of operating savings when compared to manned aircraft providing the same mission effects. The NOV of $137 million represents the value of this option in excess of what they invested.

The option value of roughly $560 million (i.e., the R&D investment plus the NOV) represents over two-thirds (i.e., 560/(750 − 72)) of the net present difference between the expected cost savings from exercising the option and the investment required to exercise it, despite the returns occurring in a similar >10-year time frame. Why is this ratio 1/3 for the semiconductor memory investment but 2/3 for the unmanned air vehicle technology investment?

The answer may be obtained by examining the quotient of expected profit (or cost savings) divided by the investment required to exercise the option. This ratio is quite different for these two examples. This quotient is roughly 2.0 (i.e., 3500/1700) for the semiconductor memory option and 10.0 (i.e., 750/72) for unmanned air vehicle technology investment. Thus, the likelihood of the option being "in the money" is significantly higher for the latter. This is why the option value is one-third of the net present difference for semiconductor memory and two-thirds for unmanned air vehicle technology.

9.6 VALUE-CENTERED R&D

Thinking in terms of options can have an enormous impact on how an enterprise approaches and formulates its overall strategy. This is particularly true for research and development. Listed below are 10 principles for value-centered R&D (Rouse and Boff, 2004). Several, but not all, of these principles relate directly to the real option model, methods, and tools discussed thus far in this chapter.

Principles 1 to 3 focus on characterizing value. These principles argue that the overarching purpose of the R&D function is to provide the right portfolio of technology options. Ideally, these options will both be aligned with business unit aspirations and have NOVs that far exceed the enterprise's R&D budget.

1. Value is created in R&D organizations by providing "technology options" for meeting contingent needs of the enterprise.
2. R&D organizations provide a primary means for enterprises to manage uncertainty by generating options for addressing contingent needs.

3. A central challenge for R&D organizations is to create a portfolio of viable options; whether or not options are exercised is an enterprise challenge.
4. Value streams, or value networks, provide a means for representing value flow and assessing the OV created.
5. Valuation of R&D investments can be addressed by assessing the OV created in the value network.
6. Decision-making processes—governance—are central in managing the flow of value.
7. Organizational structure affects value flow, with significant differences between hierarchical and heterarchical structures.
8. Individual and team affiliations and identities affect value flow; dovetailing processes with disciplines is essential.
9. Champions play important, yet subtle, roles in value flow; supporting champions is necessary but not sufficient for success.
10. Incentives and rewards affect value flow; aligning these systems with value maximization is critical.

Principles 4 and 5 concern assessing value. This involves understanding how the organization creates and deploys value. Identification of value streams and the networks they form often results in streamlining these streams and networks. The assessed values of the options flowing through these streams and networks can help to decide where the value is added—or not added—and how it can be enhanced.

Principles 6 to 10 emphasize managing value. The technical constructs of value and options are necessary for success, but not sufficient. Governance, structure, affiliation, champions, and incentives and rewards are key elements of success with value-centered R&D. Unfortunately, elaboration of these elements of success is beyond the scope of this book.

9.7 SUMMARY

This chapter has shown that it is pretty straightforward to formulate option pricing models and create associated option calculators. What is not straightforward is framing the options appropriately. There are some characteristics of well-framed options. First of all, "purchasing" an option typically involves a *relatively* small investment now to purchase the right to "exercise" the option later if the conditions at that time make sense. Exercising an option usually requires a significantly larger investment than purchasing the option. One purchases an option to hedge against the downside possibility that one may not choose to exercise the option at a later point when the true nature of the market opportunity is much clearer. Thus, one is making a two-stage (or multistage) investment decision. In contrast, if the lion's share, or perhaps all, of the investment is up front, one is making a single-stage investment and the real options construct in this chapter will not yield economic valuations much different than traditional financial analyses discussed in Chapter 8.

Why would one make an up-front, albeit relatively small, investment in an option that one might not exercise? The primary motivation is to avoid making a very large investment in an attractive future that may not materialize. Perhaps people will not want to avail themselves of the functionality possible with the technology in which one has invested. One buys an option because one wants the right to the future *if* it emerges, but one does not want to commit to it now. Further, because options require much smaller initial investments, one can bet on several alternative futures rather than just one or two. This makes it much more likely that at least one of the bets will have been right.

A second characteristic of a well-framed option is that one can see how a small investment now to purchase an option can gain the right to later exercise the option and secure a large cash flow, large being relative to the costs of purchasing and exercising the option. Typically, this large cash flow will be a new cash flow because other firms usually already own existing large cash flows. Of course, competitors' existing cash flows may be replaced by their securing the new cash flows. Consequently, one needs to make sure that one has some reasonable chance of winning this competition.

A third characteristic of interest is situations where existing owners of cash flows—the incumbents—are unlikely to be dominant competitors in the future. For example, incumbents may be unlikely to afford the scale necessary, or may not have the agility to compete in fast cycle time markets. Of course, there is also the possibility that the emerging markets will be so new that there will be no incumbents. This still raises the questions of why one will compete successfully.

With these three characteristics of well-formed options in mind, one needs to think about capabilities and markets that will yield the desired large cash flows. One approach to this is to start with particular ideas, capabilities, or technologies and ask where and how they might be transformed from being inventions to being market innovations. An alternative approach is to work backwards. What capabilities will people expect and be willing to pay for, even if they are not sure how to create these capabilities? If the current market vision becomes a reality, what capabilities are consumers and providers going to want provided in some way?

Next, one needs to think about what one would need to do to gain the option to provide these capabilities in the future. Perhaps R&D would be needed. One or more alliances might be the key. Possibly, implementing a new technology in an existing line of business would provide the competency to employ this technology in a new line of business in the future. Overall, the question is what relatively small investment one should make to purchase an option that will enable one to later make a much larger investment if market opportunities have evolved in a manner that makes exercising this option attractive.

Thinking in terms of real options not only results in different economic assessments, that is, different numbers, but can also lead to new ways of thinking about one's enterprise. Companies with large portfolios of viable and attractive options have considerable leverage to take advantage of the future regardless of the substantial uncertainties and risks associated with the future. Their options provide the means to success no matter how the contingencies play out.

BIBLIOGRAPHY AND REFERENCES

Amram M, Kulatilaka N. Real options: managing strategic investment in an uncertain world. Boston: Harvard Business School Press; 1999.

Black F, Scholes M. The pricing of options and corporate liabilities. J Political Econ 1973;81:637–659.

Boer FP. Traps, pitfalls, and snares in the valuation of technology. Res Technol Manag 1998;41 (5):45–54.

Boer FP. The valuation of technology: business and financial issues in R&D. New York: Wiley; 1999.

Dixit AK, Pindyck RS. Investment under uncertainty. Princeton, NJ: Princeton University Press; 1994.

Luenberger DG. Investment science. Oxford, UK: Oxford University Press; 1997.

Merton RC. Theory of rational option pricing. Bell J Econ Manag Sci 1973;4(1):141–183.

Pennock MJ, Rouse WB, Kollar DL. Transforming the acquisition enterprise: a framework for analysis and a case study of ship acquisition. Syst Eng 2007;10(2):99–117.

Rouse WB, Boff KR. Value-centered R&D organizations: ten principles for characterizing, assessing & managing value. Syst Eng 2004;7(2):167–185.

Rouse WB, Howard CW, Carns WE, Prendergast EJ. Technology investment advisor: an options-based approach to technology strategy. Inf Knowl Syst Manag 2000;2(1):63–81.

Smithson CW. Managing financial risk: a guide to derivative products, financial engineering, and value maximization. New York: McGraw-Hill; 1998.

Trigeorgis L. Real options: managerial flexibility and strategy in resource allocation. Cambridge, MA: MIT Press; 1996.

CHAPTER 10

CONTEMPORARY PERSPECTIVES

10.1 INTRODUCTION

There are a number of contemporary issues that we have not really discussed despite the size of this book. This is only due in part to a lack of space. Indeed, many of these issues are such that the best economic concepts and practices are only just emerging and many ideas are still subject to substantive debates. Nevertheless, it is important to round out this book by outlining these contemporary issues, as well as current thinking on how best to address them.

We provide an overview of some of these issues in this brief concluding chapter. Among the subjects that we briefly discuss are the following:

- evolutionary economics;
- path dependence and network effects;
- intellectual capital;
- value of information;
- investing in humans.

10.2 EVOLUTIONARY ECONOMICS

Evolutionary economics is basically concerned with situations where the economics of a situation evolves over time generally in a complex and adaptive fashion. If we were to examine evolutionary economics here, we would be particularly concerned with guidelines for success in industries subject to these characteristics. We would be especially concerned with such information and knowledge network issues as compatibility, interconnection, and interoperability, and how the influences of these issues, as well as coordination of pricing and quality of service, lead to emergence of a network of networks. A relatively good introductory treatment of the basic features on complex adaptive systems is contained in Chapter 30 of the *Handbook of Systems Engineering and*

Economic Systems Analysis and Assessment,
by Andrew P. Sage and William B. Rouse

Management (Sage and Rouse, 2009). Also highly relevant is Chapter 34 in this book on information and knowledge management.

Good examples of industries where these phenomena are prevalent are telecommunications, energy, and healthcare. Convergence in telecommunications among landlines, mobile lines, Internet, and television has resulted in rapidly evolving networks of supplier relationships, alliances and joint ventures, and mergers and acquisitions. Optimal pricing and quality of service have become quite complicated, with new approaches quickly becoming outdated by the evolving network of relationships. The emergence of Smart Grid in the energy industry has resulted in the rapid proliferation of new entrants with offerings in intelligent sensing and control of generation, transmission, distribution, and consumption of energy. This has made competitive analysis very difficult. Forecasting revenue for yet-to-exist market segments by, as yet, unknown suppliers and customers makes planning quite difficult. The healthcare industry is undergoing a slow and painful transformation from a federation of millions of entrepreneurs with no one in charge to an integrated delivery system to provide quality affordable care for everyone. The uncertainties associated with this transformation, in both magnitude and timing, make planning and economic analysis quite problematic and very needed. All three of these examples illustrate how evolving economies require creative application of the concepts, principles, methods, and tools presented in this book.

10.3 PATH DEPENDENCE AND NETWORK EFFECTS

An important aspect of economic systems is path dependence (Arthur, 1994). The essence of this phenomenon begins with a supposedly minor advantage or inconsequential head start in the marketplace for some technology, product, or standard. This minor advantage can have important and irreversible influences on the ultimate market allocation of resources even if market participants make voluntary decisions and attempt to maximize their individual benefits. Such a result is not plausible with classic economic models that assume that the maximization of individual gain leads to market optimization unless the market is imperfect due to the existence of such effects as monopolies. Path dependence is a failure of traditional market mechanisms and suggests that users are "locked" into a suboptimal product, even though they are aware of the situation and may know that there is a superior alternative.

This type of path lock-in is generally attributed to two underlying drivers: (1) network effects and (2) increasing returns of scale. Both of these drivers produce the same result, namely that the value of a product increases with the number of users. Network effects, or "network externalities," occur because the value of a product for an individual consumer may increase with increased adoption of that product by other consumers. This, in turn, raises the potential value for additional users. An example is the telephone, which is only useful if at least one other person has one as well, and becomes increasingly beneficial as the number of potential users of the telephone increases.

Increasing returns of scale imply that the average cost of a product decreases as higher volumes are manufactured. This effect is a feature of many knowledge-based products where high initial development costs dominate low marginal production and distribution costs. Thus, the average cost per unit decreases as the sales volume increases and the producing company is able to continuously reduce the price of the product. The increasing returns to scale, associated with high initial development costs and the decreasing sales price, create barriers against market entry by new potential competitors, even though they may have a superior product. If there is no competition, this phenomenon results in increasing profits due to the lack of incentives to decrease prices.

The controversy in the late 1990s over the integration of the Microsoft Internet Explorer with the Windows operating system may be regarded as a potential example of path dependence, and appropriate models of this phenomenon can potentially be developed using complexity theory. These would allow exploration of whether network effects and increasing returns of scale can potentially reinforce the market dominance of an established but inferior product in the face of other superior products, or whether a given product is successful because its engineers have carefully and foresightedly integrated it with associated products such as to provide a seamless interface between several applications. To some extent, the more recent success of Google and social networking websites indicates how technology changes and market forces eventually emerge in such situations.

10.4 INTELLECTUAL CAPITAL

A major determinant of organizational abilities is the extent to which an organization possesses intellectual capital, or knowledge capital, such that it can create and use innovative ideas to produce productive results. The concept of intellectual capital has been defined in various ways (Rouse and Sage, 2009; Rouse, 2010). We would add communications to the formulation of Ulrich (1998) representing intellectual capital to yield

$$\text{Intellectual captial} = \text{Competence} \times \text{Commitment} \times \text{Communications}$$

One could argue that other important terms, such as collaboration and courage, could be added to this equation.

Loosely structured organizations and the speed, flexibility, and discretion they engender in managing intellectual capital fundamentally affect knowledge management (Klein, 1998; Rouse and Sage, 2009). Knowledge workers are not captive, and hence know-how is not "owned" by the organization. Patents are of much less value, for example, as evidenced by the substantial decline in use of this mechanism. Instead, what matters is the ability to make sense of market and technology trends, quickly decide how to take advantage of these trends, and act faster than other players. Sustaining competitive advantage requires redefining market-driven value propositions and quickly leading in providing value in appropriate new ways. Accomplishing this in an increasingly information-rich environment is a major challenge, both for organizations experiencing these environments and for those who devise and provide systems engineering and management methods and tools for supporting these new ways of doing business. There is a major interaction involving knowledge work and intellectual capital, and the communications-driven information and knowledge revolution that suggests many and profound complex adaptive system like changes in the economy of this century (Shapiro and Varian, 1998; Kelly, 1998).

What are the likely returns on capabilities gained from human capital investments? Tangible assets and financial assets usually yield returns that are important elements of a company's overall earnings. It is often the case, however, that earnings far exceed what might be expected from these "hard" assets. For example, companies in the software, biotechnology, and pharmaceutical industries typically have much higher earnings than companies with similar hard assets in the aerospace, appliance, and automobile industries, to name just a few. It can be argued that these higher earnings are due to greater human capital among software companies, etc. However, since human capital does not appear on financial statements, it is very difficult to identify and, better yet, project knowledge earnings.

Mintz (1998) summarizes a method developed by Baruch Lev for estimating what he terms knowledge capital—what could be argued to be a surrogate for human capital. The key, he argues, is to partition earnings into knowledge earnings and hard asset earnings. This is accomplished by first projecting normalized annual earnings from an average of three past years as well as estimates for three future years. Earnings from tangible and financial assets were calculated from reported asset values using industry averages of 7% and 4.5% for tangible and financial assets, respectively. Knowledge capital was then estimated by dividing knowledge earnings by a knowledge capital discount rate. Based on an analysis of several knowledge-intensive industries, 10.5% was used for this discount rate.

Using this approach to calculating knowledge capital, Mintz compares 20 pharmaceutical companies to 27 chemical companies. He determines, for example, a knowledge capital to book value ratio of 2.45 for pharmaceutical companies and 1.42 for chemical companies. Similarly the market value to book value ratio is 8.85 for pharmaceutical companies and 3.53 for chemical companies.

The key issue within this overall approach is being able to partition earnings. While earnings from financial assets should be readily identifiable, the distinction between tangible and knowledge assets is problematic. Further, using industry average return rates to attribute earnings to tangible assets does not allow for the significant possibility of tangible assets having little or no earnings potential. Finally, of course, simply attributing all earnings "leftover" to knowledge assets amounts to giving knowledge assets credit for everything that cannot be explained by traditional financial methods. Nevertheless, this approach does provide insights into an important aspect of human capital—human skills and knowledge.

10.5 VALUE OF INFORMATION

Networking internally and externally provides information about your operations, your suppliers, and perhaps your competitors. A wide range of databases provide information on consumer characteristics, behaviors, and preferences. All things, it seems, are possible within the new economics of information and its strategic management (Evans and Wurster, 1997; Sveiby, 1997).

A central economic issue in the new information economy concerns how to attach economic value to information. Shapiro and Varian (1998) provide an excellent treatment of this topic. They argue, quite convincingly, that fundamental principles of economics still apply in the realm of networks and information.

The first principle is that the selling price of any product or service tends to the marginal cost of production and distribution. In competitive markets, the players will continually push marginal costs down—thus, prices tend to go down. For information products distributed over the Internet, the marginal costs are zero! Consequently, companies will tend to give their information products away—and customers will expect information products to be free.

This principle explains why so many Internet businesses have focused on making money via advertising. They trade free content for people's willingness to put up with banners and blinkers proclaiming the wonders of everything from security software to simulated sex. As irritating as this can be, people have long demonstrated their willingness to be manipulated by such messages via television.

Their second principle is that differentiation can help one to escape the fate of having to give away information products. They suggest one do this by selling customers personalized products at personalized prices. This requires in-depth understanding of customers' needs and values so that one knows what to put in the package and which things can command higher prices. For example, some customers value time much more than others.

Put simply, the idea is to sell roughly the same things to different people for different prices. Sounds great, but the network economy enables everyone to know the lowest price for anything. Shapiro and Varian suggest that one can avoid this with versioning. With a modular design, based on a common platform, one can create different versions of products tailored to the desires of different market segments. While one wants to avoid blunders, such as putting Cadillac badges on Chevrolets (Hanawalt and Rouse, 2010), this principle can help provide the differentiation one needs at costs that can be endured.

Another principle focuses on lock-in. The essence of lock-in is that customers' future options are constrained by the choices they make now. Once a customer commits to particular information products, invests in gaining competence in using these products, and becomes dependent on the tailored information they provide, it will be expensive for them to change providers. The switching costs are likely to be too high. Such customers are locked in. An installed base of locked-in customers can be a company's most valuable asset.

Yet another principle concerns network externalities and positive feedback. As noted earlier, the value of some information products, for instance telephones, is much greater if many people use these products. The more people in the network, the better. In this way, larger networks get larger—this is called positive feedback. For obvious reasons, therefore, one typically wants to grow the network of users of a company's information products.

These principles are particularly relevent to current trends toward "open access" of publications in science, technology, and medicine (Beaudouin-Lafon, 2010). The notion is that the content of research publications that was funded by the federal government should be free to the public who paid the taxes to fund the research. This has the potential to completely undermine the business models of the publishers and professional societies in these areas.

The key, assuming that these providers follow Shapiro and Varian's principles, is to provide value-added services that justify—from the consumer's point of view—charging for content that is otherwise free. A good example of this is census data, which is also free. However, most people do not buy the data directly. Instead, they buy value-added tools that enable manipulating this data and, in the process, receive the data for free. In fact, most major publishers and professional societies are pursuing this strategy (IEEE, 2010).

10.6 INVESTING IN HUMANS

Human capital and the value information are central aspects of the contemporary economy. How can we economically assess the value of investments in training and education, safety and health, and work productivity? There is a long and rich history, and many successes, associated with effective integration of human behavior and performance into complex systems such as aircraft, automobiles, factories, process plants, and, more recently, service systems. Human Systems Integration (HSI)—as well as human-centered design—is now a well-articulated and supported endeavor. We have accumulated much knowledge and the skills needed to enhance human abilities, overcome human limitations, and foster human acceptance (Rouse, 2007).

However, as with any engineering activity, there are costs associated with HSI or human-centered design. Most would argue that these costs are actually investments in increased performance, higher quality, and lower operating costs. The economics of HSI addresses the question of whether such investments are worth it (Rouse, 2010). In particular, what are the likely monetary returns on such investments and do these returns justify these investments?

Of course, nonmonetary returns are often also of interest. However, the focus here is solely on getting the economics right. Admittedly, the numbers are not all that counts. But, we need to count the numbers correctly. Then we can trade off economic attributes versus noneconomic attributes.

Understanding the economic attributes of HSI investments is not quite as straightforward as it may seem. First of all, there are several levels of costs. At the lowest level, there are the labor and material costs of the personnel who do HSI. Their efforts usually result in recommendations for improving the system of interest. These recommendations often involve second-level costs that are much larger than those associated with those doing HSI. At the third level, there are the costs associated with operating the system after the HSI-oriented recommendations have been implemented.

From an investment perspective, we would hope that the third-level costs are decreased by having incurred the first- and second-level costs. (Some HSI practitioners characterize these savings as "cost avoidance.") These reductions represent returns on having made the lower-level investments. There may be additional returns associated with selling more units of a well-designed system, such as we have seen of late with Apple's iPhone. This increased demand can lead to greater production efficiencies and thereby increase profits per unit, creating a third source of return on investment.

The investment situation just outlined is summarized in Fig. 10.1. There are time series of upstream costs—or investments—and then time series of downstream returns. Standard discounted cash flow analysis as discussed in Chapters 6 and 8 can be used to determine whether or not expected returns justify the proposed investments. However, it is not at all a straightforward effort (Rouse, 2010).

One problem is that it is difficult to estimate the upstream and downstream time series of investments and costs. Point estimates will not suffice, as there is much uncertainty. Thus, we need probability distributions, not just expected values. For all but the most sophisticated enterprises, this poses data collection problems. Quite simply, while most enterprises understand their overall costs as seen on their income statements, most cannot attribute these costs to particular activities such as operations and maintenance of the systems they operate.

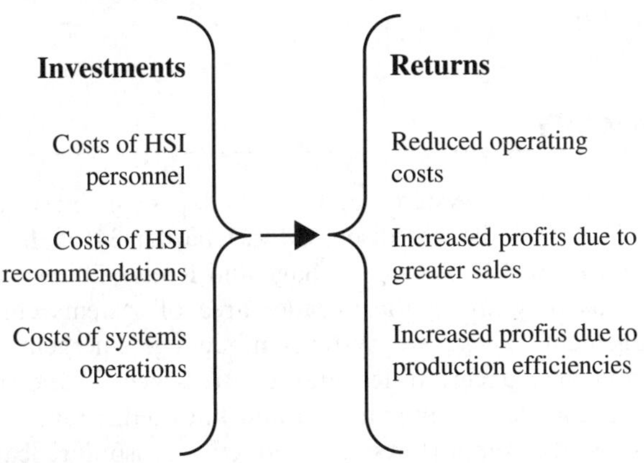

Figure 10.1. Investments and Returns for Human Systems Integration.

There are also uncertainties associated with what recommendations will emerge, which ones will be chosen for implementation, and whether the actual operating environment of the system once deployed will encounter operational demands that take advantage of the enhanced system functionality that was recommended by the HSI personnel. Consequently, the decision to invest in HSI is really a multistage decision as discussed in Chapter 9. Traditional discounted cash flow analyses substantially underestimate the value of multistage investments. While we have the analytic machinery to address these types of investments, many decision makers find this level of uncertainty daunting.

Beyond these technical and practical difficulties, there is often an enormous behavioral and social difficulty associated with the simple fact that different people and organizations make the investments and then see the returns. The organization developing or procuring a system is usually quite remote from the organization gaining the returns, both spatially and temporally. For example, engineering and manufacturing may incur the costs while marketing and sales see the returns. Further, the costs may be incurred today while the returns are not seen until years from now.

This spatial and temporal separation is less difficult for highly integrated enterprises such as companies operating in the private sector. In contrast, for government agencies and companies operating in the public sector, there may be no one who "owns the future." In these situations, investments are treated as costs. While these expenditures may yield assets that can provide future returns, government agencies—and Congress—have no balance sheet on which to tally the value of these assets. Thus, no value is explicitly attached to the future.

As formidable as this litany of difficulties may seem, we still make investments in training and education, health and safety, and performance enhancements. Culturally at least, we value a healthy, educated, productive, and competitive workforce. We have the right inclinations. However, we have not had the right data, methods, and tools to make stronger economic arguments for investing in people. Fortunately, this situation is rapidly improving due to both the increasing availability of data and the development of easily accessible and usable tools. Perhaps in the not too distant future, human capital will make it onto the balance sheet (Rouse, 2010).

10.7 SUMMARY

The subject of economic systems analysis and assessment is a continually evolving one. The 11 challenges discussed in Chapter 34 of the *Handbook of Systems Engineering and Management* (Sage and Rouse, 2009) are generally as applicable here as they are in the broader area of systems engineering and systems management. These are systems modeling, emergent and complex adaptive phenomena, uncertainties and control, access and utilization of information and knowledge, information and knowledge requirements, information and knowledge support systems, inductive reasoning, learning organizations, planning and design, optimization versus agility, and measurement and evaluation. Thus, we see that economic systems analysis and assessment is one of the core and central subject areas of systems engineering and management.

BIBLIOGRAPHY AND REFERENCES

Arthur WB. Increasing returns and path dependence in the economy. Ann Arbor: University of Michigan Press; 1994.

Beaudouin-Lafon M. Open access to scientific publications: the good, the bad, and the ugly. Commun ACM 2010;53(2):33–34.

Evans PB, Wurster TS. Strategy and the new economics of information. Harv Business Rev 1997; September–October;75(5):71–82.

Hanawalt E, Rouse WB. Car wars: factors underlying the success or failure of new car programs. Syst Eng 2010;13(4):389–404.

IEEE. IEEE Workshop on the Future of Information; 2010 May 24–26; Washington, DC: National Academy of Engineering.

Kelly K. New rules for the new economy. New York: Viking; 1998.

Klein DA. The strategic management of intellectual capital. Boston: Butterworth-Heinemann; 1998.

Mintz SL. A better approach to estimating knowledge capital. CFO 1998;February; 14(2):29–37.

Rouse WB. People and organizations: explorations of human-centered design. Hoboken, NJ: Wiley; 2007.

Rouse WB, editor. The economics of human systems integration: valuation of investments in people's training and education, safety and health, and work productivity. Hoboken, NJ: Wiley; 2010.

Rouse WB, Sage AP. Information technology and knowledge management. In Sage AP, Rouse WB, editors, Handbook of Systems Engineering and Management (Chap. 34). 2nd ed. Hoboken, NJ: Wiley; 2009.

Sage AP, Rouse WB, editors. Handbook of systems engineering and management. 2nd ed. Hoboken, NJ: Wiley; 2009.

Shapiro C, Varian HR. Information rules: a strategic guide to the network economy. Boston: Harvard Business School Press; 1998.

Sveiby KE. The new organizational wealth: managing and measuring knowledge based assets. San Francisco: Berrett-Koehler Publishers; 1997.

Ulrich D. Intellectual capital = Competence × Commitment. Sloan Manag Rev 1998;39(2):15–26.

INDEX

WILEY SERIES IN SYSTEMS ENGINEERING AND MANAGEMENT

Andrew P. Sage, Editor

ANDREW P. SAGE and JAMES D. PALMER
Software Systems Engineering

WILLIAM B. ROUSE
Design for Success: A Human-Centered Approach to Designing Successful Products and Systems

LEONARD ADELMAN
Evaluating Decision Support and Expert System Technology

ANDREW P. SAGE
Decision Support Systems Engineering

YEFIM FASSER and DONALD BRETTNER
Process Improvement in the Electronics Industry, Second Edition

WILLIAM B. ROUSE
Strategies for Innovation: Creating Successful Products, Systems and Organizations

ANDREW P. SAGE
Systems Engineering

HORST TEMPELMEIER and HEINRICH KUHN
Flexible Manufacturing Systems: Decision Support for Design and Operation

WILLIAM B. ROUSE
Catalysts for Change: Concepts and Principles for Enabling Innovation

LIPING FANG, KEITH W. HIPEL, and D. MARC KILGOUR
Interactive Decision Making: The Graph Model for Conflict Resolution

DAVID A. SCHUM
Evidential Foundations of Probabilistic Reasoning

JENS RASMUSSEN, ANNELISE MARK PEJTERSEN, and LEONARD P. GOODSTEIN
Cognitive Systems Engineering

ANDREW P. SAGE
Systems Management for Information Technology and Software Engineering

ALPHONSE CHAPANIS
Human Factors in Systems Engineering

YACOV Y. HAIMES
Risk Modeling, Assessment, and Management, Third Edition

DENNIS M. BUEDE
The Engineering Design of Systems: Models and Methods, Second Edition

ANDREW P. SAGE and JAMES E. ARMSTRONG, Jr.
Introduction to Systems Engineering

WILLIAM B. ROUSE
Essential Challenges of Strategic Management

YEFIM FASSER and DONALD BRETTNER
Management for Quality in High-Technology Enterprises

THOMAS B. SHERIDAN
Humans and Automation: System Design and Research Issues

ALEXANDER KOSSIAKOFF and WILLIAM N. SWEET
Systems Engineering Principles and Practice

HAROLD R. BOOHER
Handbook of Human Systems Integration

JEFFREY T. POLLOCK and RALPH HODGSON
Adaptive Information: Improving Business Through Semantic Interoperability, Grid Computing, and Enterprise Integration

ALAN L. PORTER and SCOTT W. CUNNINGHAM
Tech Mining: Exploiting New Technologies for Competitive Advantage

REX BROWN
Rational Choice and Judgment: Decision Analysis for the Decider

WILLIAM B. ROUSE and KENNETH R. BOFF (editors)
Organizational Simulation

HOWARD EISNER
Managing Complex Systems: Thinking Outside the Box

STEVE BELL
Lean Enterprise Systems: Using IT for Continuous Improvement

J. JERRY KAUFMAN and ROY WOODHEAD
Stimulating Innovation in Products and Services: With Function Analysis and Mapping

WILLIAM B. ROUSE (editor)
Enterprise Tranformation: Understanding and Enabling Fundamental Change

JOHN E. GIBSON, WILLIAM T. SCHERER, and WILLAM F. GIBSON
How to Do Systems Analysis

WILLIAM F. CHRISTOPHER
Holistic Management: Managing What Matters for Company Success

WILLIAM B. ROUSE
People and Organizations: Explorations of Human-Centered Design

MO JAMSHIDI
System of Systems Engineering: Innovations for the Twenty-First Century

ANDREW P. SAGE and WILLIAM B. ROUSE
Handbook of Systems Engineering and Management, Second Edition

JOHN R. CLYMER
Simulation-Based Engineering of Complex Systems, Second Edition

KRAG BROTBY
Information Security Governance: A Practical Development and Implementation Approach

JULIAN TALBOT and MILES JAKEMAN
Security Risk Management Body of Knowledge

SCOTT JACKSON
Architecting Resilient Systems: Accident Avoidance and Survival and Recovery from Disruptions

JAMES A. GEORGE and JAMES A. RODGER
Smart Data: Enterprise Performance Optimization Strategy

YORAM KOREN
The Global Manufacturing Revolution: Product-Process-Business Integration and Reconfigurable Systems

AVNER ENGEL
Verification, Validation, and Testing of Engineered Systems

WILLIAM B. ROUSE (editor)
The Economics of Human Systems Integration: Valuation of Investments in People's Training and Education, Safety and Health, and Work Productivity

ALEXANDER KOSSIAKOFF, WILLIAM N. SWEET, SAM SEYMOUR, and STEVEN M. BIEMER
Systems Engineering Principles and Practice, Second Edition

GREGORY S. PARNELL, PATRICK J. DRISCOLL, and DALE L. HENDERSON (editors)
Decision Making in Systems Engineering and Management, Second Edition

ANDREW P. SAGE and WILLIAM B. ROUSE
Economic Systems Analysis and Assessment: Cost, Value, and Competition in Information and Knowledge Intensive Systems, Organizations, and Enterprises